THE CITY

PLANNING AND PROFIT IN THE URBAN ECONOMY

URBAN ECONOMICS

PLANNING AND PROFIT IN THE URBAN ECONOMY

T.A. BROADBENT

LONDON AND NEW YORK

First published in 1977

This edition published in 2007
Routledge
2 Park Square, Milton Park, Abingdon, Oxon, OX14 4RN

Simultaneously published in the USA and Canada by Routledge
711 Third Avenue, New York, NY 10017

Routledge is an imprint of Taylor & Francis Group, an informa business

Transferred to Digital Printing 2007

First issued in paperback 2013

The publishers have made every effort to contact authors and copyright holders of the works reprinted in the *The City* series. This has not been possible in every case, however, and we would welcome correspondence from those individuals or organisations we have been unable to trace.

These reprints are taken from original copies of each book. In many cases the condition of these originals is not perfect. The publisher has gone to great lengths to ensure the quality of these reprints, but wishes to point out that certain characteristics of the original copies will, of necessity, be apparent in reprints thereof.

British Library Cataloguing in Publication Data
A CIP catalogue record for this book
is available from the British Library

Planning and Profit in the Urban Economy

ISBN13: 978-0-415-41766-2 (volume hbk)
ISBN13: 978-0-415-41932-1 (subset)
ISBN13: 978-0-415-41318-3 (set)
ISBN13: 978-0-415-86042-0 (volume pbk)

Routledge Library Editions: The City

Planning and profit in the urban economy

T. A. Broadbent

METHUEN & CO LTD

First published in 1977
by Methuen & Co Ltd
11 New Fetter Lane, London EC4P 4EE

Typeset in Great Britain by
Preface Graphics Ltd., Salisbury, Wilts
and printed in Great Britain by
Richard Clay (The Chaucer Press) Ltd.,
Bungay, Suffolk

ISBN 0 416 563201 *hardback*
ISBN 0 416 563309 *paperback*

Contents

Introduction xi

1 The overdeveloped economy 1
1 Introduction 1
2 The symptoms 2
Britain in the world economy: the decline 2
Standard of living: falling behind 4
Growth: the lagging economy 5
3 The market economy 6
4 Barriers to development 13
Growth: extensive and intensive 13
The agricultural barrier 15
Other barriers 16
5 Some results 16
Investment and growth 16
Competition, concentration and the multinational companies 18
Decline of manufacturing? 21
Size of plant and specialization 22
The population and inequality 23
The public sector 24
6 Some effects at the urban level 25
Urbanization 25
Investment in urban development 26
7 The economy as a framework for urban development and planning 27
Summary 28

2 The state: short-term myopia and the cleft stick 29
1 Introduction: the state — limited or unlimited? 29
2 Mainstream economics and the state 31
Introduction 31
Neoclassical welfare theory 32
Keynes: macro-economic management 35

3 The limits of state intervention 39
Introduction 39
Bearing the costs of production 39
The cleft stick 40
4 The state: other views 45
Convergence theories 45
The state as oppressor 46
5 The state in practice 48
Steering the economy 48
Public spending 55
The composition of public spending 58
6 The history and effects of economic controls 60
7 The 'planning' experiment 67
8 Regional planning and policy 73
The instruments of policy 73
The effects and weaknesses of regional policy 77
9 Appendix: the three estates — an activity-analysis approach 79
Unscrambling the national accounts 79
Activity analysis of the national economy 82
Systems analysis and control 86
Summary 88

3 The urban system: labour and production 90
1 Introduction 90
2 The UK: an urban economy 92
Introduction 92
Superficial trends: urban growth and decentralization 94
Overall pattern 94
Effect of urban size on growth and decline 97
Migration and natural increase 99
Centralization and decentralization 100
Broad conclusions 102
3 Towards an explanation of the urban hierarchy 103
Introduction 103
Size and function 103
Organization structure 105
Office space and office rent 107
Summary: the city and the economy 109
4 The city as a pool of labour: the first urban process 110
Introduction 110
The activity of people in cities 112
The city as a system of production 116
Households: a production activity? 117
5 Competition for space and location: the second urban process 120
Summary 127

4 The local state: urban government and urban planning 128
1 Introduction 128
2 Urban government: increasing size, decreasing power 129
Growth 129
Finance 132
Central controls 134
The local state and the urban system 136
3 The growth of urban planning 141
Early history 141
The inter-war period 150
Wartime reports and the birth of post-war urban planning 151
4 The post-war planning system: plans without powers 153
Unforeseen change 153
The 1947 system: negative control 155
The 1968 and 1971 acts: revision but no change 158
Effects and operation of the system 162
5 Land: the planner's resource? 165
Summary 170

5 Current theories of urban development 172
1 How ideas are used and produced: the function of theories in planning 172
2 Conventional economic theory 174
Introduction 174
Fundamentals: the sovereign consumer, utility, the margin and the market 174
Introduction 174
Utility 176
The margin 177
Supply and demand 178
Equilibrium 180
Elaboration 182
What is the margin? 182
The two estates: wages and profits — 'what they deserve' 183
Big firms: unfair profits 184
Many markets: general equilibrium 186
Summary of the market mechanism 187
The staying power of neoclassical economics 188
Justification 189
Explanation and description 190
Planning techniques 190
3 The urban economy: making conventional theory spatial 191
Introduction 191
Urban micro-economics: assumptions breaking down 292
Inferior economics? 192
Externalities 193

Lumpiness 195
The sticky market 195
The location problem 195
Trade 196
Monopoly again 196
Central places 197
Describing the city: zones and rings 198
Geographic margin and economic margin 200
Urban macro-economics — the multiplier again: the economic base 201
Assessment: description without explanation? 202
4 Pluralism: a market theory? 204
5 Ultra-radicals: planners as oppressors 207
6 Systems and cybernetics: technique or social theory? 210
Summary 211

6 Planning practice 212
1 Introduction: what planners actually do 212
2 The technical take-off: US transportation imports in the 1960s 214
Proving the demand for roads 214
Central control: imposing a method 218
3 Planning expertise: 'Development control' 220
4 Preparing a structure plan 223
5 The use and misuse of planning technology 228
US gigantism and its requiem 228
A land-use model in the UK 230
6 Cost-benefit analysis: the limits to micro-economics 235
Summary 238

7 Conclusion 240
1 The central dilemma 240
2 The main themes summarized 243
3 Positive suggestions? 245

Bibliography 251

Index 266

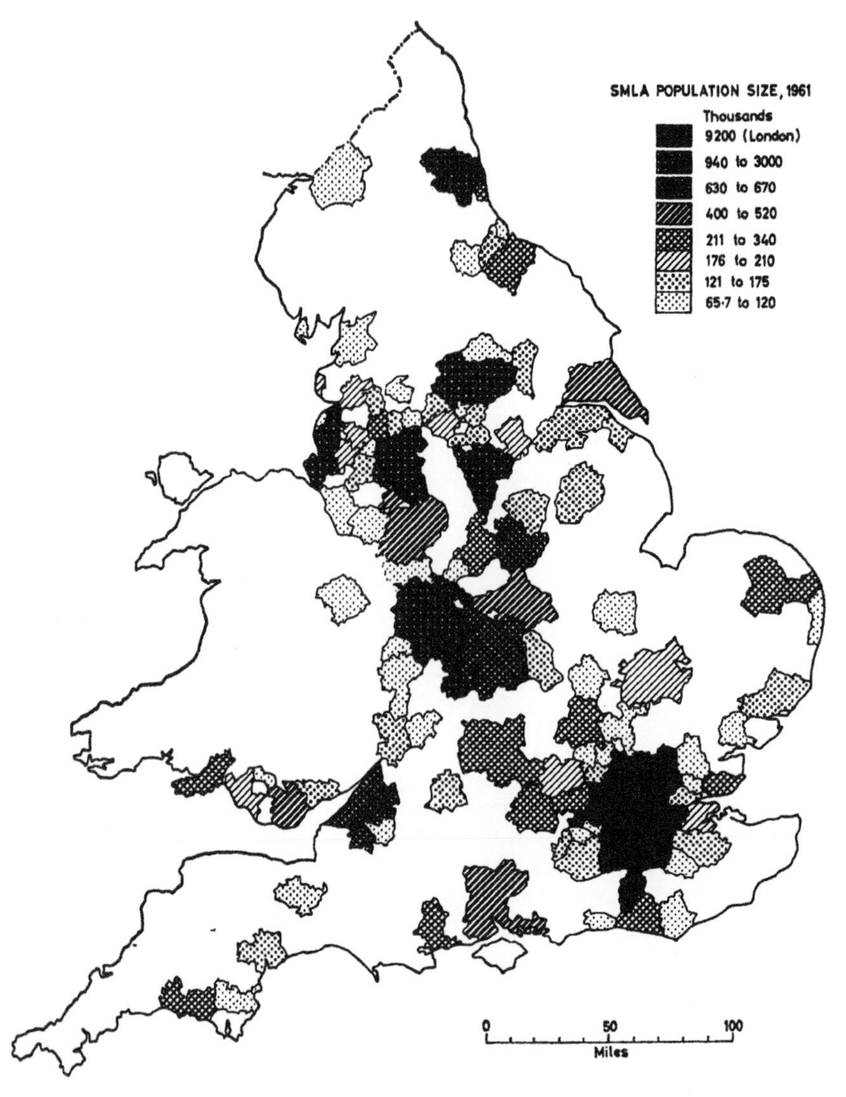

Labour pools in England. Standard Metropolitan Labour Areas (SMLA's) after Hall *et al.* (1973)

Introduction

This book is an essay — a point of view, critical but not iconoclastic. It is not a piece of research, in that the suggestions and concepts are not proved in any formal, scientific sense. The hope is that the picture presented is relatively complete and coherent, and reasonably supported by the evidence.

The book tries to answer some of the questions posed in the introduction to the British edition of *After the Planners*: what is the relationship between government and industry, and what is the role of planning within this relationship? Since the discussion is wide ranging, there is a danger that some of the different subject areas may be too superficially treated or unbalanced. The evidence has to be selective — in many of these areas there is little agreement about what the facts really are, let alone how they might be interpreted. The hope is that there is at least a certain coherence in the interpretation and explanation of evidence from the different aspects of the theory and practice of national, regional and urban planning. There is no implied judgement of works not cited; further references are given to allow the reader to follow up some of the subjects in greater detail.

It has been suggested that this kind of book is almost impossible to write — at least as a short-term project. It probably falls between every stool in sight. Each field covered could and should be considered in greater depth and more rigorously. A reader specializing in any one of the separate topics covered will be most dissatisfied with the superficial treatment of his or her topic. Inevitably there is too much detail in some areas, too little in others.

There is a concept in systems analysis that the behaviour of the whole system is greater than the sum of the separate parts. It is to be hoped, therefore, that a reader quite rightly exasperated at the sketchy treatment of individual topics might acknowledge that the whole argument does succeed at least in establishing an agenda of issues and ideas about urban development and planning which justify further more rigorous inquiry.

This book is probably just critical enough of the important institutions, individuals and schools of thought in the urban planning field to alienate them all. Professional institutions are criticized for restrictive practices, status quo academics for being too superficial, radical researchers for being nihilistic, planners for obscuring the issues. Government departments will find the description of practices and powers crude and oversimplified. Nevertheless, the viewpoint propounded is intended to be ultimately positive.

While it would be naive to assume that there are no limits to the gradual improvement or evolution of urban planning (or indeed the wider system of national state planning of which it is a part), nevertheless the existing institutions and arrangements for planning do offer potential for constructive and possibly more fundamental changes in the future.

Urban development and planning are so bound up with the very heartland of the mixed economy — the allocation and control of resources, with all the controversy and conflict that implies — that it is remarkable how many authors are able to present the subject in a relatively neutral, bland and uncontroversial way.

On the other hand, there is a new school of geographers and planners who are indeed critical and who tend to blame the capitalist system for our urban ills. I try to show here that, while the controversy and debate is welcome, much of this work tends to avoid the central issue — what do we actually *do* about it? These authors are often iconoclastic, almost nihilistic. We are all against the bureaucrats, but these authors tend to view planners largely as oppressors or as agents of the oppressors (big business or property companies). This seems to be not only opportunistic, but wrong — the state is not benign, but neither can it be cast purely in the role of the devil. Its role is essentially contradictory but has potential, and understanding this role is probably the key to understanding the future of the UK economy.

The book is also born of frustration generated by questions of the sort: why bother with 'good planning'? 'It's all politics and wheeler-dealing, so why bother with planning the economy and the cities? It always comes down to the same old political decision in the end.' But is this really so? Why is it that economic policy seems so constrained, and that politicians with vastly different rhetorics tend to come up with the same policies in practice. Pure political dealing would surely lead to a much more random set of decisions. Is there an invisible hand at work? Can it be 'the system'?

This book tries to set what are often treated as separate subjects in a wider context, and thereby aims to provide some new insight. Many accounts of geography and economics, especially urban economics, tend to see planning largely as a matter of choice, without considering whether or not the economy itself might impose limits on the state's freedom to intervene. Many modern accounts of urban development and planning do not set the city in a national context or connect the forces of competition in the city with competition in the economy at large. Again, planning techniques are often presented in the abstract, are not related to the theory that underlies them, nor the planning context in which they are used. Conventional economic theory also often tends to ignore some of the basic 'facts' of the modern economy — such as the effect of the large 'monopoly' firm on the national economy or the large factory on the urban area.

The argument begins in Chapter 1 with the UK economy (which, it is argued, is uniquely 'overdeveloped') and its decline in the world economy, looking particularly at the forces of competition and concentration.

State management and planning appears to be the logical road to regeneration, but Chapter 2 examines the limitations on the state's ability to control the mixed economy, imposed on the state by the economy itself, and cites the increasing difficulty of 'steering' the UK economy over the last thirty years. It is suggested that there are indeed rules governing the operation of the state and in particular that there is a limit cobarrier to its growth.

Chapter 3 comes down to the city or urban level and looks more closely at the forces of competition and concentration as they appear at the local level. The city is viewed as a 'system of production', with households also regarded as producers — 'small factories'. The only truly local thing about the city is its pool of labour, and its most significant development force is the competition for locations, for space and for land. This in turn is driven by the search for rent, and by the different land-using productive activities' ability to pay rent.

Chapter 4 shows how the rules and limitations on the state take on a more extreme form at the local level. Urban planning is a specialized activity which shows clearly some key aspects of the state's contradictory role — its lack of powers and its 'indicative planning' task. In spite of the increase in urban planning over the last hundred years, and more recently in the scope and content of the new structure plans, planning power and resources are not sufficient to implement development. They have proved inadequate in city centres and other areas where there is intense competition for land between different activities; at best they have succeeded in influencing or containing development at the margin.

In Chapter 5 the main ideas that influence this field are discussed. Economics — 'the queen of the social sciences' — contains the only coherent theories with associated models, statistical techniques and a corporate expertise in the social sciences. Other schools of thought — sociological or geographical — often reflect economic ideas (e.g. 'the margin'). They also often give undue importance to some particular aspect of the real world: if everything is related to everything else then any one factor can appear to 'explain' the rest. Neoclassical economics itself owes a debt to urban analysis; an early application of its key concept, 'the margin', was to the geographical margin around a city.

Having looked at the dominating ideas and the contradictions in the role planning is called upon to play, in Chapter 6 we can see what planners are actually doing — and why and how the new techniques are often used to mystify and to justify policies in the new UK structure plans.

Finally, Chapter 7 suggests how state planning and particularly urban planning can and should develop. It is suggested that planning should attempt to harness economic forces and channel them more directly to meet the needs of the city. A key requirement is to close the gap between 'plans' and 'powers' which has pervaded all attempts at state planning in the UK for the past fifteen years, to gain greater control at the urban level of the planners' basic resource — land — and by this and other means to give the public sector the resources needed to implement urban development.

Such a strengthening of urban planning could well be consistent with the continuation of a market-led mixed economy, but the extension of planning and public-sector control at the national level could raise much more serious questions.

The chapters are structured into sections and subsections so that the reader not conversant with or interested in the details of a particular topic may read through the introductory sections and the main sections, omitting the more detailed subsections for a later reading.

I should like to thank Alan Evans, David Eversley, David Donnison, Richard Barras, Derek Palmer, Brian McLoughlin, William Morrison and others in the Centre for Environmental Studies and the Planning Research Applications Group for commenting on parts or all of the manuscript. I alone, of course, am responsible for any errors. I should like to thank Frances Richards and Mary Moody for helping with the bibliography. I should also like to thank Pauline Benington for deciphering and typing the various drafts.

1 The overdeveloped economy

1 Introduction

This first chapter describes some of the main features of the UK national economy. It is suggested that the UK is the most developed of all the Western economies — indeed, that it is 'overdeveloped'. That the UK has declined over the last fifteen years is undeniable, and Section 2 points to some of the most obvious indications of this decline in relation to other countries — the falling behind of living standards and the slow growth of the economy.

Section 3 then describes in broad terms the 'mechanics' or processes by which the UK economy actually works. It is suggested that the UK is still essentially a market or capitalist economy. The driving force is competition in the private sector, and the struggle to maintain profits. The way this is done is to try continually to reduce the costs of production of goods and services. It is suggested that the market pervades nearly every corner of the national system and that, notwithstanding the increasing size of the public sector, it is to market processes — and the causes and effects of competition — that we must look first in order to understand the way any part of it develops, and this includes any urban areas or cities within the economy.

It is then argued in Section 4 that there are 'barriers' which impinge on the UK economy, and which confine and inhibit its development and the direction in which it can move. Some of these barriers are external (e.g. overseas competition) and others internal (e.g. few new resources to exploit, no residual semi-rural agricultural population). Britain is uniquely overdeveloped in that the market economy has developed as far as, or further, than it has in any other country. A greater proportion of the active population earn their keep by selling their labour for wages or salaries than in any other comparable country in the Western world. And this itself represents another 'barrier' or 'limit' on the development of the economy: there are no reserves of non-wage and salary earners to be drawn into the labour force. The labour force, which is highly unionized, has the power to resist attempts to reduce its living standards — the wages and salaries which are part of industry's costs.

There is an increasing concentration and centralization of the economy, where large firms have come to dominate the market and the production of specific goods and services. There is also an increasing interdependence and centralization of all the different branches of the national economy.

Industries are dependent for some of their finance on banks and other lending institutions, and often on the state. All this has effects at the urban level, where the size of manufacturing plants and office establishments is increasing, and an individual city becomes more and more dependent on national institutions — on large firms to provide employment and on finance companies and property companies to finance urban development. The final section shows briefly some of the other ways in which the national economy affects urban areas.

2 The symptoms

Britain in the world economy: the decline

Predicting decline or catastrophe for the UK has become a major industry in recent years. Nearly everyone must be aware by now that the country is, and has been for some time, suffering from an 'economic malaise'. From official sources (Organisation of Ecomomic Cooperation and Development 1974) academic institutions (National Institute of Economic and Social Research 1974; Cambridge Political Economy Group 1974), pressure groups (Hudson Institute 1974) or individual authors (Glyn and Sutcliffe 1972) the message is basically the same: Britain is lagging behind her 'foreign competitors' and is set fair to fall away even further.

The evidence for this virtually unanimous opinion about Britain has been complicated by recent crises in the economy of the Western world as a whole; world inflation, currency speculation, energy shortages, and the like, have all served to increase the feeling of mounting instability, if not impending doom for the UK. This is notwithstanding the existence of 'Micawberish' ('something will turn up') notions (Cambridge Political Economy Group 1974)[1] which rely on bonanzas such as North Sea oil to save the UK.

Now, while it is not possible to examine here in detail the overall validity of these assessments, nor the arguments underlying them, any study of planning at national, regional or local level must begin by at least identifying the main features of the national economy. Indeed, there are some more or less unique aspects to the British economy, and these are the very things which affect national and local planning most directly. While many authors on urban development and planning point out that the British town planning system is one of the most highly developed of all the industrial nations (e.g. Hall 1974; Cherry 1974a and 1974b; Cullingworth 1972), they generally fail to say why in any convincing way. They do not relate town planning to the wider issue of state planning in the UK economy.

Urban problems, and the attempts by planning to solve or regulate them, can be explained in terms of the development of the British economy. It is suggested that what is unique about the British economy is that it is the most 'mature' economy in the Western world. It is the most developed economy,

1 The Cambridge Group note that the very characteristics of the mature economy which are discussed in this section will deny the possibility of any such resources being fully utilized by the UK economy.

indeed it is overdeveloped — in the sense outlined below. An extensive urban planning system is partly the result of this and is in fact an essential component of this unique economy.

'Maturity' is sometimes used to characterize *all* developed industrial market economies (e.g. Keynes 1936) or even any industrial economy (market or planned) at a certain stage of development. According to Rostow (1971) the UK economy was 'mature' in 1850, but here we shall argue that the UK is more mature than other market economies. But it is not suggested that other countries will necessarily pass through this particular stage of over-maturity.

Let us be quite clear at the outset that what is of interest is not the immediate 'crisis' that the country appears to be facing. Short-term obsessions with the balance of payments, the 'defence of the pound', wage restraint or other crises have dominated the headlines for twenty-five years, and are symptoms of longer-term underlying causes. Certainly the crises do draw attention to these causes but they are often used as propaganda for far-out solutions or 'predictions', usually of impending doom, rather than to shed light on fundamental issues. More recently the 'failure to invest in our future' has become a more insistent and more constructive theme in popular debate.[1]

Short-term crises, and the action governments take to remedy them, can have and, it will be argued later, have had a cumulative effect over periods of ten to twenty years and more. These cumulative effects eventually do indeed change the structure of the national economy. But they do not change what one might call the fundamental character of the system, although they may point towards it. The UK remains a market-dominated and essentially non-planned economy, notwithstanding the ever-increasing role of government. Nevertheless the balance of forces in the economy is being progressively shifted, especially in respect of the terms on which the private sector, the wage- and salary-earning labour force and the state interrelate, compete and contradict one another (see Chapter 2). It will be argued that national economic planning in the sense of a medium- or long-term strategy is almost entirely absent in the UK. But, nevertheless, as the economy develops, planning (or at least 'management') by governments, even in a 'mixed' economy, becomes more and more necessary, and also more inevitable.

Planning is required not only by governments but by the private sector itself. Peculiarly, perhaps, longer-term planning in the UK has been most institutionalized at the local level rather than at regional or national level. All major local authorities in Britain are now required to produce written 'structure plans' for their areas.

Structure plans are statements of strategy for the development of an area, for up to ten to twenty years ahead, prepared by a local authority in relation to its own very limited direct and indirect powers to influence this

1 'Doom' merchants have proliferated in the continuing climate of crisis. They tend to predict an 'end' to everything — down to the last living lichen, compounding despondency on despondency (Meadows 1972). There are other schools which suggest that somehow we can reject or confront contemporary trends merely as a matter of choice, for instance by choosing small-scale enterprise (Schumacher 1974), or rejecting 'growth' (Mishan 1969).

development. The plans are concerned mainly with the land-use aspects, of economic and social development — spatial arrangements — and as such they are submitted to the 'land-use planning ministry', currently the Department of the Environment. Although limited and ineffective in many ways these plans nevertheless do provide a pointer to the increasingly pervasive role planning will be called to play in future at all levels of the economy.

Standard of living: falling behind

How to describe the UK economy? It is easiest to begin with the most obvious measures, and then proceed to those which point more towards the underlying causes.

The first most visible measure, the one most directly felt by the population, is the standard of living. Nearly all observers agree that this has been declining in the UK in relation to other Western industrial countries over a very long period, and that this decline has increased in the last fifteen years. Now the most direct indicator of standard of living is usually taken as the 'real disposable income' per head of population. This is meant to show how many goods and services can be acquired by the average citizen. There are severe difficulties in measuring this. Habits, tastes and customs vary from country to country, so different 'baskets of goods' tend to be demanded. Again, and partly because of this, the prices of the same items will vary tremendously — especially for items of food and clothing. It is no use comparing the price and consumption of wine in France, where it is part of the staple diet, with that in the UK, where it is still a relatively rare item.

Any measure of real income must try to take account of these differences. However, the second and even greater difficulty is that some goods and services are provided by the state free of charge, or below cost. Education and social services are sometimes referred to as the 'social wage'. They are not obtained by spending personal income but must be regarded as one component contributing to the overall standard of living (see chapter 2). It is the level of the social wage which also makes comparison of standard of living with the centrally planned economies so difficult, since this is so much more important in these countries. In other countries, many incomes are not paid in money (e.g. peasants) and are difficult to measure.

The final reservation on a blanket measure of standard of living is, of course, the very wide variations within a particular country. In two countries with similar real income per head most of the population of the first might be either very rich or very poor, whereas most people in the second might be of average means.

With all these reservations in mind, the standard of living in the UK shows a progressive decline in relation to that of other comparable countires. The Hudson Report (Hudson Institute 1974) shows that real wages of the working population have grown less than in nearly all other comparable countries (except the USA) between 1960 and 1973. To convert this into a real measure of standard of living we must also note that the labour force has increased at different rates in different countries. This is largely on account of the movement of labour from agriculture in such countries as Italy and France into the industrial sector, and is also confused by increases in female participation in the labour force and by international migration. The Hudson report here seems confused and over-anxious to show a decline in the UK standard of living, and does not clarify these factors; nor does it show how the *growth* in real wages relates to the *existing level* of wages. One would expect Italy, Spain and France to have larger growth of wages, if only by virtue of the transfer of labour from low-wage agriculture to high-wage industries. But the fact that real wages and incomes have declined relative to other countries does not mean that the *share* of overall national income going to wages (as opposed to profit) has not increased. Indeed this is the principal argument put forward by Glyn and Sutcliffe (1972) as one of the reasons for the crisis in the private sector of the UK economy.

Other favourite measures of the standard of living are what the Hudson report terms 'quality of life' indicators such as the numbers of television sets, cars, phones, fridges, etc. Most of these do not point to any startling conclusions, with the UK position varying from above to below average. However, according to what might be termed the more 'social' of these indicators — the number of doctors and the number of houses constructed — the UK comes in a poor third-from-last on the OECD list.

To summarize, the standard of living in the UK is declining relative to other Western countires, and the rate and scale of the decline is progressive and persistent but not at present catastrophic. A recent estimate based on a United Nations study of consumption and prices suggest that the 1975 UK standard of living was still more or less equal to the EEC average. The important qualifications to this is that significant sections of the population have suffered a much larger decline (see below). It is also worth noting the gross inequality in personal wealth and income (see section 5).

Growth: the lagging economy

The next most visible manifestation of the relative position and performance of the UK is much easier to measure, and there is correspondingly less debate about what it means. The total level of economic activity in a country can be measured as the gross national product (GNP). The OECD (1974) data show that the UK growth of GNP has averaged 2.8 per cent from 1965 to 1973, compared with over twice this figure for France, Italy, Spain and West Germany. Figure 1 shows that the UK is well and truly at the bottom of the growth league. At the time of writing the new 'South American' levels of inflation of 20-30 per cent have been evident for a year or so, and indeed the rate of inflation in the UK is higher than in comparable countires. This is being seen by some as another indicator of the decline of the UK. This country currently has the highest rate of inflation of all developed Western countries, and therefore GNP must be calculated on the basis of a fixed set of prices with allowance made for inflation, so that the real growth in output can be distinguished from that due solely to the increase in prices.

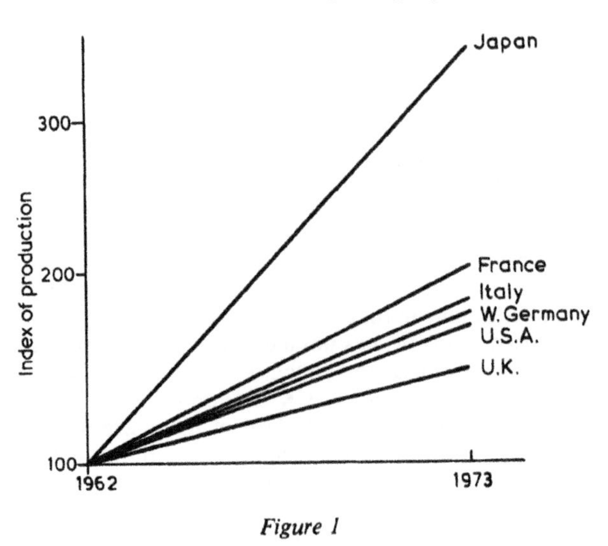

Figure 1

Even allowing for inflation, the UK economy declined from 120 per cent the size of that of France in 1963 to about 80 per cent in 1972; from 100 per cent of the West German economy to 60 per cent in 1972. The growth of GNP per head of population can be taken as an alternative, but more generalized and not so direct measure of the total standard of living. The OECD figures for 1972 show the UK per capita GNP as one of the lowest of the comparable countries. Clearly the interpretation of this measure depends very much on how the GNP is distributed between different sectors (for instance, between industry, agriculture, the government and the population at large).

Over periods of ten and fifteen years, then, the cumulative effects of different growth rates make for very major changes indeed in the total size of national economies. Looked at in this way, Britain is becoming very much an offshore island and one that is shrinking rapidly in relation to Western Europe as a whole.

Before arriving at an effective description of the national economy, we need to look a little more deeply at the reasons that underly the low growth of the UK. It is difficult to explore this in any depth without running into controversies about the causes of economic growth. On the other hand we do require some prior idea of how the economy actually 'ticks', and this is described in the next section.

3 The market economy

Of course not all the economic factors affecting urban development in Britain are unique to this economy; many of them are common to all industrial economies. Not all of Britain's recent economic history was inevitable.

Hobsbawm (1969) argues that the 'first industrial country' could have renewed its industries much more than in fact it did. We shall briefly note aspects of the economy which most affect urban development, but the actual role played by urban areas in the national economy will be spelled out later.

Although it is the *recent* decline in the UK economy in comparison with other similar countries that is of most immediate interest, there are many conventional works of economic history which point out that there has been a long-term decline in the competitive position of the UK beginning in the later nineteenth century. These works look in vain for overall explanations. They examine the relationship between investment and profitability at home and overseas — expecting the one to decline as the other increases. They may seek in the lower wages in the UK, as compared with the USA, a reduced incentive to invest. But usually they find so many counter-examples (high profits at home coupled with low investment, or low wage rates in Germany associated with high capital investment) that the conclusions are generally of the 'everything is related to everything else' type. The message tends to be that history 'defies a single unitary explanation, but explanations in their diversity relate to the whole dynamics of industrial growth rates. They cannot be explained just in simple terms of economic hypotheses such as wage rates, shifting terms of trade, or deteriorating national resources' (Mathias 1969). Deane and Cole (1969) are also reluctant to give an overall explanation of the decline of the UK.

In contrast, Glyn and Sutcliffe (1972) boldly try to tackle the problem head on. They pick out the recurring theme of international competitiveness. They point out that the reason for decline is low investment, which in turn is caused by low profitability. They isolate the separation of and contradictions between the bankers and financial interests on the one hand, and industry on the other — with the latter often preferring free trade when industry needed protection, or a high exchange rate when industry required a lower rate. In Germany and the USA finance capital traditionally provided more of the investment for industry, which in Britain had to rely on its own resources. They cite Deane and Cole's findings of a long-term increase in the share of national income going to wages and salaries. Finally, they seek to demonstrate a quite catastrophic fall in profits in the UK in the late 1960s. This fall they attributed to a joint squeeze: a strongly unionized wage push on the one side and foreign competition (which stops UK prices from going up to the point where profits could be maintained over the wage increases) on the other. Whether this analysis is correct depends very much on their method of calculating profits, which may not take account of real (hidden) profits, especially in big companies. They may also be overstressing the short-term profit squeeze in the late sixties. King (1975) shows that their results are probably distorted by the fact that 1971 was an atypical year and that the share of profits in the national income rose in 1972 and 1973. Indeed, King shows that taxes on companies have been reduced over the years and, once this is taken into account, the share of profits seems hardly to have declined at all. Nevertheless their book does serve to illustrate the key point about the modern UK economy — namely that it is a *market economy.* In spite of the enormous growth in the public sector during this century, the motor or driving force — the mechanism that makes the whole thing tick — is the market mechanism. It is competition, profits and profitability which can alone keep the 'mixed' economy going.

It is important to understand that in the 'mixed' economy the whole operation, both public and private, is circumscribed by the market. The public sector operates *within* the market economy. Holland (1975b) among others argues that the economy may be 'mixed', but it is unfairly mixed in that the private sector operates in profitable areas — producing goods and services which have a market — whereas the public sector provides the underpinning of basic utilities and social provision (see Chapter 2). Whether or not there can be a really significant extension of public ownership into the profitable parts of industry without fundamentally changing the whole economic system will be briefly examined in Chapter 2, but at least at this stage we should be clear that the public sector obeys market relationships — paying wages that are in some sense competitive with market wage rates, buying products at market prices, borrowing and lending at market rates of interest. In these many senses both sectors obey the same rules — in no sense does the public sector operate as a *deus ex mac-ina*, outside the market system.

But can we actually 'prove' that the economy is a market system, beyond saying that the state's operations are circumscribed by the market in some sense? Well, the fact that profits have to be made if a private-sector firm is to stay in business is a matter of definition. Of course, individual firms can and do make losses for several years; the value of their assets might carry them through, and those who invest in the firm may continue to do so in anticipation of better profits in the longer term. But this process can only go on for so long, and sooner or later if profits are not made the situation becomes untenable. As the recent history of industrial collapses shows, this can affect even the largest firms.

A second pointer to the market basis of the economy is that much of the growth of the state is more apparent than real (see also Chapter 2). Although the money which passes through the hands of the state has increased considerably, much of this does not actually represent real resources (i.e. goods and services actually bought or produced by the state): much of it is merely a reallocation of money from one part of the private sector to another. The state only takes up about one-third of resources in the economy, as measured in the national accounts. This again points to the dominance of the private sector. We shall further see in Chapter 2 that the state operates essentially as a complement to the private market.

Chapters 2 and 5 describe the 'ideal' market process developed in neoclassical economics, to deal with a perfect market when all firms are very small entities. (Now, of course, firms can be very large indeed.) But it is worth at this point briefly sketching out in broad terms what makes the modern market economy actually tick.

Figure 2 shows the process as it affects an individual firm. It is a cyclic process which begins with the production of goods and services for sale.

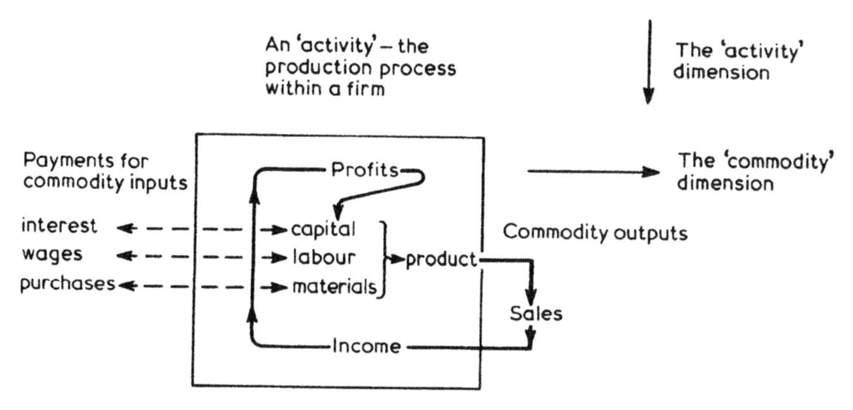

Figure 2 The development process of a single firm

There must be a 'demand' for the product in the first place; it must therefore have some 'use' in society. The *production stage* means employing resources (or 'factors') — especially labour (workers) and machines and buildings (physical capital) — together with input materials and using them and combining them to produce the product. These resources have to be paid for, and the price to be paid is determined in the long run by the costs of producing them in the economy as a whole — this is discussed later. It will be argued at various times in this book that understanding the *process of production* is crucial to understanding the economy as a whole and urban development in particular. (Not the least of reasons for this is that more of the economy now lies 'inside' firms rather than 'outside' them, since firms are now so large.)

Once the product is produced it is sold, and the resulting income, after the costs of production, wages and materials have been met, produces a profit. This profit can then be used to invest in new plant and machines to make the production process more efficient or to produce new products. This capital investment process determines the way the firm (and ultimately the whole economy) develops over a time period.

When all the firms in the economy are considered together, we arrive at a total flow of national output and income, and this is discussed in Chapter 2. It is these firms who, taken together, ultimately determine how the economy develops — through their combined decisions about how much they will produce, how many people they employ, what buildings and machines they construct, and where production takes place. But how do the firms interact with each other? The external input-output linkages of the firm shown in the figure provide the clues. On the right-hand or output side there will usually be other firms producing the same or similar products. The 'sales' stage is the point of competition with other firms. There is much theory which explains how this competition works, in different conditions (e.g. whether there are many firms or only a few competing; see Chapter 5). Firms in general will try to gain as much income as possible in relation to their costs of production — payments made for labour and capital. Depending on market conditions, they will do this by raising or lowering prices or by obtaining a greater share of the market.

But whatever conditions they are operating in, and whether they are giant corporations making hundreds of products in subsidiary firms, branch plants, subplants, divisions, factories and research establishments throughout the world, or whether they are a five-person business operating in a single town or city, the ultimate competitiveness depends on the *cost of production*. Underlying all competition, then, will be the attempt to reduce production costs. Firms thus always have an immediate interest in reducing or limiting wage payments, but over a longer period they can reduce costs by investing in more and more capital equipment: this is the way the production process itself is changed over time. The combination of output products can be changed, as can the mix of input materials and the way they are organized and brought together in the production process.

Economists talk about economies (or 'diseconomies') of scale, where the cost of production might be decreased by producing large quantities in a mass production process. This may allow the firm to benefit from increased 'division of labour' within it, the different stages of production being separated one from the other and each one simplified and made a separate operation possibly carried out in a different place.

But if this is the mechanics of the development process of a single firm, we still do not have a picture of the development of the whole economy. Again the external input—output links of the firm provide a clue. We could look along the commodity output dimension. The product might be used by households or by firms to make other products; we would be interested in the total amount of the product produced, what it was used for, and by whom — other industries, households or the government. This would give some idea of how dependent or independent the different parts of the economy were, with respect to this product. We would also want to know how the production process for the 'industry' as a whole is organized, how many firms are making the product, and whether a single firm was dominant. But above all we would need to know the *total cost* of producing the product in the economy as a whole, that is, how much of society's resources of labour — skill and effort — and capital were devoted to this activity or group of activities.

The development of the production process within the single firm is also reflected in the economy at large. New firms and new sectors might arise, each concentrating on the specialized manufacture of some special service or product which was previously made on an *ad hoc* basis within several different firms as a part of one stage in the manufacture of several other products. The product may then become cheaper, fewer total resources (labour and capital) being needed to produce it. The economy as a whole is then more efficient. Thus the production of more and more goods and services becomes socialized: they are produced for the whole economy and for *the market* as a whole. This is an 'economy of scale' and a division of labour which comes about in the economy as a whole but which is 'external' to any given firm.

It is the effort to economize on costs of production which is the driving force — the 'motor-generator' behind development of the economy, and which results in the growth of more and more specialized activities. But all these activities become more and more dependent on each other. They are tightly bound together through the single market for each good and service. This socialization of production is thus advancing through the extension of the market principle into more and more of the corners of the economy. It will be argued in Chapter 2 that the key aggregate relationships in the economy are those relating to the three estates — the private sector (firms), the state and the wage-and salary-earning households.

Figure 16 shows the relationship between these three activities, expressed in terms of the production, consumption and circulation of two aggregate commodities — labour (wages) and profit (see Chapter 2). The view of the economy as a production system involving production activities which produce a set of output commodities from a given set of input commodities is a concept known as 'activity analysis', and this again is discussed in Chapter 2.

A key aspect of the process of economic development is the circulation of the 'commodity' capital. More and more of the capital for individual firms (even large ones) is being provided by outside agencies, especially by specialist 'finance' companies,[1] which provide the capital for new equipment and are paid interest in return. So even though the very largest multinationals still raise most of their own capital internally through the reinvestment of profits, there is increasingly a 'single market' for capital (or 'finance') which now circulates between firms and promotes the restructuring process and redevelopment of the economy, since this capital does not have to be reinvested in the sector of the economy in which it was generated, but is much freer to seek new, more profitable outlets. The increasing importance of this 'finance' activity has helped to promote investment in urban development — especially the construction of new commercial developments in city centres.

This development process amounts to a comprehensive and all-embracing system for deciding how the resources in the economy are used. Now it is well known that the state has an increasingly important role in the modern economy, and the state in principle takes decisions on the use of resources which are not (on the surface, at least) necessarily determined by market criteria — by profitability. The question therefore arises as to the relationship between the market system of allocating resources (i.e. 'profits') and the other system — 'planning' by the state. We have argued above that the UK economy is above all still a market economy; but, given that state planning is needed at national level, are we free to choose *any* desired mix for planning and profits in the allocation of a commodity or resource or are there some limits to the encroachment of planning on the private sector? What are

1 See Figure 10 for the overall circulation process in the national economy.

the specific activities, commodities and resources which are important at the urban level — and how does urban planning reflect the general principles governing the relationship between planning and profits?

The private sector does not develop and grow through some comprehensive plan but rather because the various firms it comprises can make enough return from the sale of their products to make it possible and worthwhile for them to continue to produce goods, and to reinvest in plant and machinery. By the very nature of the private sector, firms are in competition with each other. This means that they try to expand the market for their own goods against those of other firms by reducing the costs of production. And the way to do this is continually to reinvest in more efficient machines and plant, often in larger-scale plants, and to increase the amount of output that can be produced by a given worker. The fact that a smaller and smaller number of larger and larger firms are coming to dominate not only the UK economy, but also the world economy, does not decrease the drive to reduce costs and to compete. Competition, under the so-called late 'monopoly capitalism' (Baran and Sweezy 1968) is no less ferocious than under early periods of 'competitive capitalism' when firms were smaller and no single firm had a significant effect on the price of a product.

Competition takes firms other than direct price competition, especially the attempt to expand markets, but underneath this the larger firms are still investing to reduce production costs. In fact there is a stark contrast between this ruthless process of modernization and restructuring within industry, and the overall poor performance of the economy as a whole. Now, as the economy develops the production of goods and services which are actually needed by the population gets overlaid by an increasing superstructure of industrial and commercial activity which by many definitions may not be strictly 'necessary', but which is required to keep the private market economy going. There is a great deal of argument about this,[1] and, of course, it is almost impossible to measure what is 'necessary' and what is not; all we can do here is to note the growth of 'services' generally and financial institutions in particular, both of which have very specific effects on urban development.

It is often said that the UK is stagnating (Shanks 1972), not only economically but in many other ways. The rapid population growth of 'developing' countries, the explosive growth of their cities, mass movements of population, and so on, are clearly absent. So is the less spectacular economic growth of other industrial countries, as we have seen. But this is not the complete story; some things are changing, and these again are largely as a result of the 'overdevelopment' of the economy, and also of the action of major world economic forces on it. These are again often the things that most affect urban development and planning.

When the total size of the economy is more or less static, then an increase in one sector implies a decrease in another. We are suggesting that the market mechanism has permeated to every corner of the mature UK economy so that

1 For one experiment in measuring waste production in a non-growth, overdeveloped market economy, see Baran and Sweezy (1968).

there are no new frontiers left where it might expand. We shall try to argue that in this situation new types of conflicts appear: the fight for profits is maintained, so firms continue to try to reduce costs and to expand markets; but if the total market is static, some firms lose the fight and industry gets restructured — firms get larger, small firms disappear or are absorbed into larger ones. Similarly there is a fight between sectors: profitable, buoyant sectors grow at the expense of the less profitable within the total fixed size of the economy, and this is discussed in the next section.

4 Barriers to development

Growth: extensive and intensive

Adelman (1961) gives a useful account of some of the different theories of economic growth. It is a wide-ranging survey, covering not only the current received 'conventional wisdom' developed as an extension of Keynesian economics, but also the older 'classical' theories of Marx and others, some aspects of which are now increasingly being brought to bear on modern economic problems (Hunt and Schwartz 1972). Indeed much of the critique of the conventional wisdom is based on ground first laid down by these earlier economists. They tried to look beneath the surface of economic phenomena, beneath the monetary transactions and the flows and exchange of goods and money which are the main concerns of modern conventional economics. They were concerned with theories of 'value' and with the interrelationship between social institutions, social classes, political struggles and the development of the economy. Conventional theories stress politically neutral factors such as the rate of 'technical progress' as the main factor determining economic growth. Marx, on the other hand, tried to explore the reasons for changes in technology, and suggested that it comes about by the needs of competing capitalists continually to invest in new plant and machinery to replace labour in a situation where competition is continually increasing and it is harder to maintain the profits on which the system depends to survive.

However, in restricting attention as far as possible to the facts rather than the 'pure' theories, why do growth rates differ? *Why Growth Rates Differ* was the title of a still definitive (but much criticized) comparative study by Denison (1967), although it only covers the period 1955-62. First of all, Denison suggested that total level of growth must be due (apart from short-term irregularities) either (1) to increases in the total means of production (what economists call 'factors'), e.g. the total number of workers, land and capital employed: or (2) to changes in how much can be produced by this total deployment, i.e. changes in *productivity*.

The first type of change is an 'extensive' change, where new resources become available and can be exploited and turned into industrial output without any change in the techniques of production or in the way production is organized in the economy. The second type of change is more traumatic, and requires that better use be made of the resources already in existence. This implies some kind of reorganization of existing arrangements either within organizations or in the balance between different types of production within the economy. The latter change may thus involve some sections of the economy or groups in the population gaining income at the expense of others. It seems that there are now so many 'barriers' to economic growth in the UK that nearly all change is 'intensive' with all that this means for internal conflicts for scarce resources between sectors and between groups. In other words, better prospects can only come about through some kind of restructuring or reorganizing of the economy. Much reorganizing has taken place in recent years, largely as a result of market forces, but the state is constantly playing a more active role. There is a continuing debate that the underlying structure of the UK can be improved only by conscious reorganization of priorities through direct intervention by the state (see Chapter 2).

Extensive and intensive growth also has implications at the urban level. Consider an isolated city. If it expands its boundaries and if the intensity of the use of land within the city (e.g. the population density) remains the same, then the city has grown *extensively*. If, however, the boundary stays where it is but the population density increases, then we have an *intensive* change — or a change in productivity.

Productivity changes appeared to be much the most important reasons for changes in economic growth. Denison then asked what caused this change in productivity, i.e. in the amount of output per person employed. Not unexpectedly, the easiest and most direct way to change productivity without having to invest in all kinds of new plant and equipment, and invest in new industries, is to take the existing industries and expand those which are already more productive, thus making a 'reallocation of resources'. (This is what one might call a 'first-order' change; changing the techniques of production and developing new types of industry would be a 'second-order' effect.) Now by far and away the largest low-productivity sector in most advanced countries is still agriculture, and Denison found that for nearly all these countries the contraction of agriculture was by far the most important element in reallocating resources to the higher-productivity sector of industry and commerce. This movement of labour out of agriculture also had a compound effect, since the new industrial and 'urban' population spends much of its higher earnings on consumer goods and the industries producing them make profit from increased production runs, more than in proportion to the increase in output ('increasing returns to scale'). Denison's study also showed up this effect; as a mechanism for growth it is stressed by Kaldor (1966).[1] This is our first, highly tentative definition of 'urban' development; the modern, capitalist industrial economy is the 'urban' part of the national economy — a well-known result. Traditional agriculture is therefore the 'non-urban' part of the economy, but this preliminary definition of 'urban' will be severely modified later.

1 Denison's results must be qualified as Peaker (1974) notes: there are many pragmatic judgements built into the analysis and none of the factors is conclusively identified as a *cause*.

The agricultural barrier

Now what is important about the Denison mechanism of growth is that it has much in common with some 'classical' theories of development. The early development of industrial capitalism involved uprooting the old feudal system of production based on agriculture. In feudalism the market for goods was not the general, all-permeating, national and international, integrated system it is now. The population did not exist by selling their labour in order that they could purchase consumer goods, but rather they remained tied to a particular rural village and largely produced their own means of subsistence. Industrialization required the destruction of these feudal arrangements; labour became a commodity to be bought by industrialists and employed at one (urban) place in large-scale industrial processes. The new industrial population also served to create a market for the new goods and services being produced. This process (which never happened in this straightforward simple manner) has been documented in many studies; it happened first, of course, in the UK in the early nineteenth century, and in the USA and Germany much later in the century (Hobsbawm 1969). Now this historical legacy, as Hobsbawm points out, is the basic key to the unique aspects of the UK economy that underlies such comparisons as those documented by OECD reports, the Hudson Institute and others, but which is not brought out in those publications.

Are we saying that in other countries industry is still in a process of early development, whereas in Britain this has long since ended? Certainly, if we look at the structure of these economies, and divide the economy into six sectors (agriculture, manufacture, construction, transport, commerce and services), it is clear that the UK has much the smallest proportion of agricultural workers, although this is low in the USA too. The UK has less agricultural land available and must import food, so even if agriculture was equally developed in all countries the UK would still tend to have a lower proportion employed. Second, and more important, agriculture in all these countries is penetrated by the industrial sector itself. Denison points out that this has gone furthest in the UK where there is more capital equipment per person employed than in other comparable countries. But are we going to exclude modern agriculture from our formal definition of the urban part of the economy? It may be industrialized and capitalist but it is not urban. Certainly all other countries have more peasants, unpaid workers, small shopkeepers and artisans than the UK — and countries like Italy with large agricultural labour forces have the fewest workers who are paid employees. Peaker points out that neither the Denison nor Beckerman (Beckerman *et al.* 1965) studies proves conclusively that there is a simple link between economic growth and large agricultural labour forces. Decline in agriculture may be a *symptom* of high growth, rather than a necessary precondition. There may be other low-productivity sectors we could replace by new industries. (But such replacements would be 'second-order' effects, having a much less immediate and dramatic effect — being a transfer within an already industrialized system — than the incoporation of new, pre-industrial population into the industrial system.)

Other barriers

We have now arrived at the beginnings of a more precise characterization of the UK economy. In their different ways Hobsbawm (1969), Keynes (1936), Morishima and Murata (1972) and other authors agree that the UK economy is the most developed capitalist economy. Morishima and Murata consider that the most immediate measure (that almost 95 per cent of the labour force is in paid employment, compared with France, 75 per cent, and West Germany, 80 per cent) shows that the UK remains the most developed capitalist country, just as it was in Marx's time 100 years ago. Hobsbawm (p. 317) notes that it is the most 'urbanized, industrialized and proletarianized state in Europe, with a relatively simple, two-class social system, and an unusually important role for the industrial working class in politics'. We now look in more detail at some aspects of this mature economy which most affect urban development and planning.

We tried to show above how the market economy develops — and specifically how the whole national economy tends to become one single market and how it seeks to draw into the market the production and consumption of commodities which were previously outside it. The fact that so few people are still outside the labour market, and that there are few immigrants to expand both the labour force and the market for goods, brings the system up against a barrier. But the economy is also faced with a barrier due to the fact that workers are organized and combined into trade unions. They can apply leverage to maintain their living standards, and can therefore stop the costs of industry's inputs being forced down.

Other barriers are fairly well recognized and include the lack of resources, especially agricultural land. (There is a prospect that the energy constraint might ease off; the UK is likely to be the only West European country self-sufficient in oil.) Overseas competition — as manifested in the balance of payments — forms another barrier or constraint, already mentioned in connection with the Glyn and Sutcliffe profits squeeze.

We shall look more closely at some aspects of overseas competition below — especially the multinational companies — and the general idea of barriers or constraints will assume even greater significance in Chapter 2, in the discussion of the state and possible limits or barriers to its growth. However, the whole highly constrained situation of the UK economy points to the need for more restructuring, rather than more extensive growth.

5 Some results

Investment and growth

This section tries to show how the existence of barriers to development, combined with the mechanism of the market economy outlined above, explains some of the recent changes in the structure of the economy. These changes are a response of the economy, an attempt to reorganize, to become more 'competitive' and to reduce costs; but this attempt has not been entirely successful. It has not solved any of the underlying problems of the economy, and in many ways it has made them worse.

Both radical and conventional theories of economic growth stress the importance of investment in new plant, factories and equipment. This new capital has to be produced out of industrial profits or from other surpluses over the production needed for current consumption. As Glyn and Sutcliffe have pointed out, manufacturers in the UK are no longer as able to finance new investment out of their own profits as they used to be. Capital funds generated internally by UK firms fell from around 70 per cent in the mid 1960s to around 50 per cent in the early 1970s. Profits are not large enough, and modern production processes often require huge plants costing massive sums. This has meant that the stock market, banks, insurance companies and other financial institutions or the state have had to lend a proportion of the new capital required — something they have traditionally been very reluctant to do. In Japan, Sweden and Germany banks have (in cooperation with the state) a stronger tradition of lending for industrial investment than in the UK. However, financial institutions are concerned with the return on their money first and foremost, not with production, and are as likely to prefer lending to property companies or to invest overseas as to help to provide the funds necessary to develop an up-to-date car industry which now requires a comprehensive plan involving massive investments over several years.

As noted above, overall explanations of the long-term failure of UK industry to invest are hard to come by, but everyone agrees that relative profitability, wage rates, the structure of finance capital and the state (taken in comparison with similar factors in competing countries) are all relevant factors. According to Deane and Cole (1969), investment rose to a peak around 1870, declined towards the 1900s and climbed again to exceed its earlier peak before the first World War. Between the wars capital formation hit a new low in relation to national income, equivalent to its level in the first quarter of the nineteenth century. After the second World War there was a revival in investment in absolute terms, especially in the 1950s, even if more recently it has been decreasing again both absolutely and relative to competing countries.

A concrete statistical proof of the link between high investment and high growth is hard to come by, but there is no doubt that Britain's position at the bottom of the OECD growth league is matched by an equally abysmal position in the investment league. True, the proportion of gross domestic product devoted to new capital formation rose from 20.3 per cent to 21.9 per cent between 1964 and 1974, but only the USA had a lower rate of investment than the UK.

This is one reason why the state, although always involved in industry to some extent, has increased its intervention recently (almost 50 per cent of new capital passes through its hands), supporting lame ducks, providing grants and loans, nationalizing whole sectors or, most recently, actually beginning to play an active part in industry through the National Enterprise Board. The role of the state is dealt with more fully in Chapter 2.

Urban areas will be crucially affected by the uncertainties and fluctuations in industrial and service activity resulting from the basic reliance on the market demand to sustain the production of goods and services and thence to provide new industrial capital (see pages 105-7 for some examples). In addition to short-term uncertainties, the future of many urban areas will virtually depend on particular industries, on their position in the productivity league and on their fate in the competitive race.

Levinson (1971) points out that far and away the most important source of new capital is still generated from retained profits within industrial firms, representing roughly 70 per cent of new capital. In his argument, which cannot be pursued here, the declared 'profit' gives very little indication of the cash actually available within the firm for investment. He suggests that the drive to reduce labour costs and invest in new plant in the largest and most modern firms overrides many conventional barriers such as the short-term reductions in demand for the product (e.g. fall in demand for cars) and is a major factor causing inflation. Firms raise prices to generate more capital.

One thing that stands out clearly from all the attempts to explain low British investment the legacy of the empire, the trading interests and the financial sectors which are peculiarly important in the UK. It seems fairly natural that investment funds should go abroad, through the City's multitude of interests and connections overseas where profits may be larger. Hence Glyn and Sutcliffe's (1972) concentration on the different and often conflicting interests of industrialists and financiers — and the resulting large British assets overseas which, if realized, the Cambridge Political Economy Group argue, could provide one escape from the short-term straitjacket which restricts British economic planning.

Competition, concentration and the multinational companies

It is part of received wisdom that Marx's prediction of ever-increasing concentration of wealth in fewer and fewer hands together with the increasing misery of the industrial workers has not been fulfilled. Recent arguments have suggested from different points of view (Glyn and Sutcliffe 1972; Baran and Sweezy 1968; Mattick 1969) that this process is indeed taking place but on a world scale with many of its effects masked and cushioned for the time being, at least in the industrial world. Whether or not Glyn and Sutcliffe are correct in detecting a fall in profits, there is no doubt that a major restructuring of the economy *is* taking place — the concentration of production in large firms, the increasing importance of finance and the increasing role of the state are all consistent with increasingly severe competition, whether or not it is driven by an overall decline in profit.

In 1950 the largest 100 companies accounted for 20 per cent of UK national output; in 1970 this had risen to 50 per cent (Holland 1975b). An official report (Bolton Committee 1971) points out that at the present rate (another 'doom' prediction) there would be few small businesses left in the UK by the end of the century. Although there is some evidence that the mature economy of the UK is concentrated into fewer firms than in the USA (e.g. Florence 1972; Shepherd 1961), this is not conclusive, and it is even less so in comparison with Europe. These measures depend very much on how one defines a 'sector' of the economy, but Peaker's (1974) comment that concentration is not increasing very rapidly, and at least not enough to influence the rate of growth, seems somewhat complacent in the light of Holland's evidence cited above.

Even in the 1950s, in cement, petrol refining and explosives the largest three firms controlled over 90 per cent of the output (Bain 1966). In electric lamps, distilled liquors, aircraft plastics and flour the three largest firms produced over 50 per cent of the output. The takeover boom is a startling illustration of how, while the economy may appear static and sluggish in growth terms, the forces of competition are generating very large-scale changes in structure: 10 per cent of manufacturing assets changed hands in 1967-8 (Donaldson 1974). Concentration of production increased especially in food, vehicles, textiles, leather and clothing. By 1968 much broader industrial groups were dominated by a few large firms, all food and drink, all chemicals, all metal production, all electrical engineering and all vehicles each having more than 70 per cent of output produced by the five largest firms in their respective industries (George 1975). In fact, even in 1968, the *average* proportion of output produced by the top five firms in each of the thirteen main manufacturing groups was 70 per cent.

Some economists divide the private sector into two distinct parts — the large-firm sector (sometimes called the 'monopoly' sector) and the small-firm sector (sometimes called the 'competitive' sector). There are still thousands of small firms — even a single city like Liverpool has upwards of 20,000 — but they are producing a smaller and smaller proportion of total output (see Figure 3).

A new feature of the mature economy is its dependence on multinational companies. These firms control a growing proportion of UK production, and much of this is due to British multinationals, like ICI, Shell, Bowater and

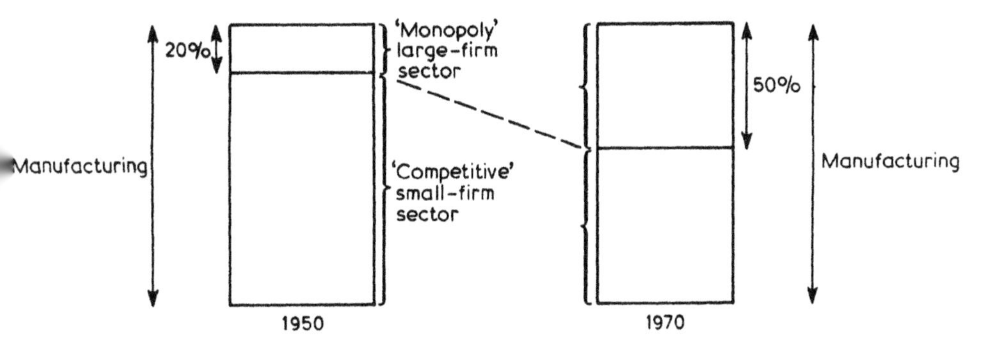

Figure 3 The growth of big firms 1950-1970

Reed.[1] US multinationals, however, tend to dominate the newer and more productive sectors, and Dunning (1975) estimates that between one-fifth and one-quarter of British industry will be owned by the USA by 1981. It has been argued that these companies can effectively circumvent and undermine such national planning as exists by artificially buying and selling to themselves across national boundaries and by moving profits and capital around from one country to another.

It is clear that key decisions affecting particular factories and particular employment centres in particular urban areas are taken by large organizations in places far removed from the local effect. We are thus presented with a clear picture of the dependence not only of urban areas but also of national governments on these large economic units, with the decision making receding even more rapidly in recent years from the local company to the multinational enterprise.

Levinson (1971), in a rather knockabout and sensational, but very relevant presentation, states that between 1000 and 2000 firms produce 75 per cent of world output, 50 firms accounting for 75 per cent of UK exports.[2] General Motors produces more than Belgium, Argentina, Switzerland or South Africa and is a quarter the size of the UK; Ford is bigger than Denmark. Some projections show that by 1985 perhaps 300 firms will control 80 per cent of the West's productive assets. Larger firms can more easily reduce costs (see, in the USA, National Industrial Conference Board 1969); they can produce more goods more cheaply and thereby make more profit than smaller firms (who have to sell at the same price to compete); thus they have more capital to invest, to reduce costs still further, produce more and grow even larger.

There is also a growing gap between different types of industry. In the USA and also probably the UK, food processing, chemicals, petrol and plastics produced nearly four times more 'value' per employee than the least productive industries like textiles. This puts certain industries, and therefore certain factories in particular urban areas, at a severe competitive disadvantage, with a major gap opening up between the group of ultra-modern, capital-intensive sectors, and the sluggish, slower-moving labour-intensive industries.

Rowthorn and Hymes (1971) in a more conventionally scholarly work and also Levinson (1971) show how the big firms, although each unambiguously based in and owned in *one* country, are interpenetrating each other's national economies to create a new, genuinely global economy. The USA, being 'too wise to try to govern the world — we shall merely own it', could own 35 per cent of the West's economy in 1973 (Levinson, p. 71). Sixty-five of the top 400 UK companies are controlled from abroad, and the UK private-sector overseas assets amounted to £61,000 million in 1974 (Bank of England 1974), with an excess over liabilities of £6700 million.

1 The number of British firms having at least six foreign subsidiaries has risen from 30 to 50 of the top 100 UK manufacturers.

2 Six Japanese firms produce 50 per cent of the total output.

Decline of manufacturing?

The process of competition and economic development outlined above provides some explanation of several recent trends. Between 1966 and 1974 manufacturing employment in the UK declined from 8.6 million (37 per cent of total employment) to 7.9 million (35 per cent) while at the same time the number of those employed in financial and professional services rose by nearly a million. The latter was the only major sector to increase; there were declines in the distributive trades (including retailing), construction, gas, electricity, water, transport and mining. Catering and hotels increased as did central and local government employment. Changes in employment cannot be related directly to the importance of the industry in the economy since, as we have seen above, the opportunities for shedding labour and increasing output vary widely between industries and there is evidence that it is increasing. Over the same period (1966-74) manufacturing output declined from 33.2 per cent of the total national product to 28.4 per cent. It is clear that in Britain manufacturing is more important in employment terms than it is in output terms. In comparable countries, the reverse is true. Nevertheless manufacturing output has a continued ability to replace labour with machines, and over the longer period 1950-73 Prest and Coppock (1974) note a significant rise in the proportion of GNP due to manufacture, with a decline in the primary industries of agriculture and mining.

We have noted that the relative backwardness of manufacturing is often explained by the fact that the financial sector is so important in the UK. The 'tertiary' sector as a whole (which includes banking, insurance, property companies, professions and personal services) is larger in employment and output terms than it is in comparable countries.

Looking at individual manufacturing industries, the fastest growing in output terms are the familiar high-profit-large-firm-dominated sectors we have already come across, with chemicals and light engineering showing particularly high growth. But the theme which appears most insistent, and most important for urban development, is the 'restructuring' theme. We shall see in Chapter 3 that the 'labour' effect is by far the most important link a large firm or factory has with its own urban area. So as far as an individual city is concerned it does not matter if a local petrochemicals complex produces a quarter of the nation's output if it only employs a few hundred personnel. Consequently, the shedding of labour from manufacturing is crucial to urban development whether or not it occurs in an industry in absolute and relative decline or in a capital-intensive, leading multinational firm which is replacing people with machines. We have here one crude key to the difference between the *national* level on the one hand, where a major company is judged mainly in terms of its total output, and the *urban* level where its employment of labour is the key measure. Some of the top ten leading national companies (in output terms) are a long way down the employment league, and employ very few workers in total, highlighting Levinson's case that there are massive differences in capital employed between different industries.

Size of plant and specialization

Another aspect arising from competition and the concentration of economic power as industries strive to reduce production costs is the economies made in the cost of production by large-scale production units, by the division of labour and by the specialization of industrial plants producing different components of a single complex product. Many of the largest firms may have small plants (such as research centres, laboratories, and the like), but they also have large and increasingly larger, single factories or service establishments — office blocks, superstores, etc. Bain (1966) indicated that while on average the UK had smaller factories than the USA (by an average 20 per cent), they were on average twice as large as those in Europe. It may be, of course, that the very largest plants in the most important or most productive and most competitive industries are to be found abroad. However, there is some indication in the figures that the economy is up against another barrier here — European industry can still benefit more from building large plants in the future than can the UK — but the evidence here is rather shaky.

The proportion of large plants has been growing: 15 per cent of the workforce were employed in plants with over 1500 workers in 1935; this grew to nearly 30 per cent in 1968. Nearly 55 per cent of total manufacturing output was then produced in plants employing more than 500. But this is not as dramatic as the growth in big firms, so there has obviously been a considerable growth in multi-plant firms — even if part of the purpose of mergers between firms is to rationalize production and close smaller, inefficient plants. This not unexpectedly leads to the multi-urban firm with its multiplicity of plants situated in different cities.

Again, the implications for urban development are fairly clear. Many people are likely to be working in a relatively few large plants in any given urban area; this may account for up to half the total local employment, making relatively specialized products for the national or international markets. On the other hand, there will still also be a large number of tiny firms employing a significant proportion of people. There will be a high degree of specialization, giving very large differences between different urban areas, with extreme cases, like Teesside, offering almost nothing but chemicals, and engineering and steel (North 1975). This specialization, together with the decline in manufacturing, produces further massive differences between different regions and between different urban areas, between areas specializing in old, declining industries (the North), those specializing in more modern manufacturing (West Midlands) and those specializing in services and finance (the South-East). Competition, concentration and specialization will create large differences between areas, reflecting the competition between industries. Some areas will be dependent on others both through the flows of commodities between areas specializing in different products, and also because of the dependence of some industries (e.g. manufacturing) on others (the financial sector) — the North is dependent on the South-East in this sense.

Specialization within regions and cities is not new; it has been recognized since the earliest days of town planning. The Barlow Report, the precursor of the existing UK planning system, discussed it in some detail (see Chapter 4). But now that firms and plants are larger and the state intervenes more and more, and with the growth of the tertiary sector, the phenomenon is being re-expressed in new forms. Both regions and cities are becoming more subordinate to the national and international plans of major firms and to national government's own policy response.

The population and inequality

We are understressing the 'social' aspects of the UK economy that so interest many writers in this field (e.g. Simmie 1974, Pahl 1975, Donnison *et al.* 1975, Eversley 1973). Surely we need to say more than that the mature economy has a 'simple class structure'; and that the vast majority of its population are 'proletarianized' — wage and salary earners. There exists a welter of literature on poverty, inequality, underprivilege and social malaise, which will be touched on later.

Suffice it to say at this stage that the distribution of personal income is massively skew: most people receive far less than the average income because a very few receive far more. This is dramatically illustrated in Pen's 'parade of dwarfs'. All the people in a typical Western country are made to march past within one hour. Their heights are proportional to their incomes, and, rather surprisingly, most of the people seem to be shorter than average height — they have less than the average incomes. In fact it is only ten minutes before the end of the hour that people of average height appear. Many people are poorer than average because a few people are very much richer.

The Labour Research Department (1973) and Atkinson (1973) document the many types of wealth, income and educational inequality in Britain. The former estimate that 5 per cent of the population still hold 75 per cent of the wealth of the country. The definition and measurement of wealth is difficult — the first report on income and wealth produced by the Diamond Commission (1975) suggests that the top 5 per cent only owned some 57 per cent of the wealth by 1973. They also report a considerable decrease in inequality. The same report put the share of total income taken by the top 5 per cent as only 17 per cent.

Inequality and poverty is indeed a central issue in urban development, it has been accepted as such across a whole range of central government policy, and there is a great deal of 'poverty' rhetoric. There are educational priority areas, inner-city priority areas, housing action areas, and many other policies which purport to direct help to 'where it is most needed'. But the poverty and inequality rhetoric may actually serve to obscure the real causes of inequality and the real economic mechanisms which underlie these problems. Thus, for instance, the idea of concentrating help where it is most needed carries with it an implied notion of means-testing, a poor-law approach which perpetuate 'two-nation' philosophy whereby those who are capable of providing for themselves are the mainstream of society and those who cannot are 'supported' by the rest of us through the state — with a further implication that public provision is substandard, or at least at a minimum standard, rather than at the 'average' enjoyed by the majority.

A further implication of the 'poverty' approach is that it sees the recipients of public provision as essentially passive, unable to exercise any leverage on society, unable to take their rights rather than being *provided* with them. The poverty view therefore divides social groups from each other. It also perpetuates a false analysis of the way the economy works. The idea is that there is a limited 'cake' available to income earners and that if those on higher incomes forgo their increases there will be more available for low wage earners, or for more government spending on social provision. The fact is that wages and salaries forgone in the big companies go to company profits or to new company investment, not to low wage earners. The further fact is that there is a labour market by which wage increases in one sector can push up wages throughout the economy, especially when there is full employment. So it may be that the size of the cake available to wage and salary earners as a whole increases — as Glyn and Sutcliffe have argued. (This is not automatic however, firms may be able to rise prices to offset such increases.)

The public sector

Public expenditure increases year by year, under both Conservative and Labour governments, and, although this has led some to believe (erroneously) that this has somehow changed the basic nature of the economy away from reliance on the market, it is nevertheless a most significant development — one which does indeed presage new possibilities for the future. The state appears to be becoming more pervasive: the total spending of central and local government rose from 41 per cent of GNP in 1964 to 52.7 per cent in 1968 — a slow but progressive increase.[1] Of all capital expenditure, nearly 50 per cent passes through the hands of the state. The significance of this for urban development and planning will be outlined in more detail in Chapter 2. Local authority spending has been growing even faster than other public spending, although much of this is really no more than central government money channelled through local authorities as agents. Most current public spending goes towards maintaining services. Apart from the fact that public authorities are a major employer in many urban areas, and provide a channel for much of the new physical capital created in urban development, the state's general role in the economy (of which central and local planning is one aspect) is particularly crucial, and will be developed in the next Chapter. The state is called on to perform many functions, but since Keynes (1936) probably its most important role, and the one coming more into question, is the short-term regulation of the demand

1 This increased to 60 per cent by 1974, but some of this increase is more apparent than real (see Chapter 2).

for goods and services, to maintain full employment while at the same time seeing that other constraints, like the balance of payments, are met. This is done by increasing or decreasing public expenditure, by manipulation of money policies and taxation.

At first sight (Holland 1975b), state intervention appears to be the one means of creating the conditions whereby the UK economy could, through the planning of investment, break out of its difficulties. But, as we shall see, there are strict limits in both conceptual and practical terms as to how far public expenditure and public control, and the central and local planning which goes with it, can be used to solve the problems of the overdeveloped economy.

6 Some effects at the urban level

Urbanization

If we take the most rudimentary definition of 'urban' — i.e. the non-agricultural sectors — the UK is almost entirely urbanized. Almost the entire population depends on non-agricultural, economic activity for its income. Within this urban economy, manufacturing and service dominate, with service growing markedly, and service and commerce now together being larger than manufacturing industry. What effect this has on particular urban areas we shall see later. If most of the population depends on the industrial/commercial economy, and since such activities are essentially 'social' — i.e. cooperative activities involving large numbers of people simultaneously performing complementary tasks — then they must all live in large concentrations, or at least near to good transport links to industrial centres. Hall *et al.* 1973 notes (p. 57) that the main 'urban' part of England which he terms megalopolis (see the map on page 00) is smaller in extent than megalopolii in other parts of the world, although he doesn't in fact draw out the most important conclusion from this, namely, that the UK megalopolis is actually the most intense, the most geographically concentrated urban area of its type (e.g. compared with the North-East United States Megalopolis of Clawson (1971) which has roughly the same population). The combination of a small total land area, the fact that Britain was the first industrial country, and much urbanization occurred before there was effective urban transport, and the increasing scale of many activities and competition for markets has resulted in a very intensive urban land use. Densities are relatively high, and there are generally many potential uses for a given piece of urban land.

Within this very high overall intensity of land use, there is competition for locations and for land in urban areas, between different sectors of economic activity. This is a reflection of the general competition taking place between different sectors of the economy (see page 124). Land is the urban planner's resource, and the intensity of development and competition for land play a major role in urban development as we shall see later.

Investment in urban development

The immediate physical appearance of urban areas is as a conglomeration of buildings and other physical assets such as roads and industrial plant, together with movables like vehicles. These are all relatively long-lasting assets compared to the goods and services which are produced and consumed within the space of a year or two by industry and by the population. Although ignoring the source of investment, Solesbury (1974) gives a useful table showing that buildings are the most important capital asset, being two-thirds of the total assets in the economy, with industrial plant making up most of the rest and vehicles being only 7 per cent of the total.

'Valuing' long-life assets like buildings in this way can be misleading. Property companies in particular have been able to mushroom partly on the basis of some rather specialized techniques for revaluing property assets — which may have exaggerated the real growth in office space. But the provision of office space has grown faster than the economy generally, and much faster than the capital invested in manufacturing equipment and factories (i.e. a 3 per cent p.a. growth rate through the sixties, and, although direct figures are not available, this increased even more during the 'Treasury' boom in the early seventies).

Some investment goes to replace 'worn-out' stock: 7 per cent in vehicles (giving an average life of fifteen years); plant at 3.5 per cent replacement rate is replaced every thirty years (a woefully high figure — most industrial plant in the UK is far older than in other countries); buildings at 1.3 per cent replacement, have a life of eighty years.

In our own research (Broadbent 1971a, 1973; Barras and Broadbent 1975), we have been careful always to distinguish 'objects' such as buildings and vehicles (which we call 'commodities') from the human social and economic 'activities' which consume and produce them. If we look at the main *activity* groups, Solesbury shows that, not surprisingly, manufacturing activity gets most of the plant, residential activity most of the buildings (although, not unexpectedly, in view of the decline in manufacture, services get more new buildings than manufacturing) and transport most of the vehicles.

We thus obtain a picture of relatively slow *physical* (i.e. commodity) change in the urban economy. As noted above, however, there have been marked changes in the scale and organization of economic activity. The comparatively static state of the buildings masks much larger changes in the structure of activities in the urban economy.

We may suggest that this very slowly changing physical stock represents another barrier to change in the mature economy. Almost a third of the housing in the UK was built before 1914 and occurs largely in central urban areas. It is probable that 20 per cent of all houses are unfit, or lack a basic amenity. Recently the stock of substandard housing has been reduced by large-scale demolition in 'comprehensive development areas'; this is now criticized as unnecessarily destructive both to the houses and to the people in them. Nevertheless, the replacement or improvement of *existing* substandard social capital, like housing, could pre-empt a significant amount of investment which might otherwise have been used to make additions to the housing stock, to meet rising standards and falling family size, or to make new industrial plant.

The picture of the stock of buildings in major UK cities is thus one of a still large Victorian legacy, surrounded by the sprawling inter-war and post-war housing, but now recently punctuated (and often completely disrupted) at key sensitive locations like city centres by a combination of large-scale clearance policy and the invasion of the financial sectors of the economy. These operate through property companies, banks and insurance houses, and deposit large-scale central area development schemes for shops and offices into the heart of the city (see Chapter 3).

We described in Section 3 above how the 'financial' sectors of the economy are becoming more important in channelling profits made from the sale of goods and services, back into profitable investments elsewhere in the economy, that is, in firms other than those where the profits were first generated. The financial sector deals purely in 'financial assets', not in any physical goods and services, nor in production. It comprises a group of private companies who deal in stocks and shares, government bonds, insurance, pensions, mortgages, currencies, etc. (Prest and Coppock 1974).

Chapter 3 will outline the process by which property assets, i.e. buildings (especially commercial property such as office blocks), acquire a very high value. They generate rents from activities which use them, and property assets often appear more valuable than industrial assets — single office blocks in London can be worth more than large industrial companies. It is not surprising, therefore, that there has grown up an intimate relationship between the financial companies and property development, especially since throughout the 1960s and early 1970s office values were rising faster than other assets, such as shares in companies. Much of the new property development has been undertaken by specialist property companies, who have become more and more absorbed into the financial sector because they are more and more dependent on banks, insurance companies and other financial institutions for loans and other forms of capital finance (Barras and Catalano 1975; Ambrose and Colenutt 1975). During the short-term boom in the money supply between 1971 and 1973, much investment went into property instead of into industry.

7 The economy as a framework for urban development and planning

So what is the UK economy, and what kind of framework does it provide for looking at urban development (which we have as yet barely defined)? What does it help us to say about other works on urban development and planning? Much of what has been sketched out is fairly generally recognized. There is frequent press comment on factory closures, on redevelopment schemes at a local level, on the 'national crisis', public spending and taxation, 'budget specials', and the like. The fact that most of this popular concern with the economy concentrates on short-term issues, such as the tax rates which mostly directly affect personal spending, does not really explain why there is so little economic background in the standard works on city development and urban planning in the UK. True, many of the histories of town planning such as Ashworth (1954), Cullingworth (1972), Cherry (1974), Hall (1974) and others pay at least some attention to the economic roots of planning in the later stages of the nineteenth century industrial revolution. They stress the link between urbanization and the rapid growth of the working class in the early industrial cities. They highlight the concern for public health which motivated the early town planners; they also note that, although essentially having its justification in Victorian upper-class altruism ('cleaning up the poor'), local planning made good economic sense. Public health and safety became increasingly essential to the success of industry. A regular, healthy labour force helped to guarantee uninterrupted production in industry. At the same time, the gradual assumption by the state of the control of building, sanitation and other standards relieved industry of a burden of cost, and was ultimately a more efficient way of sustaining the labour force.

Now although much of this is indeed explained in the standard works, the nearer they get to the present day the less and less explanation is there about the *current* links between the development of the economy and urban development and planning. Of course, it is almost impossible to ignore the decline of manufacturing, the low growth rate and the rise of services — the so-called tertiary sectors. But in accounting for these trends as they affect cities, most of these studies seem to shy away from drawing together a coherent picture of the way the economy works and the role of urban areas within it.

Summary

Any discussion about urban development or planning must begin with an account of the national economy. The UK economy is in decline relative to other Western industrial countries; it has a legacy of outdated industrial stock and outworn social capital. Yet the economy remains a basically capitalist market structure, dependent on profitability and the generation of surpluses for new investment to survive in a world of increasingly severe competition — and multinational companies. Indeed Britain is *the* most developed market economy, more people are wage and salary earning than anywhere else, and it has few or no new sources of new extensive growth (it is overdeveloped). Its only hope, therefore, is to restructure itself, to reorganize its industries and investment pattern; but the restructuring which has taken place has been largely in response to increasing competition (possibly due to declining profits), the economy has become more dominated by big firms, and the finance and service sectors have grown, taking a larger portion of GNP than in other countries: all this has not halted the decline. In this type of restructuring some sectors grow at the expense of others within the overall relatively static economy, without improving the overall competitive position — the economy seems to feed off itself. The only other possible salvation for the UK economy appears to be the State; the public sector has been slowly growing in size. Chapter 2 examines the possibilities for and limits of the state's ability to perform this rescue.

2 The state: short-term myopia and the cleft stick

1 Introduction: the state — limited or unlimited?

This chapter tries to pin down the role and function of the state in the modern UK economy (the 'state' includes central and local government and also the nationalized industries). Urban planning is undertaken by the state, and to understand the how and why of urban planning it is necessary to take a wider perspective, to understand the wider aims of the state and particularly its attempts to manage and plan the economy.

The main point of this chapter is to suggest that the role of the state and its scope to act are strictly limited by the market economy itself. The economy is not a mixture of two entirely different elements — a 'private' and a 'public' sector — but rather the public sector is really part and parcel of the private market system, which embraces all the three 'great estates' of the economy: the private sector, the public sector and the 'workforce' or the population at large. It is the balance of forces between these three which determines the future of the UK economy, and there is a constant struggle. The UK market economy has to maintain the profits of the private sector if it is to survive. It tries to do this by reducing costs of production (including the cost of wages) and by restructuring the economy as described in Chapter 1. The state tries to help in this process by steering or stabilizing the economy, and by providing a basic underpinning of economic and social infrastructure, running key utilities and services for business ('social investment') and providing social services for the population ('social consumption'). Since Keynes and the great depressions of the 1930s, the state has acted to control unemployment and to keep up 'demand' for goods and services produced — often by spending more than it receives in taxes. (Keynes discredited the neoclassical perfect-market argument that 'supply creates its own demand', and that the state only needs to balance its own budget, intervening to run the few services which could not be provided by the market — a situation which was supposed to result in a full-employment equilibrium.)

However, the key question remains — can the state keep on doing this indefinitely — and what if, by doing so, the state is growing larger? There is an argument (outlined in Section 3) to say that as the state's command over real resources increases and, in so far as it does not *sell* goods and services but provides them free, it is reducing the number of goods and services provided by the market, that is, goods which can be sold and which therefore do contribute to the profits of the private sector.

Therefore, if profits are to be maintained, there is a constant tendency to let the costs and taxes of state spending fall more on the workforce and less on industry. In the long run this is the only way the state can grow — and it results in a 'cleft stick' or 'catch 22' dilemma for the state. Both industry and the population continually require *more* state intervention each on their own, ultimately contradictory terms — the one to maintain profits and reduce costs and the other to improve living standards and hence increase private-sector costs. But the overriding constraint is that the state should not encroach on the operation of the private sector by making too many goods and services available 'free' through the state rather than being *sold*. Rising inflation and rising unemployment are recent manifestations of this cleft-stick dilemma.

This idea of a 'limit' or barrier to the growth of the state is given at least superficial support by studies which show that even in the most 'socialist' of Western economies the size of the state is far lower than in the most 'capitalist' of the planned economies. The gap to be bridged between these two systems is far wider than the differences which exist *within* the systems, and this symbolizes the 'untouchable' area into which the state cannot expand, without changing the whole nature of the market economy.

Section 4 contrasts this cleft-stick view with some other views of the state — those who view the state and the private sector as growing along in harmony, and those who view the state as an 'oppressor' (see also Chapter 5).

The remainder of the chapter describes in more detail the specific way in which the state has acted in the UK in recent years, pointing out some of the peculiarities — the overwhelming concern with the short term and the plethora of short-term controls available to that 'uniquely powerful' finance ministry, the Treasury. The overall picture is one of increasing difficulty — a picture which, while in no way proving the cleft-stick view, is at least consistent with it. The effort put into state intervention has increased and there have been attempts to make planning more effective, by promoting increased cooperation between the three estates. But the so-called 'separation of powers' remains. The public sector is unwilling to take the initiative in developing the economy or to intervene selectively in the private sector. Within the public sector itself the separation between 'plans' and 'powers' remains complete, with all the major executive powers and resources of the state harnessed within the Treasury to a short-term response to immediate difficulties — 'steering' the economy. Planning is left in a largely exhortatory or advisory role on the sidelines, with few or no resources to command directly.

One very clear illustration of the way state intervention has changed is the growth of 'transfer payments' (the direct handover of cash without getting any goods and services in return). Regional policy has been one instrument for this, where cash grants have been provided for investment by private industry in certain deprived regions. This policy has been an extension of, and a more detailed application of, Treasury-type passive 'response' measures, relying on the private sector to take the initiative — and this is where its drawbacks lie.

2 Mainstream economics and the state

Introduction

'Nationalization', state monopoly', or 'state bureaucracy' has been the emotive core of most general elections since the war. The bureaucrat makes an easy and popular target — an anonymous figure in the town hall or in the central government machine. Nobody loves the 'state', and at first sight 'we all' suffer — the would-be council tenant waiting years for a house, the squatter waiting for eviction or the national supermarket chain waiting for planning permission. Indeed, it sometimes seems remarkable that the Labour Party, so often given a 'state control' label by the popular media, has achieved power at all. It will be argued here that the role of the state, of which city planning is one aspect, is the key issue in the future economic and political development of the UK economy. To this extent, the continuing political debate about nationalization, and the underlying incipient conflict over the nature of the state, does indeed reflect, although often in a distorted way, the real political and economic choice before the country.

Many social theories see no particular problem with regard to the role of the state. Social democratic reformism (or 'pluralism'), on which much of Labour Party ideology is based, sees the state as a reforming instrument which, provided the will exists, can ameliorate the problems of industrial society — and which is subject to democratic control by the electorate. The early public health movement of which town planning was a branch reflected almost precisely this position. But in those days the public sector had a relatively small role to play. To many it has now become far too pervasive — even dominant.

Opposition to state bureaucracy and state powers comes not only from those on the Right who see the public sector as a fetter on the working of the free market economy (Friedmann 1959; papers of the Institute of Economic Affairs 1974). Nor does it stem just from liberal nostalgia for the much more 'humane' paternalism of the past. It comes also from the newer, what I have termed (Broadbent 1975d) 'ultra-radical' movements, who are most vehemently opposed to bureaucracy in all its forms and see the state largely as an instrument of oppression set over and above 'the people'. The ultra-radicals regard spontaneous grassroots actions of tenant groups or factory work-ins as much more valid and hopeful manifestations than, say, the nationalization of a major industry. Indeed, in their view, extension of state control may become counter-productive, by fostering the illusion that politicians and bureaucrats can be relied on to provide for the social and economic needs of the masses.

To Marx, of course, the state under capitalism is merely a 'committee for managing the common affairs of the whole bourgeoisie'. But this statement, if taken at face value, begs the question of *how* the state is able to do this; what are the constraints, if any, limiting its powers to maintain the hegemony of the ruling class? For clearly, if there are no limits, then there is no reason why the state should not merely continue operating in collusion with the private sector. In this situation it could operate to keep 'everybody happy' by increasing profits (increasing the demand for the goods produced by the private sector), and, at the same time, it could provide welfare services, housing and other support for the working class to keep it 'content', or else maintain the working class in a docile condition by political and bureaucratic coercion. Interestingly, protagonists on both sides of the argument, whether they see the state as an instrument for maintaining the status quo (whether it be a state of harmony and progress or suppression and exploitation) or as an instrument for change (by reform or revolution), often coincide in their view on the limits of state power — namely that it is unlimited. At least they very rarely try to explore either the limits of state power in a modern mixed economy or the way in which state spending and state influence is determined. According to this view, state power can be brought to bear largely as a matter of choice by whichever interest group or class is held to control the state.

Neoclassical welfare theory

In conventional economics there are two main areas of theory which relate to the state; these represent the two sides of the input—output coin. First the output side: how much state activity should there be, what is it for, what should it do? Second, on the input side: how can the money and resources be raised in order to undertake the task identified? Again it is essentially a 'choice' approach: decide what is needed and then raise the resources, bearing in mind the effects and especially the possible inequities of taxation on various social groups. It does not address centrally the possible limits to the state's ability to restructure or stabilize the economy.

First, then, the output side — the conventional theory is based on the 'micro-economic' theory and its development, welfare economics. The term ('economic welfare' was invented by Pigou in 1912.) This involves the idea that the state intervenes in the private market economy only when the latter is not functioning properly, that is, when competition is not working to perfection. According to the original version of the theory, developed in the nineteenth century, perfect competition would result in a kind of ideal situation — a 'welfare optimum'. This ideal state, the *Pareto criterion* (Pigou 1920), is one where the total quantities of goods and services produced by all the competing industries, and consumed by individuals or firms, cannot be changed in any way without making some individual in the economy 'worse off'. It is not surprising that this has been attacked as the philosophy of the status quo: it might be used to justify *any* existing situation. It also appears of little use in practice, since real decisions will nearly always involve discrimination in favour of one group against another. A further development to make this idea more useful and meaningful was the attempt to devise a *compensation principle*, whereby an economic change was defined as an improvement if those who benefited by the change could fully compensate those who were made worse off, without the former losing the whole of their gain — the so called Hicks-Kaldor criterion.

We shall show in Chapter 5 some of the key concepts in mainstream economics, especially those which are relevant to the study of urban areas. But it is convenient to explain briefly the ideal micro-optimizing market of perfect competition at this point. As we shall see, conventional economics focuses on exchange processes ('the market') between the *producers* of goods and services, who bring them to the market, and the *consumers* who buy them. Perfect competition is said to occur when various conditions and assumptions about producers and consumers are fulfilled. These conditions are as follows (McCormick *et al.* 1974):

1 Producers aim to maximize their profits, and consumers aim to maximize their utility. (Utility is a measure of the satisfaction gained by consumers in making a purchase of a good — it is a subjective criterion.)
2 There are a large number of potential buyers and sellers, each of them being infinitesimally small in relation to the size of the market as a whole. That is, the individual producers have to take prices which are fixed by the market as a whole; they are 'price-takers' and maximize profits in relation to these given prices.
3 All potential buyers and sellers have perfect knowledge of opportunities to buy and sell (i.e. all the prices and quantities of goods on offer throughout the economy).
4 Individual consumer tastes may differ, but consumers see all units of a commodity to be equally worth buying, i.e. *all units of a given commodity are the same.*
5 There is *perfectly mobility* of the basic resources which underpin the whole economy. (These are land, labour and capital, the so-called 'factors of production'; see Chapter 5.) This means they can be easily and costlessly transferred to their most efficient use.
6 Any production process can be perfectly divided. The costs of production remain the same, however large or small the firm, factory or plant undertaking the production. This is the assumption of constant returns to scale.
7 The goods produced are *pure private goods*, i.e. all goods are privately owned and produced and consumed. That is, they can be sold direct to an *individual*, who consumes them in their entirety, and there are no spillover effects.

The interaction between producers and consumers in the market place results in the perfect welfare optimum situation described by Pareto. When all those involved have maximized their profits and their utility to the best of their ability, the result is a stable equilibrium with a certain total quantity of goods and services produced. They are *all* sold, and at prices determined by the push and the pull of the combined action of the producers and the consumers. This elegant idea of an infinity of tiny producers and consumers mutually adjusting their behaviour until equilibrium is reached, under an ideal set of conditions, bears much affinity to the way classical physics studied the statistical behaviour of small particles. The analogy with 'statistical mechanics' was quite explicit in the works of nineteenth-century economists such as Jevons (Dobb 1973). We shall see in Chapter 6 that similar analogies are still being used to derive theories about urban areas — and especially travel behaviour of large groups of people.

Dobb pinpoints the crucial assumption about the behaviour of the economy embodied in this equilibrium view: *all factors of production are used to their full* (see Chapter 5). The obsession with the conditions of a static equilibrium involved an assumption that prices and production always mutually readjusted so that there were no unused resources and, above all, there was no unemployment.

Disregarding for the moment the manifold ramifications of the profit/competition assumptions, the main point is that welfare economics justifies state intervention when these assumptions break down. According to this theory, the state is then called in to correct or replace parts of the market, so that the ideal equilibrium welfare optimum can be restored.

O'Connor (1969) lists the instances recognized by micro-economics when the assumptions of perfect competition break down. First there are the monopoly situations, where large firms can keep prices of goods and services artificially high. Second, there may be 'externalities' of production, where the costs facing an individual firm depend on outside factors facing the industry as a whole.[1] Thus, in the case of the 'industries' of education and public transport, their outputs contribute more to the economy than is represented by the costs of the industry — and they are therefore subsidized or nationalized. Third, if there are increasing returns to scale, so that the more an industry produces, the less is the average cost of production, the state should subsidize these industries to maintain a high output and low overall costs. 'Public goods' (the opposite of pure private goods) are the extreme case of increasing returns, where there are no additional costs, however many more people consume the goods (examples being lighthouses, television programmes). If these goods are sold at a price higher than zero, then we again have an imperfect situation because total consumption would be restricted; more people would be worse off and no one made better off by this restriction. Fourth, there are regulations and taxes to control 'spillover' or 'neighbourhood' effects, such as pollution, changes in property values, working hours and town-planning regulations. These effects are not accounted for directly in the immediate purchase and sale of goods between the individual producer and consumer. State intervention to correct all these real-life divergencies from the optimum welfare situation can be justified by the theory.

But many of the modern conventional economists who criticize the traditional welfare economics do so not only because the assumptions of perfect competition are so often not met, but also because the approach

1 Externalities are the situation *par excellence* where planning at local level comes into its own; see Chapter 5.

contains inconsistencies. By developing a formal analysis of the economy as a whole, quite unexpected state policies can be justified, e.g. large-scale subsidies or nationalization. This is especially so when the external effects ('externalities') are strong. Also the theory of consumer behaviour has been criticized by orthodox economists (e.g. Little 1967). Consumers may be inconsistent, so it is not possible to say when they are 'better off'. Preferences or aspirations are often determined collectively, by social groups rather than individually; furthermore, over time, one set of consumers is replaced by another who may have different utilities. Again, there are many situations where the purchase of one good at one price affects the purchase of other goods. Cost-benefit analysis (COBA), which is often used to decide between different public works and urban development projects (e.g. roads), also contains inconsistencies, based as it is on micro-economic welfare theory (see Chapters 5 and 6).

But there are important areas of state policy which welfare economics leaves to 'value judgements', and which it does not purport to be able to analyse by reference to the welfare criterion. The most important of these is the *distribution of income*. Conventional economics is indeed able to explain in its own terms the way income is distributed between the factors of production (see Chapter 5); but generally it does not argue that this distribution in itself is optimal in social terms. Samuelson has argued that according to this theory, 99 per cent of national income could be paid as profits, dividends or rents, rather than wages, which would be 'exploitation enough in the eyes of radical agitators'. What Dobb refers to as the watershed between the old and the new welfare economics was the frank admission in the 1930s that the wants (and the utilities) of two different people could not be compared — that is, that welfare economics was a more or less neutral discipline which clearly separated value judgements about income distribution from the rigorous examination of efficiency and compensation. The welfare justification for state intervention at the micro-level has developed relatively separately from the study of the behaviour and regulation of the economy *as a whole*, especially the role of the state in keeping the economy stable and growing. This is the preserve of macro-economics.

Keynes: macro-economic management

Modern macro-economic theory was not developed as a pure theory, but was rather invented as a practical series of proposals in response to the pre-war economic crises of mass unemployment. It did not try to be value-free in the same way as micro-economics, and was specifically designed to control the level of income and employment in the economy by state economic policy — that is, to keep the market economy going without the massive unemployment and inflation which may create conditions leading to its demise. But great as this turnabout in thinking was for government economic policy, according to Dobb (1973) the Keynesian revolution did not overturn or revolutionize the overall conceptual approach of mainstream economics.

Keyne's theory attacked Pigou's theory of the 'ideal state' by challenging only one crucial concept: that the economy would always come into equilibrium at the unique single point of *full employment*. The new theory stated that a situation of unemployment, far from being a temporary aberration, could itself represent an equilibrium situation which could persist over long periods, and that it was up to governments to intervene and create sufficient effective demand for goods and services so that unused capacity was taken up and full employment achieved. This was the prescription — even if government expenditure then outran its income and the state budget was unbalanced.

Keynes recognized that much of this theory was a reworking of an old debate between Ricardo on the one hand and Malthus on the other about the existence of 'gluts'. Say's Law — 'supply creates its own demand' (and this includes the demand for labour) — was developed in the early nineteenth century and echoed by Pigou (1933) in his Theory of Employment: 'in stable conditions everyone will be employed'.

The full-employment equilibrium was justified in the pre-Keynesian theory by the notion that all income received in the economy by firms and individuals should immediately be fully used for consumption or investment. All income would therefore be used to buy the output of firms which produced consumer goods or capital goods (machines, etc.) — or to buy labour. There was no way in which income could 'disappear' from the production cycle through saving, since all money saved would be invested. But why? Why should the money put into the bank by individuals and firms necessarily all be borrowed by businesses to buy new equipment? The overall interest rate in the economy moved up and down so that businesses' demand for borrowing for new investment and the rate of saving just came into balance; all savings were then invested. In other words, there was another perfect market: if saving was higher than investment the interest rate came down so that saving became less worthwhile and borrowing for investment by businesses cheaper, and vice versa.

Keynes argued that the interest rate could not be determined by the supply and demand for savings because the *total* amount saved was not only dependent on the incentive to save but also on the level of *total* income. Now this depends on how many people are employed in total and this in turn is also partly dependent on what is being produced in total. Part of this production is the production of capital goods — i.e. *investment*. Thus total saving is dependent on investment itself; they are not operating independently on two sides of a supply-and-demand relation as hitherto assumed. The interest rate could therefore not operate to bring saving and investment into balance and therefore a full-employment equilibrium could not automatically be achieved. The causal link between saving and investment is reversed (Dobb 1973). Investment is not dependent on saving, but rather helps to determine the total savings, because investment is an activity which employs people and therefore generates income, part of which is saved. The proportion of income spent depends on the community's 'propensity to consume'. Keynes argued that in mature, wealthy countries most basic needs were already satisfied, so that the propensity to consume was lower than the capacity to produce and there was thus a chronic tendency to over-production.

In Keynes's view, the rate of interest is determined not by the supply and demand for saving but by the supply of money and by the demand for money *as compared to other assets* (the liquidity preference). But Keynes argued that increasing the money supply indefinitely will not necessarily increase investment sufficiently to create full employment, because under extreme conditions there will be a cumulative incentive to hoard money — the 'liquidity trap'. This means that the onus falls firmly on the government to create sufficient income in the economy. Income could be increased directly by raising the propensity to consume, by raising investment (e.g. by subsidies to industry) by direct government spending or by tax reductions. It is therefore crucial to study the relation between any initial investment made (e.g. by the government) and the total income finally generated in the economy (or between the initial employment and the total final employment generated after all the new income has been spent or saved). This is the so-called 'multiplier'. Thus, to paraphrase Keynes, 'if the propensity to consume is such that nine-tenths of any new income is spent on consumption, then the total increase in the employment generated by increased public works is ten times the primary employment generated by the works themselves' — that is, the initial employment is 'multiplied' by ten. Keynes used this example to show how nicely the whole system comes back into balance, because, of course, part of this new total income stemming from the total employment generated is saved and in fact the new total savings are just enough to match the initial investment.

Keynes's work rekindled an interest in the theory of economic growth, so that his original study of the short-term correction of unemployment was subsequently integrated into an analysis of the relationship between investment and the growth of the labour force over long periods of time by Harrod (1948) and others. The detailed elaboration of the interrelationships between the key elements (the macro-economic variables) of Keynesian analysis — consumption, investment, saving, employment, money, interest — has continued over the last thirty years.

But O'Connor feels that the study of 'public finance' (the way the government actually raises the resources to promote the desired level of employment) often tends to be divorced from the theory of economic growth and hence from the question of whether or not the government can maintain full employment in practice over a long period, or, if it can, whether this imposes other unwanted burdens on the system (inflation, increasing debt, etc.).

Conventional works on public finance tend to try to define some optimal kind of behaviour for the state, an economic optimum which is efficient for the public authority and for the economy as a whole: thus Musgrave's (1959) optimal theory of the 'public household'. Public sector economists are concerned both with the effects of particular taxes and with the proper limits to what the government should 'take away' in taxes and borrowing. Musgrave does indeed examine the implications of government intervention on the growth of the economy, although the overall implications of a continuing increase in the public sector are not faced directly, nor is there an assessment of what the overall balance between public and private sector might be over the long term, and the resulting implications for the health and profitability of the private sector.

Musgrave relates taxation and spending, in a strictly Keynesian analysis, to aggregate quantities such as total income and total investment, and considers various types of balanced and unbalanced state budgets. From there he goes on to consider the possibilities of securing not only an adequate level of demand and hence full employment but also a desired growth rate at full employment, one which can be achieved without creating inflation (i.e. a growth rate which marries full employment to the continued increase in economic capacity resulting from the increase in investment). The crucial question of whether or not the tendency to Keynesian over-production or 'stagnation' tends to increase is indeed mentioned, but there is no overall judgement. If the tax rate (tax /income) is less than the rate of government spending (i.e. government spending/income), Musgrave declares a 'state of stagnation'. But can the government *choose* to have a state of 'stable stagnation' — which causes no problems in that there is no increase in the public debt (see below) and there is no need for increased taxes to finance interest payments? This is not made clear. The running-faster-to-stand-still syndrome does arise, in Musgrave's view, if stagnation is increasing — but is it? This seems to be the crucial unanswered question of the conventional theorists; the fact that state intervention is continually increasing at least suggests that the question should be raised.

Clearly, an adequate account of Keynesian analysis and its implications for taxation and spending is beyond this book, but at this stage it should at least be clear that some of the major implications for policy and hence for the development of the UK economy since 1945 are as follows:

1 There has been a rejection of the neoclassical micro-economic theory as the general model for the state economy.
2 A balanced budget is therefore not as essential for a state as it is for a firm.
3 There is no guarantee of full employment and the government must be responsible for increasing demand in the economy so that full employment (if possible, with growth and no inflation) is achieved — even at the cost of a budget deficit.

The potential problems raised by long-term application of these policies have not been adequately faced in the literature. If the tendency to stagnation is prolonged or because of other changes in the economy the state has continually to increase intervention, the policy might somehow be storing up trouble for the future — by increasing debt, taxes or inflation.

As for the potential limits to successful state management, the first clue, noted by Keynes, comes in *The General Theory* (1936), p. 105: 'each time we secure today's equilibrium by increasing investment we are aggravating the difficulty of securing equilibrium tomorrow'. Since 1945, the UK government has absorbed Keynes's policy message and has taken upon itself the task of regulating effective demand to maintain full employment, as we shall see below.

3 The limits of state intervention

Introduction

This section will try to examine some of the (fairly inchoate) ideas which have been put forward to question the possible limits to the state's ability to operate Keynesian policies over a long period, and which question the reasons for, and effects of, the growth of state spending. There appears to be relatively little formal analysis of this problem, and some of the arguments below, especially that of Mattick, may seem difficult. But the conclusions are clear enough: the state's growth in real terms is strictly limited; it cannot absorb more than a certain amount of real resources without threatening the profitability of the market sector and hence the viability of the market economy. Whether or not this analysis is correct, some naïve support for it is provided by the fact that even in the most socialist of market economies the public sector in real terms amounts to only a minor part of the economy, far smaller than it is in the most 'capitalist' of the planned economies of Eastern Europe.

Bearing the costs of production

Both the micro and macro arguments for state intervention in the 'mixed' economy influence current practice. Nationalization and the extension of state control are often justified because a market is held to be malfunctioning, because there are monopolies, etc. Government macro-economic management has been explicitly Keynesian for many years. But it is also worth looking at more radical (indeed, explicitly Marxist) works such as those of O'Connor (1969) or Mattick (1969) which analyse the growth of the public sector and consider the implications for the survival of the mixed economy over the longer term.

O'Connor begins by trying to examine the exact functions of state spending. There are three categories:

1 Social investment (infrastructure, etc). This type of spending increases the productivity of the economy generally and especially benefits industry and commerce. Examples are roads, airports, utilities (gas, water, electricity), industrial development, urban renewal for commercial development, capital grants and incentives to industry. Some of these investments may be essentially determined by the projects that firms in the market economy decide to undertake; they are *complementary investments* without which private capital projects would be unprofitable. *Discretionary investment*, on the other hand, is undertaken to improve the regional or national economy as a whole. Nationalized industries also appear here — often providing commodities for sale. More often than not these commodities

are the basic inputs that are used by nearly all sectors of the economy, such as coal and transport.

2 Social consumption. This state spending is focused more or less directly on the population (and the labour force especially). This includes both capital investment and current spending on income maintenance, education, health and welfare, housing, food subsidies, etc. Social services, in Marxist terminology, 'socialize the reproduction of labour power'. The cost of maintaining the labour force in housing, education and health is borne by the community as a whole through the state. This means that the cost of social provision is not borne by individuals or by firms. This is more efficient for the economy as a whole, and the reduction helps to sustain and increase the productivity of labour and capital and hence to maintain profits.

3 Military, police and other 'social expenses'. This category is said to be spending which is unavoidable if the market economy is to be protected from social and political disruption, from outside and from within. It also includes repayments of the state debts, and it is a polyglot category that does not immediately concern us here.

The first two classifications are helpful in that they distinguish between the profit-making private sector (the 'motor-generator' of the economy) and the population (the 'labour sector') which provides the skill and effort to make it work. Profits are necessary to keep the system going, and the classification helps to show that some state spending increases profitability directly, by providing cheap inputs for firms, or by providing markets, and some indirectly by bearing the costs of sustaining and reproducing the labour force. Nevertheless there is some possible confusion: spending is often a mixture of all three types: roads and electricity, for instance, are used both by industry and by the population. It should also be noted that both social investment and social consumption involve a mixture of 'capital' ('investment') spending and current spending.

We are still left with the other side of the state spending coin: where does the money come from and what is the *net* effect on profitability, taking into account not only the direct effect of the spending but also of the taxation and borrowing required to finance it? This doesn't drop out of the sky.

The cleft stick

According to Mattick, the Keynesian multiplier effect described above can be misinterpreted. It may appear to imply that, if a given quantity of new government spending is injected into the economy, the actual total effect of this expenditure once it has been realized in successive rounds of sales and purchases of new products through the economy, adds up to *more* than the original spending. According to Mattick, there is the possibility of a serious illusion here. Income cannot multiply itself merely by being passed from one firm to another. It is simply exchanged for an equivalent amount of goods and services down the line. If the government has obtained this money in taxes from industry or from individuals who have earned it by the sale of goods or labour, it merely represents a quantity of manufactured goods or labour which *already exists.* In this sense the purchase of goods and services by the government cannot confer more benefits than it has already taken away. A transfer of money from the private sector to the public may change the composition and character of the total production in the economy, but it does not necessarily enlarge it in the first instance.

Mattick is here looking at the public sector within the economy as a whole. His argument is not inconsistent with the fact that increases in public spending over the short term or in specific local instances can bring benefits to the other activities in the economy, benefits which amount to more than the initial public spending. This is especially true where there are strong externalities. Subsidizing commuter rail travel to the tune of a few hundred thousands may bring benefits to hundreds of firms in the centre of a large city, far in excess of the subsidy. Indeed, this is a central reason why 'social investment' spending on basic utilities like electricity, gas and railways takes place. Here Mattick is at his most controversial: he sees the state in the context of the economy as a whole, where it acts as a reallocator of resources, not as a creator of new wealth. His analysis really requires a more rigorous formulation. Pages 82-6 below show one way this may be done.

But how does the government get its money? First in the same way as both private firms and households do — by 'selling' goods and services and by borrowing. But the government can compulsorily transfer resources from private to public sector (by taxation) to pay for government operations, and this is not at all like pricing a good to be sold on the market. The government can, in addition, print money and (in times of emergency) can physically commandeer resources into the public sector. The practical effects and limitations of these different forms of finance are not considered at this point. In general we must regard the government sector and the private sector as essentially part of the same economy, because government raises its resources by *taking them away* from the private sector (firms) or from the population, or by borrowing overseas.

Bearing in mind what we said in Chapter 1 about the primacy of the profit motive in determining investment and growth, then it is quite clear that, if the government obtains its resources from industry by taking away profits, it can only go so far, otherwise profits will be reduced to the level where the private sector grinds even more rapidly to a halt. This Mattick sees as the key to the character of the mixed economy; it is essentially a single, unified economic system where the limits to the growth of the public sector are defined by the private sector itself. Although some sectors do come increasingly under government ownership, Mattick considers that the most usual characteristic of the modern economy is the extension of government *control.* The state is acting largely only as a *channel* for privately produced wealth, but it requires an increasing proportion of private wealth to succeed in its role of stabilizing the economy. Thus the *public sector* is still to a large extent the *private sector*; it is certainly not the autonomous, independent despot, benevolent or otherwise, that it plays in some conventional theories.

The money extracted by the government comes from that part of the income generated by private firms from the sale of their products which is left after all the costs of production have been met — that is, after all raw materials and labour have been paid for. It is thus a part of a potential surplus or profit[1] in the hands of the firms which is taken away by the government in taxes. If these profits were recycled into the private sector and simply handed back, then the overall profit would remain the same. But this, Mattick points out, is not so. They are indeed used to purchase goods and services from the private sector, but the government does not generally *sell* these goods and services for profit. There is no final sale for this type of production and therefore no final profit on these particular goods. The government spends most of its money on education, social services, defence, public administration and other things that have no market. If it did concentrate on producing profit-yielding goods, then again it would be competing with the private sector directly and reducing profits still further.

The purchases the government makes do indeed channel the tax money back to the private sector through government contracts, so that all the tax (which was originally part of the surplus made by the private sector) is now used by that same private sector as payment for new products on which, as we have just seen, there is no final profit. So this part of the total production is initially financed from the potential profits or surplus made by the private sector and represents a quantity of goods and services which, by taxation, the government actually *takes away* (expropriates) from the private sector. Rather than the term 'public sector', Mattick uses the term 'not-for-profit sector', to highlight the fact that this production takes place within the private sector, and that it is also originally financed by the private sector.

If tax money, being mostly a transfer from the private sector, does not *necessarily* increase total production, the other form of government finance — borrowing or deficit financing — definitely does. The borrowed money puts productive resources to work. The private sector will initially appear to make profits from the new government contracts, but the new production is again being paid for by the private sector itself, since it loaned the government the money in the first place. Now if this money is again being used to produce goods and services which have no market and, therefore, yield no profit in the long run, even this new, debt-financed increase in production represents an increase in 'not-for-profit' production as against the profit sector. If the government keeps on financing new production in this way, it will have to borrow more and more, thus increasing the 'national debt'. The only way this debt can be honoured by the government is to raise more taxes or to increase the supply of credit money generally. The debt can only be honoured in future by still further reductions in the future income retained in the private sector. Mattick does recognize that there are many factors complicating this picture, and that private investment and therefore profits can be increased through improved 'business confidence' and other indirect effects of government activity. However, the main conclusions are

1 A strict Marxist analysis would distinguish between these categories much more carefully.

unambiguously stated: the 'not-for-profit' sector (i.e. the public sector) is increasing in all Western industrial societies. 'The persistence of government-determined production is a sure sign of the decay of the private enterprise economy.' 'The increase in the public sector forecasts the end of private enterprise.' To Mattick, this is not apparent on the surface of events at present, because, as we have seen with Keynes's argument, the private sector does in fact require government expenditure to maintain a stable, full-employment equilibrium, and it therefore cooperates with government, and government still strives as far as possible to avoid direct competition with the private sector.

Now whether one can agree with this formulation or whether in fact the analysis is correct (or even clear), it certainly raises fundamental questions. The argument seems to begin at the point where Keynes himself expressed doubts about the long-term effects of postponing the crisis through deficit financing to maintain employment. Mattick's conclusion is that over the longer period the only chance for the private economy is to recover the national debt through additional income in the private sector (i.e. in addition to that injected into the economy by the government); unless this profitability actually materializes 'today's additional income becomes tomorrow's loss'. Conventional economists hold that there is no problem if the national debt increases, provided it does so at the same rate as the national income as a whole. But, here again, Mattick points to an 'error' — that the debt can only justifiably be related to that part of total production which is not due to government spending in the past.[1] He argues that the published figures tend to understate the real growth in the national debt, since a large part of the total GNP is due to previous government debts — debts which have been shrunk over the years by inflation.

Mattick's views are based on his own particular Marxist analysis, and, as there appear to be so few other authors with similar views and there is no space here to analyse them fully, they have to be treated with caution. Perhaps the underlying strength of his analysis is based on what we might call an 'accounting' logic. He is at pains to trace through the origin and destination of government spending, without double counting and without generating income 'out of the sky'. His ability to discern the decay of profits in the private sector stems in the first instance from his ability simply to account for the source of public spending (in taxes from profits) although there are of course many other assumptions about the origins of profit and new production — before he is able to draw his main conclusions.

It is useful to make some cautionary points here. First, Mattick's argument seems to assume that most or all taxes come from the private sector, whereas in fact an increasing proportion of tax is drawn from the population — the 'workforce' — and not from business. This may not necessarily contradict his thesis, since it may show that the state is trying to avoid taking too many resources away from the private sector and so trying to preserve profits.

1 Inflation provides one means by which all debts (including the national debt) are repudiated. (Note that Levinson (1971) points out that the US national debt is only a tiny element in the economy.)

Another point worth stressing is that the part of government spending that subvents real resources — the purchase of goods and services (including wages) — should be clearly identified. These are the state's *real* claims on the total gross national product. These resources are used to run the various social services or public utilities. The other part of government spending is only a 'transfer' from one part of the economy (e.g. taxpayers) to another (e.g. recipients of social security). These transfer payments appear as part of the government spending as presented in the UK national accounts, but they cannot legitimately be regarded as contributing to the size of the state in real terms. This distinction should be clearer in the discussion on the history of UK state controls below, but it is remarkable that the *real* growth of the state has been very slow indeed compared to the growth of transfer payments.

O'Connor's analysis (1973) is rather different. As we have already seen, he tends to view the state much more in the role of *increasing* private-sector profits by providing cheap inputs and by reducing some costs of production, including labour. This Mattick tends to understress, attaching more importance to the fact that, whether or not public-sector production helps to reduce industry's costs, no profit is made on it directly. Similarly O'Connor sees the national debt as 'tightening the grip of capital on the state', since the borrowed money is used to finance unprofitable social investment which, unlike private investment, does not generate new funds and new profits which might increase the state's ability to pay. Nevertheless O'Connor also foresees, an increase in the state debt, and shows how the financial sectors — banks, insurance houses, etc. — are reluctant, at least in the USA, to support the state by 'buying up' the debt, even when interest rates rise or the debt is shortened.

O'Connor's central thesis is that of the 'fiscal crisis' or 'structural gap' between the increasing level of state spending on the one hand and the state revenues on the other — with the former always tending to increase faster than the latter. The gap arises because more and more of the costs of production, indirect and direct, are being borne by the state, whereas the surplus and profit is still appropriated by the private sector. This seems roughly consistent with Mattick's views. But while O'Connor's economic analysis seems to point to a similar kind of cleft-stick dilemma for the state — with more and more state activity required to maintain the economy in profitable equilibrium — he doesn't seem to follow through fully the economic implications of the increasing fiscal gap in order to discover whether it might of itself lead to a profitability crisis in the private sector.

So he emphasizes the 'political struggle' to shift more of the burden of finance away from industry and finance on to the population at large, and to reduce services to the population. This forms a rather unsatisfactory conclusion to his analysis, which would perhaps be stronger if combined with the 'economic' arguments of Mattick.

There are other theories on the state — often drawing on elements of the Marxist tradition. One view — known as the 'under-consumption' thesis — reflects Keynes's concern with the low propensity to consume in the mature economy. These theories see the central problem of the modern economy to be the conflict between the increasing drive by the large corporations to produce more and more by investing in more and more productive capital, on the one hand, and, on the other, the inability of the national and international market to buy all the goods produced. This view is put forward by Baran and Sweezy (1968), who suggest that state spending is one way of absorbing this surplus production. But this seems to imply that there is no cleft-stick dilemma for the state, and, in particular, no limit on its size, since the greater the surplus business produces the larger it requires the state to be.

We have stressed Keynesianism and its radical critique, largely because it is the most immediately relevant area of economic theory to national, regional and urban planning. It is an economic theory, and equally it is most emphatically government practice, a practice which has successfully managed the mature economy for a generation — 'postponed the crisis', in Mattick's words. Mattick indicates that there are possible limits to the ability of the state to maintain the mixed economy and, therefore, that there are limits to the state's activity, of which urban planning is one aspect.

When we come to look at planning in practice, we shall be looking for some of the aspects noted above. We shall be looking for evidence of the increasing gap between spending and revenue, for the conscious avoidance of direct competition with the private sector, for the provision of direct infrastructure and other direct inputs which help to decrease production costs in the private sector, for demand-increasing purchases to maintain full employment and stability, for other regulations which help to keep the private sector running smoothly, and for the political struggles and debates over the size and composition of public-sector activity which might indicate the possible limits to the growth of the public sector in the mixed economy.

It is worth noting in passing that even the most 'socialist' of Western industrial economies has a public sector representing less than 35 per cent of total employment. The gap between the size of the public sector in the most 'socialist' market economy and the most 'capitalist' socialist country is more than twice the range of public sector size *within* these two groups (Pryor 1970), so there may be an upper limit which cannot be exceeded if the mixed economy is to retain its essential character.

4 The state: other views

Convergence theories

Of course, there are many other views of the state. Galbraith (1962, 1967) has some well-known ideas on taxing private waste production which he sees as being artificially created by private industry through advertising. He notes the perpetual shortage of public goods and services (public squalor) needed for urbanization, for roads, schools, utilities, planning, etc. Galbraith continually stresses the converging interests of large-scale corporations and the state — and the demand by the former for a large public sector. Not least of the reasons for this is that if the state is to regulate the total demand in the economy it must be large enough itself for its spending and taxation to have an appreciable effect. He is continually stressing the role of state planning as the provider of a stable, risk-free environment for business. As investment decisions become larger and larger, and as the time needed to plan major new production facilities (e.g. introducing a new make of car) becomes longer, the state is required to provide more and more certainty that the market for the product will actually materialize. This points to a crucial aspect of the role of the state but it does not allow for any contradictions in its role or any limit on its size.

Fitch (1968) in a hilarious piece characterizes Galbraith's ideology as 'monopoly without tears'. Large multinational companies are no longer motivated by competition and profit, but the managerial élite (not the owners) who now effectively run these companies are now more interested in long-term stability and planning. This points the way to a smooth convergence of the socialist and capitalist economies into a neutral, technology-dominated, benign, industrial monolith. He is in favour of a larger role for the state, to meet the demand for public services. This smooth convergence is, of course, flatly impossible for the state in the long run, if we accept the arguments that while the growth of the state is necessary to the private sector, it might also threaten its survival.

The state as oppressor

Miliband (1969) agrees with Galbraith that the state does indeed collaborate with business, but only in the sense that a ruling élite actually runs the state. He stresses the intimate personal and cultural ties which exist between top civil servants and business interests. Needless to say, being a Marxist, he doesn't view this in Galbraith's sanguine terms. He sees the cooperation and collaboration as essentially against the interests of the working class. He is attacked by Poulantzas (1975), one of the 'Althusserian French', for being too empiricist, and for describing state power in terms of the individual bureaucrats and their close links with the bourgeoisie rather than by a 'scientific' analysis of society. Poulantzas again stresses that the state largely operates in collaboration with the large firms — 'the monopoly sectors' — and helps to eliminate smaller firms. He and similar writers attach great importance to the 'relative autonomy' of the state — that is, its ability to act in a way that is partially independent of the economy and particularly of the ruling class or specific parts (or 'fractions') of it. He sees the state as a summary or 'projection' of the current balance of social forces; it is a 'sounding box' for various economic and social crises whereas in earlier periods it used to be a 'safety valve' — much more capable of smoothing over the malfunctions in the economy. This analysis has a certain force, especially if we relate it back to the discussion on the economic limits to state intervention. The real resources used by the state in running services for the population and for the private sector — 'social consumption' and 'social investment' — reduce industries' direct and indirect costs. Industry will always want more social investment, whereas once the state has provided certain minimum standards of health and education, it is at least arguable that a further increase in social consumption with its concomitant increase in real resources used by the state, while it would increase the standard of living, would not necessarily increase the ability of the labour force to produce more. Such an increase in social consumption would represent a decrease in the total resources going to profits. This is one example of how decisions on state spending represent a sounding box for the wider conflicts in society.

But the transfer payments made by the state, involving direct expropriation from one group and direct payments to another, may involve even more severe political choices. This is a real sounding-box role, and as we shall show below more and more of the burden of taxation is falling on the working population and there is every reason to expect this *visible* transfer of income to become more controversial as it grows more visible.

However, Poulantzas's analysis as presented seems incomplete. It does not discuss in concrete economic terms *how* the state operating as a sounding box is to carry out its task. He does not investigate the possible limits to its power to act, and his overall emphasis tends to be on the 'repressive' nature of the state apparatus.

There is a diverse band of writers, many of whom are 'urbanists', who in one way or another see the state fairly unequivocally as an oppressor. They do this by taking what Mattick terms the 'superficial collaboration' between central and local government on the one hand and the property speculator or big business on the other. They see the state as smoothing the process of exploitation by legalizing the ejection of people from their homes, or by erecting institutional barriers to stop them escaping from urban ghettos or imposing substandard private housing using planning laws and other regulations. This view ultimately relies on the 'unlimited' view of state power; they do not consider the means available to the state to do this, and what limits or contradictions there are in its ability to maintain oppression. It is thus very much a onesided picture which again tends to disregard the economic basis and limits to state power. If Mattick's picture is correct, the market system demands progressively more intervention to keep it going, while, at the same time, this intervention itself only gains its strength to preserve the status quo by feeding off the very system it is preserving and thus mortgages its future. Nevertheless, this view, elements of which are present in some of the writings of Harvey (1974), Pahl (1975) and Simmie (1974), has brought the notion of exploitation to the fore of thinking about cities and has had considerable influence. The challenge this work offers is to turn its pessimistic and often nihilistic message into something more constructive, something which illuminates the positive possibilities inherent in the current exercise of state power. Otherwise the ultra-radical message to 'the people' is to suffer, or else to react in blind fury and throw stones. This will be discussed more fully in chapter 4.

5 The state in practice

Steering the economy

The main point of this section is to demonstrate the nature of 'planning' as an activity of the UK government rather than to itemize in great detail all the different planning instruments, regulations and *ad hoc* institutions developed to give effect to national, regional and urban planning. This part will, therefore, be a summary sufficient only to show the broad outlines and extent of planning in the UK at the national and regional level, so that we can later trace through its effects down to the local level.

At the national level, as we get into more detail of what has happened since 1945, we can fill out the general picture of the economy outlined in Section 1 by showing in some detail how Keynesian management has been operated by the state to maintain the economy in a relatively crisis-free condition. History shows, superficially at least, that this appears to be becoming an increasingly difficult task. It is a picture which, without proving Mattick's thesis in any way, does lend it some support. It is a picture of increasing visibility and increasing size of state activity, and also increasing sophistication in the way the activity is managed politically and institutionally.

There are three main aspects to state management. The first is the 'budgetary measures'; these are the main Keynesian instruments for 'steering the economy' (Brittan 1971). A major aim of these measures is to keep the economy 'on course', on a road whose boundaries are strictly defined. The balance of external trade has to be kept within limits, unemployment and inflation kept down, and a reasonable level of growth achieved. The second aspect of state management is direct intervention in the economy through public ownership and through direct investment in industry, either through the nationalized industries or through grants and loans for investment to the private sector. The third is the planning and institutional mechanisms including all the various regulations, standards and practices promulgated by government.

The Treasury emerges as the key state institution for managing the economy. Brittan (1971) shows that from 1947 it had responsibility not only for money and credit but also for production, exports, wages, manpower and the coordination of other ministries' economic policies. Keynesian-type policies were adopted from the end of the war. First the Keynesian principles were applied 'in reverse' to suppress demand after the war, which had outstripped the ability of industry to supply goods and services. The 'planning' system was, therefore, as strong then as it has ever been since. There was a massive reconstruction programme, the nationalization of basic utilities (in accordance with both classical and Keynesian economics) and a period of harsh physical controls on consumption — rationing. When these restrictions were abolished in the early 1950s the Treasury became even more important as 'the only remaining department with economic sanctions'.

A detailed study of Treasury economic policy would take us far away from the subject in hand. We should soon be into a discussion not only of economics but of the whole institutional and social manifestation of the British establishment. Since direct planning and even indicative planning have been almost entirely lacking for long periods since the war, the Treasury's budgetary measures have been the supreme 'national planning' mechanism. Brittan shows that the Treasury's control of public spending rests on three historical rules. First, since 1713 no MP can put forward proposals for spending without government consent; second, since 1924 any suggestion for new spending cannot be circulated to Cabinet until the Treasury have discussed it; third, since 1884 Treasury consent is needed for any new item of expenditure in any government department, whether or not it means an increase in spending.

The Treasury uses a five-year rolling programme as a basis for estimating and controlling public spending. Since 1969 these have been published annually as Public Expenditure White Papers. It is usually stressed that spending up to Year 3 represents existing commitments; substantial changes can come only in subsequent years. Brittan considers these papers to be the nearest the UK comes to a 'National Plan' but, of course, this is only for the public sector, not for the economy as a whole.

The overwhelming impression of the management of the British economy since 1945 — for it cannot be called planning — is of the absolute dominance of short-term considerations, uninformed by any underlying strategy for improving the performance of the economy in the longer term.[1] *Steering the economy* is itself a question-begging title — who builds the road? who finds the route? The overall direction of the economy, rather than being consciously planned, appears to be determined as a residual — as an *outcome* of the short-term management rather than as an input to it. Yet it is possible to discern some kind of meaningful pattern to events over the years. It is one of alternate short sharp booms in production, followed by rather longer periods of instability, with production oscillating rapidly about a fairly static average (Figure 4). Thus there have been booms in 1952-4, 1958-9, 1963-4 and 1972-3, with the intervening periods having several mini-booms and slumps within them. The unemployment index follows the production index but oscillates rather less wildly.

There is an unresolved debate as to whether the Treasury action (by budgets and other measures, which are largely designed to respond to these cycles) has actually helped or hindered. There are some who argue that the method of forecasting used (again dominated by short-term considerations) has generally led to 'corrective' action being applied too late, so that it tended to reinforce a trend that was already developing, rather than to hold it back (Dow 1970; Brittan 1971). But Little (1967) queries this conclusion. Figure 4 also shows that expansionary budgets have tended to occur when the economy was already expanding, and deflationary ones when the boom was already beginning to falter.

A plethora of short-term instruments is available to the Treasury. This is another unique feature of the UK economy: far more effort has been put into

1 'The short-run obsession of British economic thinking' (Brittan 1971, p. 153).

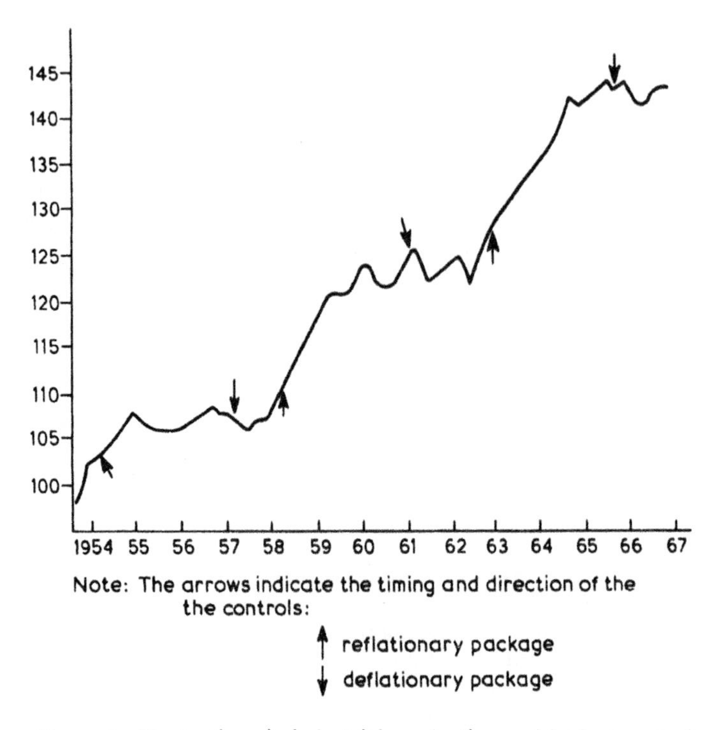

Figure 4 Fluctuations in industrial production and budget controls

stabilization policy in the UK than in other comparable countries. These instruments usually applied through annual 'budget' statements, can be divided into three broad groups:

1 Taxes
   (a) Taxes (and tax allowances) on individuals — income tax and surtax, profit tax and capital gains tax
   (b) Taxes on companies — income tax, profit tax, capital gains tax
   (c) Taxes on spending and tax allowances: general (purchase tax, value added tax); specific (drink, petrol, tobacco, etc.)
2 Money policy
   (a) Interest rates
   (b) Restriction of bank advances, special deposits
   (c) Hire-purchase restrictions
3 Public spending

These instruments have been used to keep the economy within two overriding constraints: first, to maintain the overseas trade deficit in balance and, second, to keep unemployment low. More recently more attention has been given to reducing or limiting the level of inflation and to maintaining or achieving economic growth. The continuing tendency for imports to be larger than exports cannot be examined here, but it has much to do with the weak competitive position and the particular structure of the UK economy outlined in Chapter 1. To Brittan, of the two constraints, the Treasury has the trade deficit uppermost in mind. On the other hand, it is the politicians rather than the bureaucrats who regard high unemployment as unacceptable.

The overall performance of the economy, so far as it has been affected by state management, has thus been largely determined as an outcome of the measures to satisfy these immediate goals. None of the measures was specifically directed to changing the structure of the economy outlined in Chapter 1, and only recently have these more fundamental issues come to the centre of political debate.

The emerging pattern is one of increasing difficulty in handling the short-term situation — as evidenced partly by the increasing number and sophistication of controls, and more frequent budgets and partly by the increasing levels of unemployment inflation and trade deficits. Kennedy (1974) notes a continual increase in the sophistication of the techniques of short-term management over the last thirty years. Improvements include:

1. Statistics — the Central Statistical Office now provides quarterly expenditure figures, seasonally adjusted at constant prices where appropriate, of all the major variables in the national accounts.
2. Techniques and models have been developed for forecasting the level of GDP for eighteen months ahead, and for linking changes in output to unemployment. This work has been carried out both within the Treasury and in outside bodies like the National Institute for Social and Economic Research.
3. A model for estimating the effects of budgetary measures on the GDP is also used within the Treasury.

There has been a growing awareness in the country at large that the underlying economic structure does matter, and especially as it affects our overall competitive position in the world economy. During the 1960s there were attempts at specific intervention in the economy, with the beginnings of indicative planning machinery and other selective measures directed at industry. This culminated in the National Plan of 1965, which, however, was soon abandoned in the face of an immediate crisis on one of the classic short-term aims — balance of trade and defence of the pound. The late sixties and early seventies thus saw a hiatus in 'planning' activity. Now there is renewed flirtation with the planning idea, but this history is spelled out in more detail below.

Brittan (1971, p. 456) refers to recent UK economic history as having the inevitability of a Greek tragedy and says that a certain feeling of 'fatalism' may not be misplaced — although he does think that other courses could have been adopted by the Treasury over the years. It is sometimes pointed out that fluctuations in the output of the UK economy are less than in comparable countries — but that these movements may be more serious in a stagnant economy, where growth sometimes actually halts, than it is to a rapidly growing one, where much larger fluctuations may only mean that the growth process slows down. Brittan (p. 457) points out some of the longer-term effects of the short-term Treasury measures. He considers that although the average level of demand in the economy may not have made for slower growth, the boom/slump cycles themselves and the measures taken to manage them have tended to reinforce the factors which slow down growth. Firms will not invest unless they feel that the growth of markets is going to be long-term, smooth and continuous.

In controlling any dynamic process it is important to see how large are the fluctuations to be controlled, and to know something about the processes creating them. It is particularly important to know the characteristic time periods over which different effects make themselves felt throughout the economy as a whole, and similarly to know the likely effects of government measures — the scale of their impact and the time period over which they make their full impact. The fluctuations which have actually occurred are partly the outcome of the controls that have been applied; it is difficult to establish what the economic fluctuations might have been without them.

Unemployment, until the late 1960s, fluctuated from between 1 per cent and just over 2 per cent. More recently, in the troughs of 1972 and 1975-6, it has climbed to well over 4 per cent, with between 1 million and 1.5 million unemployed — another indication of the increasing difficulty of steering the economy. To be fair, this is not only a function of the UK economy; clearly the world oil price rise has something to do with it. The fluctuations in all Western economies tend now to be synchronized, with peaks and troughs reinforcing each other rather than cancelling each other out as they sometimes have previously, because of the increasing integration of the world economy and the interpenetration of national economies by multinationals discussed in Chapter 1. The UK economy can fluctuate quite rapidly in a short time, with unemployment rates changing by 1 per cent or more within a single year and the economic cycle going from a peak to a trough within an eighteen-month to two-year period. It is important therefore to have relatively quick-acting control instruments.

Whereas it has been said that the object of Keynesian economic management is to achieve 'a little unemployment and a little inflation,' it is more recently stated that inflation is the price paid (*sic*) over the long term for a consistent application of Keynesian control to limit unemployment. But whatever the strict causal mechanism, it is clear that inflation is an increasing problem. The average annual rate of price increases was around 3 per cent in the 1950s and early 1960s. It then became 3.5 per cent from 1960 to 1965, and 4.6 per cent in 1965-70, 8.6 per cent in 1970-3 and jumped to over 25 per cent in 1974-5. Inflation naturally then became the dominant issue for the Treasury.

The controversy over the causes of inflation continue, but mainstream economics usually puts forward four possible causes:

1 Excess demand for goods and services — demand inflation.
2 Excess increases in costs, particularly wage costs (e.g. trade-union bargains irrespective of market forces) — cost inflation.
3 Imported inflation — caused by excess demand or costs overseas.
4 Monetary theories — too much money in the economy.

We have already seen that Glyn and Sutcliffe's analysis might be consistent with (2) as wage earners increase their share of the total income. Levinson would probably reject these analyses as they stand, and combine elements of (2) and (3), so that multinationals are able to raise prices throughout the world economy to maintain profit levels (and raise capital) despite wage increases produced by union bargaining.

All the first three conventional explanations of inflation are derived directly from Keynesian macro-economics. Keynes's was called a 'general' theory precisely because it brought together into a single coherent framework phenomena which had previously been considered to be independent. In pre-Keynesian neoclassical theory, real physical factors in the economy (goods, labour, etc.) were analysed separately from money phenomena. Thus the relationships between the physical factors of production were supposed to determine the overall allocation of resources in the economy (how much of which factor was used for which purpose) and hence the overall level of unemployment (the use of the factor labour). These physical relationships also determined the relative prices of the different factors. But, of course, inflation is to do with the *overall* level of prices ('absolute' prices), and neoclassical theory held that this was determined not by 'real' factors of production but by the total quantity of money in the economy — i.e. inflation and unemployment were independent, Keynes, of course, showed they were interrelated (Ball and Doyle 1969).

Thus, just as total income and expenditure in the economy could be equal at less than full employment because there was an excess of saving over investment, inflation could result from the reverse situation where at full employment there was an excess of planned expenditure over income. This 'inflationary gap' would therefore increase prices; this was the original version of 'demand inflation'. Cost inflation could be seen as a derived demand inflation, where there was excess demand in a particular factor market — e.g. excess demand for labour pushes up wages.

Keynes also showed that money wages resulted from the relationship between prices, productivity and costs, and therefore sparked another line of inflation research, based almost entirely on an analysis of costs, with excess demand considered hardly at all.

Although Keynes seemed to relegate the stock of money solely to the determination of the interest rate, the quantity theory of money has been revived since the 1950s. The argument here is that the imbalance between demand and supply which causes demand inflation is itself caused by an excess of money in the economy. Hicks has also suggested a 'money' explanation for cost inflation, whereby the fact that labour is organized in unions gives it sufficient power to be independent of market forces and push up money wages. The current consensus does not accept money as being a prime determinant but both economists and government are giving it more and more attention. Ball and Doyle consider that most economists and governments accept that avoidance of over-supply of money is a necessary but not sufficient condition to halt inflation. It is not surprising in view of our Chapter 1 argument that money phenomena (and 'financial capital') are playing a larger and larger part in economic phenomena — this is to be expected as a market economy develops. The consensus view would accept that inflation is a measure of the underlying balance of forces — as Ball and Doyle put it, between aggregate demand and supply and the 'monopoly power' of organized labour, or, as Levinson would say, the monopoly power of the corporations to raise capital by raising prices above whatever level unions manage to achieve.

We would be generally in line with the consensus if we take the increase in inflation as another rough-and-ready indicator of the continuing difficulty of the 'motor-generator' of the market economy — the private sector — to maintain profits and to reduce costs in a situation of struggle with its workforce.

Phillips (1958) developed a technique to predict changes in wage rates from data on the level and rate of change of unemployment. According to the Phillips curve, the first 1 per cent increase in unemployment reduces wages by nearly 9 per cent, but an increase in unemployment from 2.5 per cent to 3.5 per cent only reduces wages by 1 per cent. This curve had a very close correspondence with the movement of wages and unemployment over a long period (e.g. 1861-1913 and 1953-66). But from 1967 it began to underpredict wage increases rather seriously. This is one piece of evidence used by Kennedy (1974) to suggest tentatively that inflation was largely demand-led up to that date and costs were much more important thereafter. Suggested causes of the so-called 'new inflation', which began in the late 1960s, include: import prices, increased money supply, trade unions and greater all-round anticipation of future price rises by everyone. This new inflation, together with the larger-scale fluctuations in output and employment, tends to support the impression of increasing 'steering' problems for the UK economy. These latest steering problems have led to a rethinking of Keynesian control techniques. The recent Conservative government tried to control wages directly, monetarists have been coming forward with 'new' ideas about maintaining a slow, steady growth in money supply, and radicals generally have tended to proclaim 'the end of Keynes'.

Kennedy shows how some of the main aggregates in the UK economy fluctuate in relation to the total economy:

Investment. This has generally continued to increase annually, with very much bigger increases in boom periods. But in the latest 1972-3 boom this increase was very small indeed. Manufacturing investment is the most volatile; the average annual deviation has been £130 million or 7 per cent. For other non-dwellings investment, the absolute deviation has been much the same, but this only represents between 2 and 3 per cent. Housing, although varying less, is very important for urban development.

Stocks. These have fluctuated in a similar way to investment but in a more exaggerated manner. Stocks have changed by up to £900 million (at 1970 prices) over a one-year period, but the stock fluctuations are typically only 1 or 2 per cent of total stocks.

Government spending. Although this is in part a control mechanism, it is also a causal factor in the cycle. It has not moved particularly with or against the cycles, and has been more dominated by changes in social policy and defence policy than by a desire to stabilize.

Consumer spending. This has been fairly closely matched to changes in output but by no means perfectly. It is by far the largest part of total demand, and although its fluctuations are small it has a very large effect. Kennedy notes that the average propensity to consume has gradually declined over the period 1950-72, this being consistent with Keynes's 'under-consumption' thesis.

Imports. These have moved in phase with total output, but again with larger fluctuations, they are usually taken into stocks first.

To calculate the effect of government spending as a method for controlling the economy, we have to know the 'multiplier' impact and how long the effect takes to work through. On various assumptions (Kennedy 1974), if an initial government expenditure of £100 million is made, then the new income received by firms and individuals is turned into further purchases (i.e. further consumption) amounting to £31 million one-quarter of a year later — that is, a 31 per cent effect. Thus *a quarter* of a year is the time scale over which new income gets translated into expenditure. In practice this time may be drastically changed by variations in the holding of stocks in the economy. If the government continues to inject £100 million in each quarter, the total impact is fully worked through in successive rounds of income and consumption generation, so that the full impact is £144.9 million (the multiplier is 1.44). Bray (1970) notes the increasing length of time needed to plan capital spending on roads, education, etc., and the even longer period needed to complete these projects. He argues not too convincingly that increased 'cost effectiveness and flexibility' rather than longer planning time is the key to greater control over public spending. The idea is to be able to increase or reduce spending without undue waste: 'there is no point in completing a building and leaving it unused, etc. etc., but there is no reason why a school has to be approved 5 years before it is needed'.

By now we should have a rough idea of the nature of the steering task. The way it has worked out in practice is described in Section 6 below.

Public spending

The new expenditure White Papers do enable the Cabinet and parliament to obtain some kind of total picture, and hence some kind of control of the pattern of spending *as a whole*. They are a marked improvement over the old system of piecemeal interdepartmental bargaining with the Treasury. The Public Expenditure Survey Committee (PESC) tries to predict the expenditure in these categories for both existing and additional commitments, but there has been a consistent tendency to underpredict public spending. There is also an unresolved debate as to the definition of public spending.[1] Since projections are made at base-year prices, the fact that costs tend to rise more in the public sector than in the private sector means that even a constant predicted share for the public sector implies an actual rise in the public share in terms of *current* prices.[2]

Some Cambridge economists have recently argued that public spending is 'out of control' — especially spending by local authorities — but other evidence to the Layfield Committee on public spending has refuted this, saying that increases in the *prices* of public goods in a period of high inflation accounts for most of the underprediction of public spending.

Again, where physical policies are concerned, such as schools, buildings and roads, the final cost may be different from that predicted. Openended subsidies to agriculture or deficit subsidies to nationalized industries again can only be estimated. Social security — a growing component of national income — again depends on take-up rates. Although sometimes used as a regulator for the economy as a whole, changes in public capital spending can have disastrous and wasteful effects in the short term.

It is not surprising that capital expenditure, which is more 'discretionary' than much of the current spending, tends to take the brunt of government cutbacks despite the waste and disruption caused by intercepting or delaying ongoing projects. A recent classic example has been the encouragement to local authorities to buy up old property, followed by a central government cut in funds allowed for improvements — leaving many local authorities as slum landlords. It is a similar story with new building projects, with wasteful cancellations or rephasing taking place. Dow shows that, largely because of the long time lags involved, control of public investment has not proved much of an instrument for stabilizing the economy. It takes two or three years to make major changes in public investment; tax and money policies tend to be quicker acting.

Alternative definitions of the size of the public sector include:

1 *Employment* — 25 per cent of total employment is in the public sector.
2 *Total purchases* of goods and services by central and local government — this was 30 per cent of GNP in 1970, rising to 32 per cent in 1974.
3 *Current spending* by central and local government — 20 per cent of GNP in 1970, rising to 22 per cent in 1974.
4 *The Treasury* definition — a wide one covering all the flows of income passing through public hands including transfers, national insurance and capital spending of nationalized industries — 52-53 per cent of GNP in

1 Investment grants are 'spending' but investment 'allowances' are not — they are offset against revenue.

2 In 1976 cash limits were imposed.

1970, and 60 per cent in 1974. This increased from about 43 per cent in 1963.

5 *Brittan's* own definition comes to 42 per cent of GNP, and represents the total deductions in taxes, national insurance, rates, etc., from private income, needed to finance the public sector.

These definitions show quite clearly how easy it is to overstate the size and growth of the public sector. The true size is represented by the purchase of goods and services — approximately 30 per cent; this is the fraction of GNP taken up by the public sector. The Treasury definition makes the public sector look spuriously large by including all the transfer payments — money simply taken from one sector of the economy and given away to another. This is the redistribution role discussed earlier which is so subject to political controversy and political debate. It is these transfer payments which have grown faster than the real claims of the state on the GNP. To include transfer payments in a comparison between the size of the state and the total GNP is simply to be guilty of double counting — and over two-thirds of the real national product is still accounted for by the non-public sector.

The public sector is usually classified into three parts (Rees 1973):

1 Central government — employs 8 per cent of the total labour force in the country (2 million).
2 Local government — employs 10 per cent of the labour force (2.5 million) and incurs roughly one-third of state expenditure.
3 Public corporations — employs 7 per cent of the labour force.
4 State shareholdings. This interesting and little-known part of the public sector reflects the growing integration of the economy noted in Chapter 1. The mixed economy is now penetrating inside individual enterprises. There are many partially and wholly owned companies. Public control over their operations is very unclear, and scrutiny is by no means close. On the other hand, the government is less circumscribed by statute than it is with the public corporations and is free to operate them as it wishes, subject to company law.

The increase in public spending noted by Brittan would be expected from our earlier discussion of the state. He sees the 'danger point' for liberty now being reached in the increase in the public sector. This idea of the 'danger point' has been echoed recently by politicians — including members of the Labour government in the debate leading up to the public expenditure White Paper of 1976, which proposes reductions of up to £3000 million in public spending up to 1979-80. The real size of the state in terms of spending on goods and services including wages had grown to 35 per cent by 1975-6, and the aim is to cut this to 28 per cent by 1979-80. Transfer payments rose much faster (from 19 to 26 per cent from 1971-2 to 1975-6). But this 'danger point' idea begs many of the really interesting questions. Brittan, for instance, makes the size of the public sector entirely one of social 'choice' rather than a necessity for maintaining the stable economy — that is, 'which goods do we want to produce collectively, how much do we wish to spend and to whom do we want to transfer purchasing power? But it may well be that this kind of clear political choice just does not exist. The state has to be large in the modern economy. Whether or not we 'want' to produce goods collectively, it may be that we just have to, if the economy is to be kept stable. Brittan does conclude firmly that spending may not be entirely under control, especially in weapons, aircraft, space and atomic energy where there are 'complex projects and powerful commercial interests'.

It is quite remarkable how most of the standard works on state planning or public control avoid discussing the implications of the central simple fact that the public sector is growing in relation to the total economy. Is it getting too big? Does it affect profits? Does the growth seem to be approaching a limit beyond which it cannot increase without changing the character of the system as Pryor's evidence would suggest? None of these issues is addressed by Brittan (1971), Dow (1970), Meade (1975), Shonfield (1965), Livingstone (1974), Donaldson (1974), Cairncross (1970), Prest and Coppock (1974) or Bray (1970). Perhaps more surprisingly, Holland's '1975b) *Socialist Challenge* doesn't either.

This is the fundamental question about the role of planning in particular, and the state in general. The arguments outlined in Section 3 above, taken with the fact that the state seems to take up only a minor part of the GNP of all major Western countries and that the growth of the state *in real terms* has been fairly slow, do seem to be consistent with the idea of some kind of upper limit or 'barrier' to the real growth of the state. Further evidence of the difficulties in improving and extending state planning in the UK also seem broadly to support this idea, together with the associated notion of a delicate and possibly ultimately contradictory interrelationship between the 'three estates' — the state, the private sector and the wage- and salary-earning population.

The composition of public spending

'Social consumption' expenditures (O'Connor 1973) dominate public spending. The largest sector is social security, followed by education and health. Defence and debt interest ('social expenses') have typically been of the same size as education and health. Next comes capital spending by nationalized industries, then housing and industrial grants. Of these major sectors, defence has been declining in its share over recent years, while all the others (except debt interest) have increased by more than the percentage increase in total spending (3 per cent between 1968 and 1972). The implication for the urban economy will become clear later, but it is apparent that the largest sectors are also very labour-intensive and are generally geared to 'social reproduction', and the social support of the labour force and the population in general. Many of these services are specifically geared to local labour-market areas and population centres and, therefore, intimately bound up with urban development.

Social consumption by the population is therefore taking an increasing proportion of the cake. Although private consumption may tend to lag,

	Category	Proportion of total public spending (1964 in brackets)	Proportion of the category which is 'real' resources	Proportion of the category which is capital spending	Proportion of total public spending on real resources taken by the category
'Social consumption'	Social security	16% (16)	5%	0	2%
	Health and social services	11% (10)	99%	9%	20%
	Education	12% (11)	81%	10%	18%
	Housing	9% (6)	42%	42%	7%
	Environment	4% (4)	100%	43%	7%
'Social investment'	Agriculture	3% (3)	12%	1%	–
	Transport	5% (4)	61%	60%	5%
	Commerce and industry	10% (8)	40%	32%	7%
	Roads and lighting	3% (3)	100%	60%	5%
'Social expenses'	Defence, ext. relations	12% (18)	90%	1%	20%
	Police, etc.	3% (2)	100%	9%	5%
	Debt interest	9% (11)	0	0	0
	Other	3% (4)	2%	1%	4%
	Total Public Spending	100%	57%	17%	100%

Source: National Income and Expenditure 1964–74; Central Statistical Office (1975).

Figure 5 Composition of Public Spending 1974

public consumption is potentially far higher than it is now — health, education and housing are all still very overstretched services. The ratio of labour costs to other current inputs is increasing, but these services are also very capital intensive, requiring elaborate buildings and equipment. Whichever political party is in power, the general increase in the public sector continues, and the main sectors of social consumption continue to become more and more costly.

Out of a total current and capital public spending of £41,606 million in 1974, £6918 million (or 17 per cent) is capital spending, and the total is broken down as in Figure 5.

The table brings out more clearly the relation between real claims on resources and the transfer payments, which now represent 44 per cent of total public spending. Although social security takes a massive 16 per cent of the total budget it only represents 2 per cent of the total real resources commanded by the state. Conversely defence, health and education are much more prominent users of real resources, each taking up to one-fifth of total real spending. Comparing column 2 with column 3 shows the capital intensity of the category. Housing and transport, followed by industry, are the most capital-intensive activities, with defence being the least intensive. Defence is thus unique in commanding significant real resources but contributing nothing to future growth or productivity.

6 The history and effects of economic controls

This section shows how effective have been the various budgetary and other measures used to control the economy, and at the same time tries to bring out the essential difference in the impact of the different controls, especially on those aspects that most affect urban areas. We are thus discussing the general effectiveness of the various controls at the same time as giving a brief history of UK economic management since the war.

We have already seen that, on the expenditure side, the government has some difficulty in using its own expenditure to control the economy in the short term: the effect is too delayed, it is wasteful and non-productive. The income side, on the other hand, looks more promising. We have already noted briefly the main ways in which the government can raise finance and influence the economy through the budget, and the two most important are taxes and monetary measures. The government has two things in mind — first, to achieve its stabilization goals (and it is therefore interested in knowing the effect of its tax and money changes on the economy) and, second, to achieve a satisfactory balance (or desired imbalance) between its own income and receipts. It is worth bearing in mind where the government (central and local) gets its money from.

The burden of taxes has shifted more towards wage and salary earners, and away from companies. The former contribute the bulk of taxes as seen from Figure 6. King (1975) shows how the taxes on companies has decreased in recent years. An 'average' married man now pays 25 per cent of his income in tax compared with only 10 per cent in 1960-1. This shows up another

	£1 000 m	*Percentage*
Income tax on wages and salaries	7 089	25
Net expenditure taxes on consumers	6 462	23
Income tax on companies	1 940	7
National insurance payments by employees	2 019	7
National insurance payments by employers	2 733	10
Rates	2 991	10
Rent	1 958	7
Interest	830	3
Trading surplus	2 545	8
	28 557	100

Figure 6 Main sources of income and public authorities 1974

interesting sidelight on the view of the state and its struggle to maintain profits: in shifting the burden of tax it is now coming up against another 'limit' or 'barrier'. The tax 'threshold' has descended to such a low level of income that it has crossed the 'social security' line — so that people now start paying taxes at incomes lower than the supplementary benefit levels. Thus an unemployed person may lose more in income tax and loss of social security benefits when he or she takes a job than the extra income gained from employment. The marginal rate of taxation on these low incomes is thus more than 100 per cent.

By 1974-5, when corporation tax had been separated out from personal income tax, companies did appear to be contributing proportionately rather more taxes. But generally, as GDP fluctuates, profits may do so even more, and so therefore does the contribution by companies to the total tax bill. This has several times actually decreased (in money terms) between one year and the next, whereas personal tax payments never do!

Bray (1970) gives a useful checklist of the detailed economic control measures which have been used and the effectiveness and limits of each. Figure 7 lists some of the more important ones.

There is much disagreement among mainstream economists over the effectiveness of the various types of tax and money measures that have been applied since the war. The various commentators do seem to feel that economic management has been modestly successful in limiting unemployment (at or below 2 per cent until the 1970s) and inflation (again until the 1970s) and even securing a degree of expansion and growth, especially during the 1950s. Most also recognize that fluctuations have occurred and these have been costly, and that the instruments used may have been too crude or of the wrong kind. It is generally recognized that the controls themselves have probably been a factor helping to cause slow growth and investment.

We have already seen that the controls themselves *may* have been wrongly phased so that in effect the policy has often been 'destabilizing'. But there are disagreements about how the various controls interact with each other and

	Direct effect (of tightening the control)	*Delay in implementation*	*Uncertainty of total impact*	*Limitation and comments*
Control				
Tax on individual incomes	Reduces spending on durables, e.g. cars; affects industries which are rapidly growing	6 months	low	Increases wage pressure; complex effects on distribution; cannot be changed quickly (e.g. between budgets)
National insurance contributions	As above but less effect on high-priced durables	6 months	low	Not usually regarded as regulator. Bears hardest on lower incomes.
Taxes on spending	Purchase tax used to hit a narrow range of goods. Now VAT can be spread more widely	zero	low/ medium	Tends to affect lower-income groups. Not a good regulator; has been too selective in a narrow range of industries, e.g. cars, tobacco
Business taxes – i.e. corporation tax	Reduces tax on retentions of profits for investment	1 year	low	Does not encourage new firms; separates business tax from personal tax
Monetary controls				
Bank rate	Discourages loans and investment	1 month	high	Increases general level of interest rates and discourages all forms of capital spending
Building societies' borrowing rate	Reduces demand for private housing	3 months	medium	Arbitrarily links one sector of money supply with one industrial activity
Public borrowing rate	Increases cost of public investment	zero	high	A poor regulator; it does not determine total demand which is anyway directly controlled
Hire-purchase controls	Reduces spending on a narrow range of goods, esp. consumer durables	zero	medium	Easy to apply and quick acting – but affects a narrow range of industries
Bank advances	Reduces funds available for loans	3 months	low	Can be used selectively to limit different categories of borrowing, e.g. for imports

Figure 7 UK economic controls
Source: After Bray (1970)

with the economy. There are disagreements as to whether taxation or money measures have had the greatest effect, reflecting a theoretical confrontation between 'monetarists' and extreme Keynesians who believe money matters not. Kennedy, not surprisingly, opts for a consensus — that all types of control have their place.

According to Dow, between 1945 and 1960 taxes dominated economic control policy more in the UK than in other countries. After 1945 the physical rationing controls were enough to reduce demand for imports by up to 10-15 per cent. Dow argues that such controls can work only in special circumstances. The economy was really only free to fluctuate after 1951 — after these controls were removed. Up to 1960 taxes were generally reduced in the budgets. Taxes tend to affect consumption much more than they affect investment (and investment is the key to urban development), but Dow firmly argues that taxes did have more effect than monetary or other measures. Dow makes a determined attempt to identify the effects of different policy measures on the behaviour of the economy. He concludes that the measures in 1952-4 (tax cuts), 1957-9 (credit increase) and 1959 (tax cuts, the largest since the war) did have a major impact on the expansion that took place during these years. In his view it has been difficult to influence the economy using taxes on spending because these inevitably fall on a narrow range of highly taxed goods, although the introduction of VAT has spread the net wider. Public investment and public spending was in the event not (according to Dow) determined mainly by the anti-fluctuation policy. But it did have an effect on the economy, and both public spending and investment were increasing from 1948 to 1953; both declined from 1953 to 1958, partly in order to restrict demand. In 1959-66 both increased and added to the expansionary pressures.

The effect of a tax change again depends on a multiplier effect. According to Kennedy, the 3p change in tax rates in the 1974 budget raised a direct £94.2 million; the first-round indirect effect on GDP was £521 million or 0.75 per cent of GDP, and the total final effect was approximately 1 per cent of GDP. The effects of a change in taxes on spending are much harder to work out because prices change and the demands for different goods and their possible substitutes also change. A rough estimate by Kennedy of a 10 per cent change in VAT (e.g. from 10 per cent to 11 per cent) suggests an immediate effect of 0.5 per cent of GDP and a final total impact of 0.7 per cent. The importance of these figures is that an apparently severe tax change has only a relatively small effect on the economy — small, that is, in relation to the 4 per cent fluctuations that have occurred in GDP within one year. This means that considerable disruption may be borne by particular income groups or by particular industries (and hence particular areas) to not very much purpose. This in turn is a further reason why some of the more direct and quicker-acting money and credit controls are called into play. But there is general agreement with the Bray table that both taxes on spending and hire-purchase restrictions again bear selectively and unfairly on a narrow range of industries, especially cars, even though the effects do spill over into other sectors.

Some industries have suffered from the many changes in the tax and credit rates which apply to their products and therefore affect demand. 'Uncertainty' is a consistent reason given by industrialists as to why they don't invest — uncertainty often produced by rapid changes in government policy. This uncertainty probably helps to shake up and restructure industry, with the larger firms able to concentrate on what they hope are less vulnerable 'core' operations, subcontracting out the manufacture of components so they carry fewer cost commitments in the face of a change in demand. We shall see some of the effects of this at local 'urban' level in Chapter 3. The same changes in policy have also affected nationalized industries, who often complain that they are not allowed to implement sensible medium-term investment policies. They are urged alternatively to balance their books by charging higher prices and to provide a subsidized social service. There appears to be little or no overall rationale based on precise calculations as to the appropriate level of subsidy for different nationalized industries, in relation to the economic and social benefits they bring to the economy at large. As mentioned in the discussion on welfare economics, the decision to subsidize depends on the industry's contribution to the total economy in relation to its costs. This is not an easy calculation, but one which has to be performed.

The Bray table is slightly confusing in that it may seem to indicate that a change in interest rates acts rather quickly; but of course the full effect on the economy is delayed for a long time after the measure comes into operation. Indeed, Kennedy argues that, although credit restrictions and cutbacks on money supply can act fairly quickly, monetary instruments in general are more uncertain and delayed in effect and therefore will generally take second place to taxes for shorter-term control. Thus Dow (1970, p. 174): 'The British Chancellor of the Exchequer is in a position of unique strength among the finance ministers of the world. This explains why tax policy has a more central role here.'

The following are the major turning points between 1945 and 1960:

1948 Wage freeze (Cripps)
1949 Devaluation
1950 Defence spending increased by 50 per cent
1951 Butskellism — 'planning and freedom', investment allowances, appeasement on wages
1954 'Turning point' — rejection of floating pound
1955 'Error' — tax relief in middle of boom
1957 Thorneycroft, first stop/go
'Turning point' — first squeeze in a depression; missed opportunity to devalue
1959 Biggest tax handouts ever
1960 Policy reappraisal (Brittan 1971, p. 230)
Not controlling unemployment
Not boosting exports
Credit controls damaged the car industry just when demand was

> already falling. This brought industry itself to question these short-term policies. They were now seen to have shortcomings, even in their own terms.

Dow (190, p. 403) concludes that the lessons of 1945-60 are that economic strategy should be broadened with more emphasis on the longer term (prices and growth). According to him this means increasing industrial capacity *in association with* expansion of demand through short-term policies, otherwise 'price inflation will get out of control', but in general he argues for fewer tax and money changes. However, much of the argument is dominated by the 'conventional' views of inflation — 'cost push or demand pull'. Dow regards these methods of controlling the economy as an 'inheritance of an earlier age' which are effective only in a rough-and-ready way. Dow sees the key to the desired longer-term economic control to be the influence on *private* investment, again an issue which has now come to the very forefront of the political and economic debate. He tentatively suggests that fiscal measures, and particularly investment allowances or grants, are most likely to be effective, possibly with firmer controls on investment (e.g. capital issues control, building controls, etc.). But here Dow's own arguments seem to symbolize the failure of the Treasury method of management and its inability to tackle the underlying structural economic problems head on. He first of all speculates as to how effective the various modifications in short-term Treasury tax and money policy might be on long-term investment and growth, and only then is he able to look at the 'planning' option.

The early 1960s are generally supposed to represent a watershed, with the reassertion of longer-term government 'planning', and a new awareness of the ineffectiveness of short-term control. We shall discuss the planning initiatives of the 1960s after a discussion of the continued operation of the traditional controls. Indeed the two can be discussed almost independently, for it is a characteristic of all attempts at planning in the UK that they have had little direct power to affect existing arrangements for controlling and directing economic resources.

The budget saga continues, with small adjustments to correct a balance of payments deficit during the first two years of the 1960s, followed by an extremely expansionary budget in 1963 prior to the 1964 election.

Brittan characterizes Labour's economic management from 1964 to 1966 as bad in the short term (especially the failure to devalue — a self-imposed straitjacket), but possibly laying the basis for longer-term improvements. Conservative budget policies in the early sixties having failed to correct the growing overseas imbalance, Labour struggled against increasing odds to correct this until devaluation in 1967; the whole short-term control of the economy was geared to the 'defence of the pound'.

1960 Mildly expansionary budget
1961 Contractionary budget
1962 Neutral budget
1963 Very expansionary budget — first deficit for a long period
Labour elected

1964 Taxes increased, interest rate increased
1965 Tax increased
TUC agrees to wage 'norm' 3–3½ per cent
1966 Corporation tax introduced at 40 per cent; selective employment premium; squeeze on government spending and credit; wage freeze
1967 Neutral budget
Relaxation of squeeze
Devaluation
1968 Tax increases — budget surplus
1969 Tax increases — budget surplus
1970 Tax cut 'moderate' — budget surplus
Conservatives elected
1970 Conservatives cut public spending
1971 Smaller budget surplus, some debt repayment
1972 Very expansionary budget, small surplus, large borrowing requirement, increased money supply
1973 Even more expansion, despite bad trade gap, budget deficit and more borrowing
Labour elected
1974 'Reining in'
1975 'Social Contract', wage increases limited to £6 per week
1976 Continued squeeze, small tax cut in exchange for 4 per cent wage limit, attempts to control money supply

It is not worth dwelling on the details of budgetary management in this period; the point should have been made by now that not only did this ultra-short-term management fail to halt the decline of the UK or to restructure the economy, but it has also become (partly because of this very failure) an increasingly difficult task in itself. Thus in early 1974, the *National Institute Economic Review*: 'it is not often that the government finds itself confronted with the possibility of a simultaneous failure to achieve all four main policy objectives — adequate growth, full employment, a satisfactory balance of payments and stable prices.' The problem of the 'new inflation' already referred to seems to have changed the traditional trade-off relationships such that in mid-1975, with an unemployment of over a million, there was still talk of further *reducing* government spending — 'Keynes overturned'.

The increasing importance of monetary policy in this recent period highlights the advanced development of UK financial institutions and the complex and interlocking relationships between the Bank of England, discount houses, clearing banks, secondary banks, finance houses and building societies (Gibson 1974). Here again power is concentrated in a few large institutions, but there has been considerable restructuring as the sector overall has grown in importance with an especially large growth in the secondary banks. Innovations in the use of quantitative and qualitative controls for money and credit in the early 1970s introduced an element of competition between the four large clearing banks so that their interest rates could vary in order to attract money. This was coupled with some control on credit, but bank deposits did grow in the early 1970s, especially in the secondary banks.

But how does the saga fit with our previous discussion of the limits of the mixed economy? The problem of the national debt hasn't been very apparent until very recently. The state at local level (i.e. local authorities) suffers much more from the problem, and this is discussed briefly in Chapter 4. The bizarre effects of the short-term anti-fluctuation controls often bear selectively on particular industries and tend to encourage restructuring. Mergers might be regarded as a defence mechanism by private industry against the vagaries of controls. Larger firms are generally (though not necessarily always) in a better position to withstand the constant assaults on their markets, or on their sources of finance.

The property industry itself illustrates rather like a speeded-up cameo, how short-term controls help to push firms towards further dependence on financial institutions, and towards mergers. During the 1960s the supply of buildings for offices and shops became an increasingly specialized activity undertaken not by firms and industries who used these buildings, but by specialist property companies. This is the increasing division of function ('division of labour') noted in Chapter 1. However, during the boom in the money supply in the early 1970s, a large number of these companies and fringe banks made large profits out of buying and selling assets (buildings) which were rapidly appreciating. This shows first of all how such a rapid release of money could not be absorbed by industry but went into buildings or into purely monetary gains. Second, when inevitably the brakes were applied all the over-inflated parts of the system — the fringe companies — collapsed first; the few largest companies which in the early 1960s only accounted for 17 per cent of total assets of the industry had well over 50 per cent in the 1970s (Barras and Catalano 1975). At the same time these companies themselves were being absorbed into the financial sector itself (insurance companies, banks, finance houses) as the increasing size and scale of development required more capital, and the finance sector increasingly took equity stakes or otherwise took a direct interest in the assets of the companies. In this way the short-term fluctuations and government measures to counter them act as a periodic 'shaking of the sieve', exaggerating the effect of competition — speeding up the longer-term trend to concentration and increasing the role of finance in the economy.

7 The 'planning' experiment

For Smith (1975) the reappraisal of 1960 led to the only period of 'planning' (other than the immediate post-war period) in British history. He acknowledges that 'planning' was what was intended rather than what was actually achieved.

It would be fully consistent with the theoretical arguments outlined earlier, and also with Galbraith, that, while short-term economic management was becoming more necessary and more difficult, business and politicians should become more interested in longer-term planning, something which would help to sustain a smooth growth path and avoid some of the negative effects of the 'short-term' fetish. The mistakes which hit the car industry around 1960 helped to change business attitudes in favour of planning. A whole panoply of new institutions and procedures were introduced in this period. A list is given below. Beginning under the Conservatives, the style of developing new *organizations* and *procedures* was adopted by Labour — often as a substitute for effective direct control over real resources. Besides these new bodies, some of the existing short-term control procedures were reformed as we have seen. These included longer-term projection of public spending, incomes policy, new short-term 'regulator' powers, new budget accounts and new policies for the nationalized industries. The important thing is to discern where the new institutions actually had real powers to influence and control resources in the economy, either through direct intervention or through some other government body. In fact, as already stated, the striking thing about the whole exercise was its divorce from the existing arrangements for control of resources; planning was 'indicative' in the extreme, exhorting from the sidelines.

Changes within government

1961 Treasury reorganized following Plowden report; individual departments given more scope; more consideration of real resources, not only money.

1963 Board of Trade extended to cover industry and regional development.

1964 Labour government created the Department of Economic Affairs (DEA) and Ministry of Technology. BoT and Treasury functions reduced.

1965 'Government Economic Service' constituted. Regional economic planning boards set up in each region.

1968 Fulton report on civil service; Civil Service Department created.

1969 DEA abolished; planning returned to Treasury. Ministry of Technology enlarged to include Power. Housing and Transport grouped under a Minister for 'local government and regional planning'.

1970 Conservative government combined BoT and Min. Tech. into a Department of Trade and Industry (DTI); Central Policy Review Staff (CPRS) attached to Cabinet Office; Department of Environment — combined housing, planning, transport, local government.

1974 Labour elected. Split Department of Industry (DI) from Trade and set up Department of Energy.

Quasi-governed agencies

1962 National Economic Development Council (NEDC), National Economic Development Office (NEDO) and National Incomes Commission (NIC).

1964 Economic Development Committees (EDC) formed for particular industries under NEDC; Industrial Training Boards (ITB)* formed for particular industries.

1965 Regional Economic Planning Councils (REPC) formed in each region. Prices and Incomes Board (PIB) formed instead of NIC.

1966 Industrial Reorganization Corporation (IRC)*
1969 Commission on Industrial Relations formed (CIR)
1970 Conservatives abolished PIB and IRC and reduced the numbers of EDCs. Created Office of Manpower Economics (OME), National Industrial Relations Court (NIRC)*.
1972 Industrial Development Executive (IDE).
1973 Pay Board and Prices Commission created
1974 Labour abolished NIRC and Pay Board. Finance for Industry (FFI)* strengthened, proposals to establish National Enterprise Board*.

Smith points out that these institutional innovations were inspired by French 'indicative planning' presaged in the suggestions of the 1930s by Mosley, Salter, Wootton and G. D. H. Cole, and by the general disenchantment, in sections of industry, government and labour, with the failure of Keynesian management. It is significant that, initially, this move towards a more covertly 'planned' economy was introduced in many ways as an apolitical move, a drive for efficiency introduced by the Conservatives.

Shonfield's (1965) imaginative and forceful history of the British failure to plan properly for the longer term takes the view that sociological rather than economic reasons have been paramount — 'the striking thing in the British case is the extraordinary tenacity of older attitudes towards public power'. He points out that in 1948 Britain was actually *in advance* of other European countries, in having the basic elements necessary to institute effective government leadership and planning — Development Councils, modern budgeting methods and a large battery of Treasury controls; social welfare and other public spending was well advanced. The Labour government of the time talked about planning (just as they did later), but in practice even the stringent direct physical controls were used only for short-term ends, to ease shortages; indeed the latter were the sole rationale of planning. Even within government, coordination was *ad hoc*. The nationalized industries were given an independent, *laissez-faire* brief, with coal, gas and electricity authorities going their separate ways.

The public sector was regarded by both parties as essentially passive, and certainly not as an initiator. Shonfield's arguments support what was said in Chapter 1 about the 'poverty' view of social policy. In Britain public social provision was seen as a minimum absolute standard, for housing, unemployment pay (a fixed sum for everyone) or social welfare 'directly related to the traditional act of charity'. The European view tended to be that those drawing on state welfare were entitled to the *average* standard of the community as a whole, with benefits related to earnings. The only difference Shonfield can discern between Labour and Conservatives has been that Labour tended to demand less proof of 'need' than the Conservatives. There was no instinct to increase the 'depth' of public services.

* Interventionist agencies; the others — the 'planning' agencies — were advisory, or otherwise non-executive.

Both ministers and civil servants tended to avoid the responsibility for initiating anything, while industrialists were suspicious of the post-war industrial development councils as a waste of money and effort on their part. With consumption, productivity and investment all being relatively healthy during the 1950s, the rhetoric of planning was dropped. But again Shonfield stresses sociology, and speaks of a sudden 'ideological wave' sweeping Whitehall in the early 1960s — a rebellion against administration by short-term expedient. Writing in the mid 1960s, he detected a desirable shift in attitude to the use of positive public power, with outsiders brought into the civil service, and a new willingness on the part of government to *discriminate*, to be active and to encourage 'best-practice' firms. In the event this optimism was unfounded; 'planning' did not succeed in achieving an improvement in economic performance.

What Smith calls the planning agencies — NEDC and DEA — were essentially divorced from any direct control over resources (real or financial). NEDC was a tripartite body, independent of the government with participation by organized Labour (Trades Union Congress (TUC)) and the Confederation of British Industry (CBI). Its objectives were: (1) to examine the national economic performance; (2) to examine obstacles to growth and efficiency; (3) to seek agreement on ways of increasing competitive efficiency and the rate of growth. Under the Conservatives, NEDC produced a plan for a 4 per cent growth rate; the associated NEDO was the greatest 'concentration of economic expertise in the country'. There was also an industrial division concerned more with the 'micro' problems of individual industries. In 1964 the DEA became the premier centre of longer-term planning. The so-called interventionist agencies, unlike the planning agencies, did have some real functions — and Smith shows ironically that, after the National Plan was abandoned in 1966, they actually increased their activities and possibly became more effective. Yet the IRC, the most significant interventionist institution, had an initial stake of only £150 million — a derisory sum. It engaged in a game of 'monopoly' but in relation to total investment in manufacture it remained a tiny factor. It did, however, succeed in promoting several large-scale company mergers, notably British Leyland, GEC/AEI/EE and ICL.

The rationale behind all the reforms was, perhaps inevitably, never formally spelled out, but essentially the DEA had overall responsibility for producing economic plans — it took only eleven months to produce the National Plan (DEA 1965) as a longer-term backdrop to short-term Treasury management, with a 'creative tension' between the two institutions. NEDC was to develop a 'moving consensus' and act as a 'switchboard' between industry, government and labour. This set-up meant an almost complete divorce between 'planning' (viewed purely as the production of statements of policy, targets and analysis) and executive functions with real power over resources (the Treasury). No doubt there is some imaginative philosophical piece to be written on this orgy of institutional creativity and destructiveness; but it did seem to some extent that Dow's disease of too many budget changes had spread further so that Whitehall was continually preoccupied with restructuring and reorganizing, merging and splitting departments and agencies. No doubt some would say that 'creative tensions' between institutions are the product of an individualist view of social change. If personalities and people matter more than social forces and economic power, then bringing in a new 'personality' — a reorganized ministry or an IRC — may appear an effective move in itself.

There was much overlap and confusion between these bodies, partly because of the personalities involved. Some EDCs acted almost as an IRC, and the Wool Textiles EDC overlapped with the Yorkshire and Humberside REPC. The IRC was promoting industrial mergers to reap economies of scale and develop large firms capable of competing in world markets, whereas the Monopolies Commission was trying to operate as a watchdog against big firms.

It is relatively easy to say, then, that the new planning, although introducing a new rhetoric and some genuinely new perspectives on economic problems, was *in execution* much weaker even than the old 'management'. There were three attempts to produce 'national plans' integrating all aspects of the economy — the NEDC plan, the National Plan (DEA 1965) and the Task Ahead (DEA 1969a; see also Barker & Lecomber 1969). But much energy was wasted trying to coordinate and sort out the overlaps of the different bodies. The complete separation of planning from the powers of execution meant that the REPCs operated basically as pressure groups (especially in the North, North-West and South-West). Some were less visible and less effective. The central government agencies in the regions, the so-called REPBs, were, as might be expected, more passive. Regional policy is dealt with in the next section.

There was little political opposition or querying of the NEDC from either side in parliament, other than from the extreme Right. In Smith's view planning came to an end with the crisis budget measures of July 1966 followed by devaluation in 1967. In the terms of our previous discussion, this was merely a more formal reassertion of the dominance of short-term Keynesian management over longer-term economic development. 'Formal' because it had never really ceased to be the only aspect of government economic management or planning which had executive teeth. Without clearly saying so, Smith shows how the major economic trends outlined in Chapter 1 — especially the rise of the multinational company and the increasing strength of world economic pressures on the UK economy — meant an immediate retreat from the very reforms which might have put it in a better position to withstand these pressures.

The Conservatives in 1970 made an even more formal retreat and tried to reinforce a free, unfettered market. Gradually, measures were introduced to keep down the level of wages and salaries. This was operated first by using the public sector to take the lead in keeping down wages. In Glyn and Sutcliffe's terms there was a certain logic in this: it aimed to lower one of the barriers to renewal of the mature economy noted in Section 1 — the organized strength of the industrial workforce and its ability to maintain its standard of living. However, in the event this strength was sufficient to resist and finally overthrow the government after it resorted to more legalistic measures and to more direct control of wages. This led to massive industrial action, culminating in a 'who governs Britain' election, at which Labour was re-elected as a minority government.

During the Conservative 'interregnum', Labour set up an unprecedented number of study groups, began to re-cement its broken ties with the trade unions and came up with a new 'analysis' and a new approach to economic problems. In *Labour's Programme* (1973) the most important proposals were towards more direct public control over industry, for three new reasons:

1 Industry was already receiving massive aid and grants from government; those receiving these resources should be more held to account; the government should get more direct return for its outlay.
2 Many of the short-term problems of steering the economy and the other things which affect investment, such as the outflow of capital from industry into 'non-productive' activities or overseas, were directly due to the large firms, '250 of which' effectively controlled the economy. These firms should therefore be directly accountable to the wider national interest.
3 There is no reason why the public sector should not benefit from operating in profitable manufacturing industry and not only in loss-making national basic services, which in any event are subsidizing the profitable private sector by underpricing their goods and services.

There is a certain logic in the analysis: it comes face to face perhaps for the first time with the 'new' facts of concentration in the UK economy and embraces Shonfield's arguments for positive *active* discriminatory government (Holland 1975b). The aim was to 'harness' the power of the big firms in each sector by taking a leading firm into direct public ownership and influencing the rest by means of planning agreements.

The main output of this analysis was the proposal for a National Enterprise Board, which at the time of writing has just been constituted with powers to take firms into public ownership *by agreement* and to make formal 'planning agreements' on the future operations of large private firms. There is every sign of history repeating itself yet again, as the board has already been largely disarmed before it has been launched. Its initial resources are tiny (£700 to £1000 million), it will be able to acquire very few companies and it will not be able to participate in a firm's activity without that firm's consent. In fact there is at present a much greater uncoordinated, *spontaneous* state takeover of private industry on account of bankruptcy and other financial problems within the mature economy (Rolls Royce, Upper Clyde Shipbuilders, Ferranti, British Leyland), in addition to proposed further 'traditional' nationalizations such as the shipbuilding and aircraft industries. Nevertheless the NEB represents a strengthening of the executive 'interventionist' agencies, but now in the absence of an overall plan, other than a basic requirement to improve investment in the 'manufacturing industry'. However, the NEB idea, and the analysis that supports it, does appear to reflect a growing level of economic awareness in politics. While it would be too much to say that the increase in public ownership in Labour's programme 'points to the end of the private sector', it nevertheless points more directly towards the genuine alternatives facing the economy. The choices are becoming starker, and more clearly identified. The reaction of the Right is also significant; whereas here, as on the Left, there has previously been a tendency to regard government as a relatively independent force (although a malevolent one), there are signs that there is increasing recognition that both public and private sector are part of the *same economy.* So while the Left says 'public money' is going into industry, the Right now counters with the essentially correct point that 'public' money is essentially 'private' money taken in taxes or loans by the government.

The fact that any new proposals, such as that for the National Enterprise Board, are so rapidly and so completely straitjacketed back into the same traditional mould could itself be consistent with the idea that the state is 'limited'. If it did break out of its straitjacket, say in the guise of a large and active National Enterprise Board with teeth and resources, then the private sector (e.g. the stock market) might simply collapse — and the whole of British industry would fall into the lap of the state for a song. Otherwise, why the rush to disarm a relatively sophisticated instrument for improving the structure of the UK economy? It may be on the other hand that the industry cannot recognize its own best interest — that a National Enterprise Board suitably manipulated by business interests could well have been used to boost profits.

8 Regional planning and policy

The instruments of policy

We now focus on one aspect of the new economic management of the 1960s which did have some impact, where major resources were involved, and which was important for urban development. We have seen how the Regional Economic Planning Councils and Planning Boards were set up in 1965 as part and parcel of the overall attempt to plan the economy. Since the national plans had so little effect on the national economy, it is not surprising that the various regional plans had little impact; indeed they were usually modestly titled 'studies' or 'strategies', not plans. The fulfilment of the National Plan itself was dependent on the utilization of regional resources, especially labour reserves. But as we have seen, there were no direct powers available to implement national or regional plans. Consequently the various 'strategies' — 'Strategy for the South-East', 'Yorkshire and Humberside Regional Strategy', etc. — tended to be mixtures of exhortation, pleading and forecasts of the future, rather than plans for action.

But we have seen that real resources remained firmly in the control of the Treasury during the 1960s and consequently, when resources were brought to bear on the regional problem, it was largely through a refinement of the Treasury's fiscal and monetary management of the economy. From 1966, firms were eligible for grants of 40 per cent of the cost of new plant and machinery and 35 per cent of the cost of buildings within a series of broadly defined 'development areas'. Incentives and controls had been available before this, but the new measures helped to increase total government investment in regional grants from £80 million in 1966 to over £250 million in 1969. Some would say that regional policy provided a convenient rationale for the increasing scale of government assistance to industry in the late 1960s. The *raison d'être* for regional policy was that the national economic malaise was concentrated in the older industrial areas with their high unemployment and older capital stock.

However, refining our definition of the UK economy still further, it is quite clear that there is far less regional disparity in GNP per capita and in income per capita than there is in other European countries (Alden and Morgan 1974). This is a further indication of the integration and the uniformly high intensity of development in the most developed capitalist economy. Unemployment is often taken as an indicator of regional disparity, partly because reducing unemployment is a major aim of regional policy. Moore and Rhodes (1973) feel this is not a good guide to the success of the policy; unemployment may stay high because some regions have a large proportion of industries which are releasing labour. Also, new regional investment may stimulate labour to move in and keep unemployment high. Regional growth in output and investment are regarded as more valid indicators.

In addition to investment grants there are a whole battery of other regional policy measures, and McCrone (1969) reckons that the UK is again unique in having a greater variety of regional policy instruments than most other countries.

Tax policies at the regional level have always been used positively, to encourage growth in underdeveloped regions rather than to discourage growth elsewhere. Before 1960 there was some limited government assistance, but major tax incentives and grants only appeared in the 1960s. The incentive to capital investment has continued since 1966, with the rates and sizes of grants varied occasionally. The Conservatives in 1970 replaced the grants with tax allowances for capital investment in the underdeveloped regions. These were broadened to cover the whole country in 1972, and additional grants were introduced for undeveloped regions.

Moore and Rhodes show that in real terms the relative advantage these incentives gave to the development areas was progressively reduced over the 1960s, partly because the value of incentives to other areas increased. In principle, the tax allowance of 1970 restored the differential, but in order to collect the full benefit firms had first to show a profit. A matching subsidy for labour was started in 1967. The regional employment premium was paid to *all* firms in development areas, lowering wage costs to the tune of 7½ per cent — even of those firms not engaged in expansion.

A separate class of controls were the purely administrative restrictions on new development in prosperous areas — the so-called industrial development certificates (IDCs) and office development permits (ODPs). These controls are of the same type as the classic, negative land-use planning controls we shall discuss in later chapters.

The other main determinant of the absolute size and effect of these regional

	Plant and machinery		*New building*	
	Development areas	*Other areas*	*Development areas*	*Other areas*
1963				
Cash grant	10%	–	25%	–
Investment allowance	30%	30%	15%	15%
Initial allowance	Free depreciation	10%	5%	5%
1966				
Cash grant	40%	20%	25%	–
Initial allowance	–	–	15%	15%
1970				
Cash grant	–	–	35%	–
Initial allowance	Free depreciation	60%	40%	15%
1972				
Cash grant	20%	–	20%	–
Initial allowance	Free depreciation	Free depreciation	40%	15%

Figure 8 Capital grants and allowances 1963—1972

policies, is, of course, the geographical areas over which they apply. The approach has alternated between broadly defined areas, such as those now in force, covering a major part of the whole country except the Midlands and South-East, and the much more restricted areas such as the so-called development districts of the early seventies where the really severe problems are held to occur. Here again there has been a progressive refinement of policy controls, so that now there are broadly defined 'development areas', 'special development areas' and further 'intermediate areas' which are also eligible for assistance.

It is important to note that most of the initiative in the practical operation of regional policy comes from the private sector; government only encourages or discourages, it does not initiate. Both nationally and regionally in resource terms this policy completely dominated not only the planning initiatives outlined above but also the interventionist agencies such as IRC. Alden and Morgan (1974) estimate the gross money cost of the policy at £2000 million between 1967 and 1974. McCrone estimates the gross cost at £200 million per annum in the late sixties. All regions are reckoned to have benefited, with Wales coming out relatively well, and Scotland relatively worse (Moore & Rhodes 1974).

The pre-eminence of the private sector in regional policy can be highlighted further, by reference to the two cases where, atypically, the government did retain some discretionary initiative. The Highlands and Islands Development Board is an agency with that mixture of powers in which Shonfield would delight. Run by bureaucrats with only secondhand scrutiny by government, it does have a budget, initially only £1 million per annum, but it can borrow money, run its own enterprises, buy land, and so on. It did indeed initiate developments, and exercised discrimination within the private sector, helping to build up local industry in this peripheral region. Initial results were encouraging although (inevitably, as will be argued later) its policies and distribution of funds were controversial. Another national initiative has been the development of 'advance factories' in backwater regions, first in the late 1940s and again the 1960s, £21 million being spent in 1960; thirty-two factories were built between 1964 and 1967.

Public investment, being at least 45 per cent of total capital formation, clearly has a major impact on total regional investment. Not much is known about its impact, but there is a general feeling that all regions should get their fair share of roads, housing, airports, etc. Figure 9 shows that in 1973-4 public investment in new construction is distributed according to regional population. Scotland has received rather more than her share in recent years.

Region	*% pop. of UK*	*% public investment in new construction 1973–4*	*Region*	*% pop. of UK*	*% public investment in new construction 1973–4*
North	5.7	6.0	W. Midlands	9.2	8.5
Yorkshire	8.6	8.2	North-West	12.1	9.6
E. Midlands	6.2	6.2	Wales	4.8	6.3
E. Anglia	3.1	2.7	Scotland	9.3	12.7
South-East	31.0	29.8	N. Ireland	2.7	3.2
South-West	6.9	6.7			

Source: *Abstract of Regional Statistics* (Central Statistical Office, 1974).

Figure 9 Public investment in the regions

The government was spending between £50 million and £100 million per annum on labour training in the late 1960s, which also has some effect on regional growth potential. It was officially estimated by government that, of the total expenditure made on the package of policies in the late 1960s, investment grants and regional employment premiums contributed about one-third each.

The theory behind some of the regional location policies will be discussed briefly in Chapter 5. Putting investment into the regions tends to reduce production costs of industries in the needy areas. The main relevant costs are capital, labour and transport, (Chisholm and Manners 1971) feel that total transport costs in the economy would not increase in total, even with fairly major shifts in jobs and population between regions. In a multiregional system the main processes involve a mix of movements — of capital and labour (migration) — as an alternative to or in association with continued interregional trade in goods and services.

The effects and weaknesses of regional policy

Regional policy, a relatively large-scale effort, both in national economic terms and in regional terms, is nevertheless difficult to evaluate. In so far as it does not create new resources, new productive capacity or new jobs, moving industry and jobs to a depressed area clearly deprives another area of jobs and income. The Board of Trade (1968) showed that the South-East and West Midlands exported 300,000 jobs to other regions between 1945 and 1965. More recently both the GLC — in the Greater London Development Plan (GLDP, 1971) — and the West Midlands County Council (WMCC) have developed more expansionist policies and have begun to exhort central government to provide new incentives for industry and commerce to stay and develop in these more prosperous areas. They have tried to demolish shibboleths that the West Midlands is prosperous ('A time for action'), or that there is still a drift of population to the South-East.

Net migration figures (i.e. the difference between in- and out-migration from a region) hide almost as much as they reveal, and, although it is tempting to associate high net out-migration with unemployment, Cordey-Hayes and Gleave (1973) show that out-migration from Wales is no more than average, whereas in-migration is way below average. The pattern of migration in the early 1970s shows a net outflow from North Yorkshire, the South-East, the West Midlands, the North-West and Scotland, with an average annual total net movement in these regions of around 40,000 people. This is not a large figure; it is much smaller than recent losses due to international emigration. It is clear, however, that even if job creation and relocation is a major aim of regional policy labour migration continues to be a very major factor.

Whether due to regional policy or not, the disparity in unemployment rates between regions was somewhat reduced between 1964 and 1968 (McCrone 1969). Even the deep depression of 1975 has not affected these regions to the extent it would have in the early 1960s. Unemployment in Scotland, Wales and the North of England was roughly twice that of the South-East in 1975. In 1965 it was three times that of the South-East. Moore and Rhodes (1973) conclude that regional policies did indeed direct 220,000 jobs to development areas, and also helped the balance of payments (through exports and increased competitiveness); it did stimulate national output and did not strain national resources unduly. With only 20 per cent of total employment, the development areas did appear to be receiving about 50 per cent of the new employment being allotted via the development certificates. It is in line with our description of the UK economy as a highly integrated structure that, although UK regional disparities are low, regional spending was relatively high compared to that of several other countries.

Conventional criticisms of the regional policy tend to range around the inflexibility of the blanket, non-selective measures to concentrate specific help geared to the unique problems of different regions. Many critics thus argue for more selectivity and sensitivity. Bray stresses the different types of problem in different regions. Capital grants tend to go to the most productive and expansive industries, which are already the most capital-intensive, like chemicals and petroleum. These are the very sectors shown in Chapter 1 to be dominating the world economy, and who are outcompeting other industries. They are just as likely to *displace* labour with the new capital as to create new jobs. The Teesside chemical complex is an extreme example where a few workers can run a plant worth hundreds of million pounds (see Chapter 3), although the total Teesside employment in chemicals still runs into many thousands.

It has been argued that, whereas the measures of the 1940s and 1950s cost only £1000 per job created, the capital grant method cost £20,000 for each job created in the Northern development area in the 1960s. Conventional economists therefore argue for more efficiency — for better statistics, for more cost-benefit analysis and sometimes (McCrone) for strengthening market incentives even more.

It is also argued that public investment itself could be planned and coordinated as a total package aimed at the particular problems of the region rather than as fractions of separate national roads or housing cakes. Creating 'growth centres' within a region is one such suggestion — advocated, for instance, by the Hunt Committee (Department of Economic Affairs 1969). This has been implemented to a degree, especially in Central Scotland. The House of Commons Select Committee studying regional policy concluded that it was empiricism run mad, and possibly a waste of resources with no proper evaluation of the costs and benefits. They found it impossible to ascertain the local spin-off or multiplier effects of an investment in particular firms. This is not surprising. In Chapter 3 we shall argue that there are few if any such effects in the modern UK economy. It is worth mentioning a critical school of regionalists (e.g. Carney *et al.* 1974). This work tends to be 'ultra-radical'. It recognizes a system of 'total' exploitation; for example (paraphrased): 'Investment will take place in the North-East, its people and resources will be systematically exploited . . . capital drained off. . . . While . . . investment may lead to prosperity . . . this is inevitably transient . . . contains the seeds of its own destruction . . . decline, social and environmental problems and disruption.' While valuable in pointing out many flaws in the conventional wisdom, this analysis seems of little use as a guide to action or to understanding the potential for change in the existing situation — whether industry comes to the region or stays away, the system of total exploitation remains — an ultimately nihilistic analysis.

The conventional critics tend to run up against a recurrent, but often unspoken paradox. As the number and sophistication of controls and incentives needed to kick and push the market into doing what is required multiply, it may well become cheaper, simpler and more feasible for the government to intervene directly and simply build the factories, develop the industries and renew the infrastructure. In other words, if we are so sure of what we want the market to achieve for us, why not simply go ahead and do it? The possibility of putting teeth of this kind into regional planning is seldom discussed in the UK.

Current regional policy represents a mixture of the two views of the state — the 'micro-welfare optimum' and the Keynesian. It is argued on the one hand that firms do not have perfect knowledge of the costs and benefits or that the presence of regional externalities and spillovers and other imperfections means that the state has to intervene. On the other hand, the policy has been operated as a kind of regionally disaggregated Keynesianism which has injected investment and other spending into a series of regional 'states', to correct imbalances and create jobs. It has also been a vehicle for national Keynesianism, and has contributed significantly to the total increase in grants to industry. According to our working hypothesis, regional policy has helped the state to redistribute the surplus straight back into profitable industry — for, as we have seen, such grants do not represent an increase in not-for-profit production; they are a straight handover. Indeed, they help to increase profits in total since a larger portion of government income now comes from the population, not from industry.

And what of our picture of the economy outlined in Chapter 1? Since the policy depends on private-sector initiative, it is not surprising that the most dynamic parts of the economy — the big firms, the capital-intensive firms and the multinationals — have played a big part and have thereby received much state aid. To a multinational company, regional differentials in the UK are negligible; labour is far cheaper in Taiwan. Capital grants may be more significant, but firms have stated that the policy made little difference to their decisions. It has been suggested that the development areas employed about 28 per cent of American-'owned' employment in the UK, with Scotland having 40 per cent of this. It has also been suggested that 150,000 jobs were created in the development areas by foreign capital between 1945 and 1965 (Dunning 1975). Leading firms may prefer high-unemployment and low-wage areas for the obvious reasons that their total labour costs will be lower. It is clear that the regional policy must have been largely implemented by national and international firms, using capital owned from outside the region. ICI — a major UK multinational — alone received £160 million out of a total £600 million allocated through investment grants between 1966 and 1971, but in evidence to the Expenditure Committee the firm still regarded these grants as having very little effect on its decision to invest and locate in peripheral regions.

9 Appendix: the three estates — an activity-analysis approach

Unscrambling the national accounts

This section spells out in more detail the relationship between the three estates — the private sector, the state and the wage- and salary-earning labour force — as it is displayed in the official statistics, the national accounts. Activity analysis (which was introduced briefly in the introduction in Chapter 1, Section 3, to the production process in the market economy) can also be used to analyse more exactly the relationship between the three estates, using these published figures.

In the usual method of constructing a system of national accounts, the gross national product is broken down into its various components. The main divisions of interest here are, not surprisingly, the three estates. The private sector (the companies sector) is, of course, ultimately owned by a small minority of 'households', and the profits of companies ultimately belong to those individuals who own companies; but it is convenient to regard company profits as a proxy for these 'personal' profits and to keep the private profit sector quite distinct from the wage- and salary-earning sector. As the UK government publication states, incomes can be broadly divided into two types — earnings from work or from the ownership of property.

The money value of the GNP is usually measured in one of two ways (Samuelson 1973):

1 The total value of the *final* products (goods and services), made up of three parts — sales to:
  — households
  — the government
  — investment in plant and buildings

Note that consumption of goods and services by firms other than for investment in new plant and equipment is not included, because this is 'intermediate', not 'final' production. Such goods are consumed only to produce other goods and therefore adding them in would result in double counting. A bottle of whisky sold to a household includes the value of the whisky *and* the bottle. It would thus be wrong to add the value of bottles produced by the bottle manufacturers to the value of the whisky-and-bottle sold by the whisky distillers: we would be counting the bottles twice. All products sold to households include the value of all their separate components, perhaps produced in hundreds of different firms down the line. The concept of 'final' consumption tends to suggest that households are somehow outside the economy, but we shall argue in Chapter 3 that this is not the case.

2 The total *costs of production*. This includes:
  — wages and salaries
  — interest
  — rent
  — profit
  — indirect business taxes
  — depreciation of capital stock

Each contribution to the gross national product should be measured accurately, but it should be measured only once. Goods and money circulate in the economy: they are produced and consumed and do not appear and disappear out of the sky — and the use of proper accounts helps to provide the formal framework for proper rigorous measurement of the different contributions to the national output.

McCormick *et al.* (1974) use a conventional circulation diagram to show how this national income flows through the economy (Figure 10). This shows

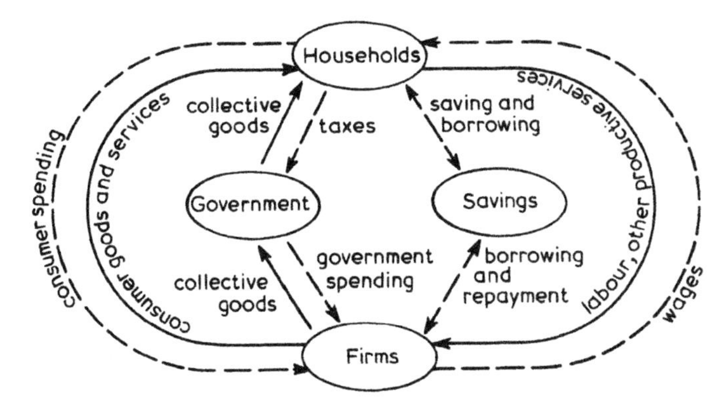

Figure 10 Conventional income — flow diagram
Source: After McCormick *et al.* (1974)

up our three estates more clearly. Households again appear as an endogenous element of the economy; their purchase of goods and services may no more be regarded as 'final' than their labour input may be regarded as 'initial'.

Figure 11 shows this circular flow of income and of goods and services as it is measured in the official national accounts statistics. We can see that the system is more or less 'closed' with income flowing round and round. The idea of 'final sales' (sales to final purchasers of commodities who do not use them

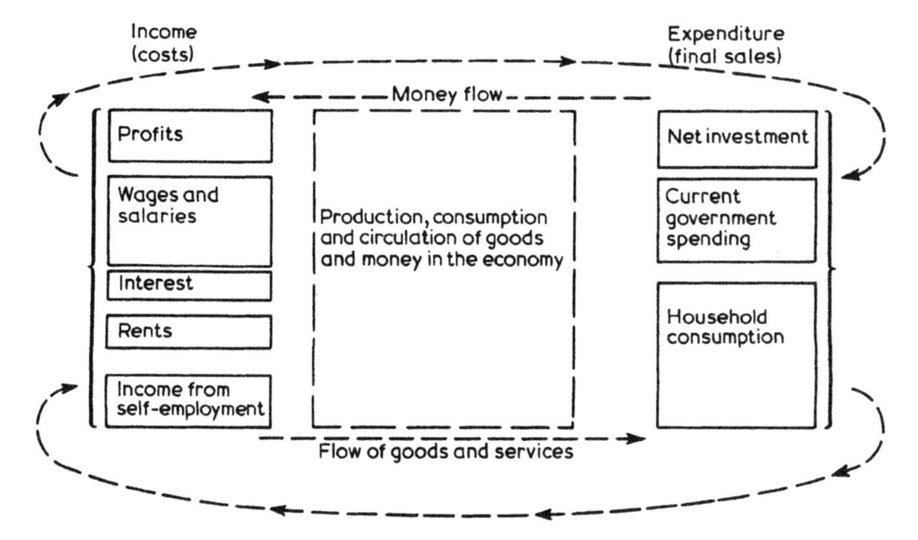

Figure 11 The income and expenditure methods of measuring GNP

Figure 12 The input–output process

to make other products and sell them again) and the notion of 'primary income', although these definitions are to some extent arbitrary, do provide a net through which the total GNP flows — and through which it can therefore be measured.

But as we shall see from Figure 14 the national accounts figures need some unscrambling if we are to calculate exactly the size of the three estates as we have defined them. Thus, on the expenditure side, some of the net investment is in the private sector but around half is also due to the state. On the income side, interest and profits are clearly part of private sector profit as we have defined it, but some of the 'rent' is imputed rent of wage earners who own their own houses. Some income from self-employment should be regarded as wages and some as 'profit'.

An attempt was made to undertake some of this unscrambling by Phillips in Baran and Sweezy (1968). This was a heroic attempt to measure the total 'surplus' production including not only profits and rents but also 'waste' and unnecessary output. Figure 14 restricts the unscrambling solely to rents and self-employment income, and undertakes some minor rearranging to bring the public corporations within the state sector.

Activity analysis of the national economy

In activity analysis introduced in Chapter 1, any production activity — whether a small part of a production process within a single firm or whether it is a whole sector of the economy — can be represented as an input–output process, a process consuming and producing commodities (Figure 12).

In the national accounts the two largest input and output flows into and out of households are consumer goods and labour (Figure 13).

Figure 14 shows how the activity-analysis approach can be used to illustrate more explicitly the relationship between the three 'estates'. This activity analysis table will need fairly careful examination and can be omitted on a first reading. The table is developed from the official figures on the gross

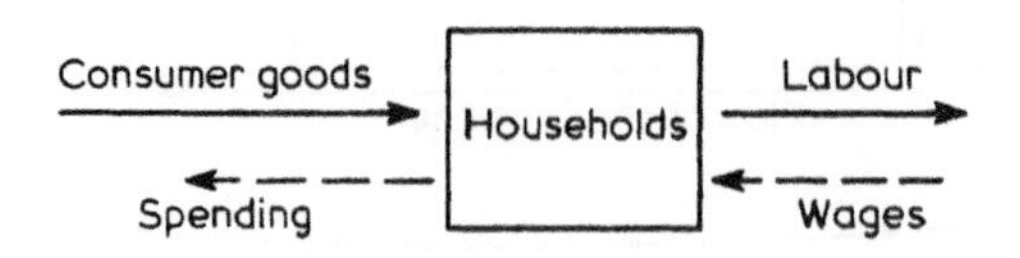

Figure 13 The input–output process in the national accounts

Commodity	Activity: Private sector	State: Central government	State: Local government	State: Public corporations	Households: Wage- and salary-earning
Private-sector current goods and services	[+54,133*]	[−4,338]	[−1,074*]	[−6,462*]	[−42,259*]
State-traded goods and services	[−3,050*]			[+12,461*]	[−9,411*]
Labour employed	[−35,302]	[−5,532]	[−5,697]	[−5,470*]	+52,001
Labour self-employed					+4,000*
Self-employed profits	+3,895				
Profits	+9,706				
State surplus on trading				+2,545	
Rent	+2,357*	+75*	+510*	+158*	+2,606*
Capital goods	[−7,020]	[−973]	[−3,280]	[−2,692]	[−2,309]
Net exports	[+3,256]				
Property abroad	+1,352				
Taxes on spending less subsidies	[+8,407]				
	[−1,082]				
Stock appreciation	−4,884			−309	−771
Error	−736				

Figure 14 Activity/commodity representation of the national economy 1974

Notes

The figure summarizes the gross national product in terms of the flow of money payments and receipts for traded goods and services. Each row represents a commodity produced (positive entry) or consumed (negative entry) by one of the three activities. Thus the right-hand entry in the first row is the flow of goods into the household sector.
Bracketed entries: Components of the gross national product expressed as *expenditure*. Unbracketed entries: Components of the gross domestic producted expressed as *income*. (The sum of the bracketed entries — The sum of the unbracketed entries.) Boxed entries: Estimated accounting quantities included to give a complete picture of the flow of goods and services through the three sectors.

Source: National Income and Expenditure (1964-74)

national product. The entries in the table deal only with transactions which are direct payments for goods and services. Except for a correction for taxes on spending, therefore, they do not show social consumption and social investment output by the state since these are not paid for directly by consumers: they have no final sale.

Each activity (i.e. each estate) is represented as a flow of inputs and outputs of commodities valued in money terms in 1974. Each column represents an *activity* and each row represents a *commodity* which is bought and sold. Down each column — for instance, the 'household' column — is displayed the set of inputs and outputs for one of the three estates. The inputs and outputs to a single activity (estate) are represented by a column of negative numbers (inputs) and positive numbers (outputs). There are three main columns corresponding to the three main estates. Each of these activities produces one main output for the economy. This is regarded as being produced *and* consumed within a single year. These three main output commodities are represented as the first three rows, above the thick line. The positive (outputs) are displayed along the diagonal. That is, the private sector produces private-sector goods and services, the state produces public goods and services, and the household sector produces labour. Each of the commodities is consumed *in its entirety* by the other two activities.

The commodities *below* the thick line, are more in the nature of 'surpluses' which are not consumed in 1974 but which are produced by the economy, surplus to requirements, in 1974 and are carried over into 1975 and later periods, or else are exported abroad.

As explained above, the gross national product is usually calculated either as the sum of all 'incomes' in the system (i.e. the sum of all unbracketed entries in the table) *or* as the sum of all final *expenditures* (the sum of all the bracketed entries). The activity-analysis table helps to show up the double-counting problem more clearly. For instance, in the third row (commodity labour), if we not only counted the total *income* to households from selling labour (£56,001 million) but also began adding the expenditure on labour by the other sectors, we would be counting the same money twice.

So, therefore, the GNP as represented in the national accounts is either the sum of the bracketed entries *or* the sum of the unbracketed entries. The state, therefore, does not appear at all on the income side (the unbracketed entries) except for the surpluses of the public corporations. The boxed entries have been added in afterwards to complete the whole flow of income around the system and are merely accounting identities. The production and consumption across each of the first three rows thus exactly balance out.

The 'commodities' below the thick line have been treated differently from the commodities above the line which are produced and consumed within a single year. Commodities above the line appear as *both* income *and* expenditure. But in this table the 'surplus' commodities below the line appear *either* as expenditure *or* as income because it is not easy to identify both the income and expenditure with transactions for a single commodity.

This activity-analysis table shows more clearly the size of the different parts of the economy in 1974 in real terms and the relative capital invested by

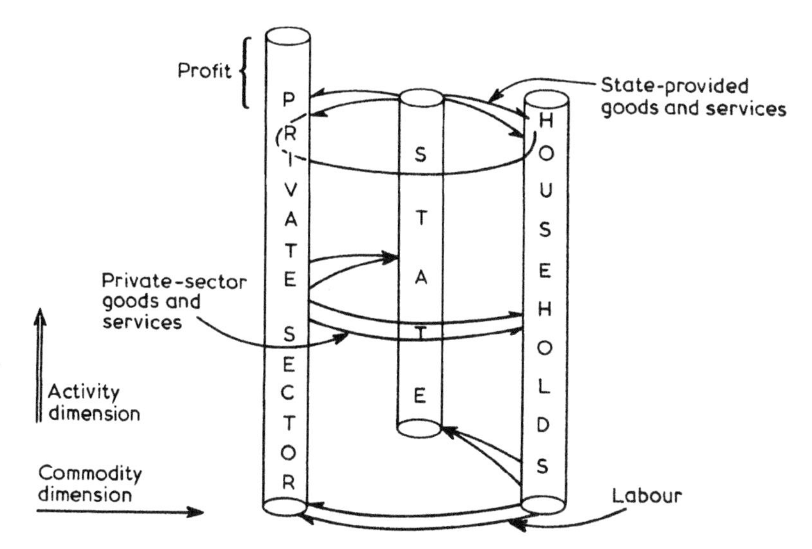

Figure 15 Activity/commodity representation of the national economy

the different sectors; it shows, for instance, how the local government part of the state undertakes more capital construction than central government. All in all, it provides a comprehensive and clear idea of who is paying whom, and for what.

The three-dimensional diagram (Figure 15) shows the full flow of goods and services in the national economy between the three 'estates'. Each horizontal level or disc corresponds to one particular *commodity* — labour at the bottom, private-sector goods and services in the centre and state-provided goods and services at the top. The three vertical pillars correspond to the three activities, each of which produces one commodity by consuming the other two. The other type of income, the 'unearned' profit income of the private sector, is also shown.

The GNP, as we have seen, is made up of measures on the income side in terms of income from work and profits, and this means taking the bottom 'disc' (labour) and the 'profits' of the private sector. Both these commodities are shown in plan-view in Figure 16 with the width of the annuli representing the size in money terms of the relevant income. Wage- and salary-earning households are shown to draw employment income from the other two estates, most of it from the private sector. The element of profit symbolizes the motor-generator part of the whole system which, as argued in this chapter, must be preserved if the system is to survive in its present form.

But if the activity-analysis table shows unambiguously the size of the national cake, in terms of traded goods and services for which payments are made, it shows nothing about the effects of the state as a reallocator of

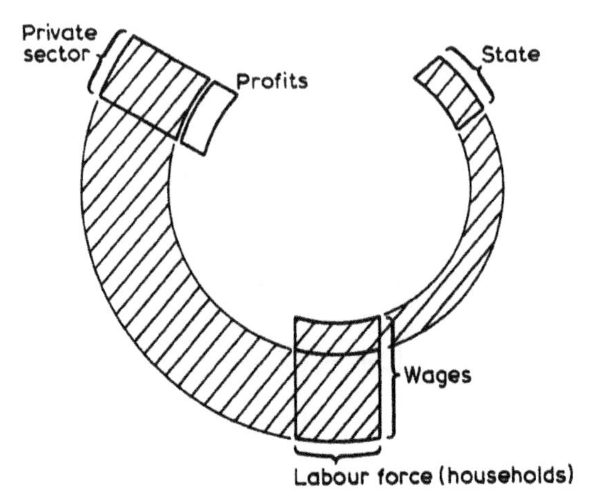

Figure 16 The three estates (The gross national product showing two commodities in section: 'labour' Wages) and 'capital' (profits))

resources through transfer payments of taxes and benefits. Nor does it show how the state affects the profitability of the private sector. This is principally because it doesn't show the *transfers* of income and spending made by the state — the taxes paid out of income and the payments received in the form of subsidies and grants. This would require a second activity-analysis table. The income from labour, then, although shown in full in Figure 15 above, does not remain in the household sector, but is transferred out in taxes, some of which comes back in social services or as direct subsidies, and some of which goes to private industry.

Systems analysis and control

In Chapters 5 and 6 we shall introduce the 'systems approach', which has been so fashionable in recent years, not least in urban planning, although some disillusion has set in recently. One branch of systems theory developed out of a concern with the *control* of a system — especially the control of an engineering process. It seems reasonable, then, that this theory should have something relevant to say about the idea of controlling the economy.

On the one hand, systems theory is supposed to be universally applicable (von Bertalanffy 1971) and to provide a way of bringing together all the different scientific disciplines. Its only *specific* subject matter is the laws of 'general' systems behaviour which can then be applied to a biological system, a physical system or an economic system. There is a continued danger that its 'laws' are so 'universal' that they also say very little, since they do not deal with the specific subject matter of the system being studied.

An early definition of systems analysis, from the engineering field, is given by Hall (1962):

> A *system* is a set of *objects* with *relationships* between the objects and between their *attributes.*
> *Objects* are 'parts' or components of a system; they can be anything from atoms, springs, wires, variables, equations, etc.
> *Attributes* are properties of objects e.g. temperature, weight, size.
> *Relationships* tie the system together.

The activity-analysis approach described above is one specific way of describing a system. Each activity is a 'system' which uses one set of input commodities to produce another set of output commodities. Each commodity is thus an 'object' within the (activity) system, and the production activity is itself a set of relationships between these commodities.

But in the real world there are systems within systems. Individual systems or activities can be combined together to form another, larger system. The urban system or the national economy is such a system, where this time the 'objects' or parts of the system are the individual production activities — households producing labour and firms producing goods and services. The relationships between the activities are then represented by the flow of commodities, from producing to consuming activities. Such a system, which is composed of smaller, interacting subsystems, is described by Lange (1965). The most fundamental law is that the behaviour of the whole system cannot be understood simply from a knowledge of the way each individual part or subsystem within it behaves. It is the pattern of relationships *between* activities which also determines the behaviour of the system: the whole is more than the sum of its parts.

Many other ideas about systems are relevant to our discussion of the national economy: the idea of the 'open' system with flows of inputs and outputs crossing its boundary, the notion of subsystems, the idea of a hierarchical system — for instance a multinational company controlled from the centre with orders going out to subsystems lower down the hierarchy. Other properties of systems include 'progressive factorization', where there is an increasing differentiation of function and new subsystems keep on appearing. This is similar to the development process of the market economy described in Chapter 1. At the same time this can occur with an apparently opposite process of 'increasing systematization'. As different subsystems arise they may also become increasingly unified and bound together, for example through the increasingly pervasive single-market economy described above. Now, vague as these 'systems' ideas might be, it is remarkable how closely they conform to the chapter picture of the market economy. Some systems are supposed to be able to 'adapt' or modify their behaviour (or even their internal structure), so that for instance they are able to survive in a changing environment. (A biological system is often cited as an example: the body can survive changes in external temperature, etc. etc.)

But the idea of controlling a system goes one further — and introduces the notion of 'feedback', whereby a system can be guided, through continued observation of its behaviour, and given new stimuli or inputs to keep it on course (Figure 17).

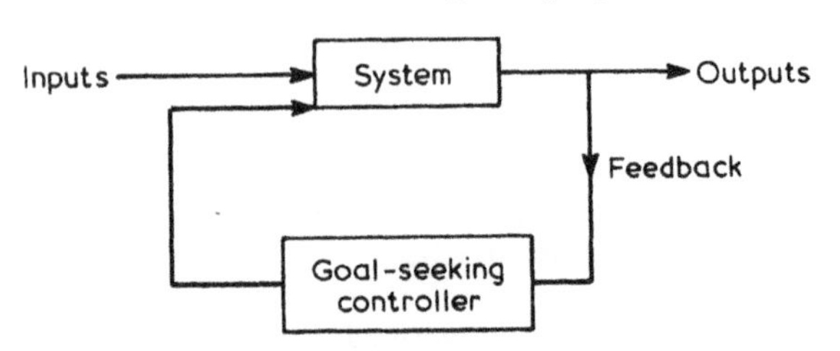

Figure 17 Controlling a system

Usually the aim is to 'correct' some malfunction, so the controller will try to provide some input which will make the system output change direction and come back on course: this is *negative* feedback. The parallel with the process of steering the economy is obvious. But the various types of formal feedback and control processes are fully described in Ashby (1956), Klir and Valach (1967), Beer (1966) and others. The only point of introducing it here is to point out how dangerous it can be to cast the state in the role of 'controller' of the economy, and planners as 'controllers' of urban development. The state is already part and parcel of the economy, it is of a significant size and its growth threatens the private sector. It cannot therefore be regarded as an external *deus ex machina* type of controller as with the classical feedback process; there is a far more complex multiplicity of interactions between the state and the economy to be taken into account.

Summary

The amount and nature of state intervention in the UK economy is a political issue, but it also colours many theories and discussions on urban development and planning. The crucial question is: can the state intervene sufficiently to revive the market economy? The answer to this lies in the relationship between the three 'great estates' in the mixed economy — the state, the private sector and the population at large — and the final effect on private-sector profits. The state helps to reduce some of the private sector's production costs by providing services, grants, etc., but part of this is financed by deductions (in taxes) from potential profits. Similarly the amount which the private sector has to pay in wages to the labour force (population) compared with the taxes contributed by the population to run the state also helps to determine the final overall profits of the private sector; the proportion contributed by the population to the state has been increasing. Some writers suggest that continued growth in the public sector — the 'not-for-profit' sector — presages the end of the market economy.

There are two conventional arguments for state intervention: (1) to correct imperfections in the market — to run monopolies and provide 'public goods', etc. (micro-welfare economics); and (2) to maintain the whole economy in a state of stability and full employment — macro-Keynesian economics. Other views on the state include 'convergence' theories (all economies are becoming alike as combined business and government operations) and the state-as-oppressor theories. In practice, UK state management has been under the control of the Treasury — a uniquely powerful executive authority in the Western World, but one which has used its enormous battery of controls only to maintain equilibrium in the *short* term. Whether or not these controls have actually helped to disturb the balance is a matter of controversy, but they have certainly done little to halt the decline which itself has made the short-term steering problem more difficult, as recent history shows.

During the 1960s there were attempts to introduce more longer-term planning, with the notion of greater cooperation between the public and private sectors on the French model, but the classic separation-of-powers in the UK meant that plans and proposals had no effective teeth, remaining largely on the sidelines as exhortations. The more important changes, not unexpectedly, have been due to Treasury-type incentives to private industry. Regional policy has had considerable effect as an extension of and more detailed application of these classic controls rather than as an exercise in direct planning — and this is where its drawbacks lie.

3 The urban system: labour and production

1 Introduction

This chapter tries to throw some light on the specific structure and function of urban areas within the national economy and the processes operating therein as outlined in the first two chapters. The aim is to give a broad general impression of the most important features of city structure and development. It does not go into great detail in describing the whole system of urban areas in the UK nor in examining specific cities in depth. The aim is to draw attention to the most important forces at work in urban areas, possibly with the risk of some oversimplification, but drawing on evidence from recent studies where appropriate. The main point is to try to identify the reasons for the existence of urban areas and for the changes taking place within and between them — that is, to identify the *processes* behind urban development. Not surprisingly, following the discussion in Chapter 1, this is largely a matter of showing how the process of development in the market economy as a whole works out at the urban level.

There are many standard texts on urban economics and urban geography which discuss (often in neoclassical economic terms — see Chapter 5) the reasons why urban areas came into existence. These arguments centre around the general issue of economic efficiency in a market system, where firms maximize profits and consumers maximize utility. According to these arguments, there is a tendency for activities to cluster together in geographical space, because it is more efficient that way. The costs of reaching markets, for instance, are less if the markets are nearby. The present discussion does not try to recapitulate these arguments but rather to throw an alternative light on urban development, one which more explicitly recognizes the obvious facts of the modern economy and which incidentally, throws some new light on such fundamental concepts as 'production' and 'consumption' which have special connotations and significance in urban development. Once the main processes behind urban development are established, we shall then be in a position in Chapter 4 to assess the role of the local state (i.e. local authorities) in attempting to control these processes.

The cornerstone of the argument is that the urban system should be first and foremost viewed from the point of view of the people who live in it, or at least from the point of view of their 'activity'. Looking at a city, almost by definition, focuses the attention (or should focus the attention) on the total activity pattern of the inhabitants. It should make the household and its links to the outside world the centre of attention, viewing the inhabitants in their combined role of consumers and producers. The conventional modern view usually separates these two roles entirely (see Chapter 5). To define human beings in terms of this 'purposive activity' (i.e. labour) is a very old concept in political economy, one which has been out of favour for a long time, but it seems to be having something of a revival. It helps to focus the attention on the way production is organized in the market economy and especially on the separation between 'home' and 'work', between the different roles people perform as producers and as consumers. The urban system is best defined in the first place as a *pool of labour* – a relatively self-contained area where a whole community can and does enter into common production and consumption processes on a daily basis. This is the *first urban development process* — the household output process (the sale of labour) and the household input process (consumption), considered together. Viewed from the household, the local urban system is a miniature representation of the national system. The important economic flows into and out of the individual household begin and end *inside* the urban area. From the household viewpoint, the urban system is relatively self-contained. The production/consumption process of the household forms the underpinning of the urban system and its relationship with the economy as a whole. The largest economic flows within the urban area go through the household.

Contrary to the popular view, there are very few other direct economic flows and links in the urban system. If we took the 'firm' as the centre of attention instead of the household, then the main link with the local urban system would be through the *employment of labour* (i.e. payment of wages); other than that, the main links would be outside the urban area through the imports of components and exports of finished products. Often the local firm will be only a branch of a large national or international firm. The firms which serve household consumption will again, in the main, be branches of national retailers and will import most of their goods from the outside world. Looked at from these firms, then, the urban system is a pool of labour and/or a pool of consumers, rather than a source of other inputs or outputs.

However, to gain a complete picture of the urban development processes within an urban area, one which will provide some explanation of the 'superficial trends' in the urban system, the hierarchy of urban sizes and the consistent pattern of growth and decentralization which is occurring (see Section 2), there is a further urban process to consider. This process involves all the activities in the urban area, both firms and households. It is the competition for space and location — competition in the land market.

All the activities that take place within the urban area can be regarded as undertaking some kind of production. Each activity takes in a series of inputs (goods and services) including a certain bundle of basic resources — land, labour and capital. It is suggested that this 'input/output' method is a very useful way of looking at households themselves, which can also be viewed as undertaking a production activity: they produce labour from a mixture of inputs — including housing, food and other goods. Households are thus 'small factories'.

Once *all* activities are looked at in this way, the process of competition in the land market can be examined. Two types of competition occur, the first between establishments producing the same or similar outputs, for instance between households or between shops. The second type of competition occurs between activities producing different outputs, that is between households and shops or offices. Production activities attempt to minimize costs of production, one of which is the cost of transport. Activities in the best locations have lower costs of production and hence make more profit than those activities located in worse locations (e.g. on the edge of the city). If land is privately owned, those who own the land can charge a rent which creams off the excess profit, leaving the activities occupying this location no worse off than those occupying worse locations. In the competition between different activities, activities making higher profits out of a given piece of land (e.g. offices and certain types of shops) will be able to pay a far higher rent and hence will tend to push out the lower-profit activities e.g. households) from the best locations. This means that competition in the land market eventually tends to separate different activities and to create mono-use *zones*. Prime locations like city centres are then restricted to high-profit activities only. 'Zoning' is thus a product of the land market and not of planning itself.

This process of competition for space and location, together with the increasing scale of activities and the increasing competition in the economy at large, goes a long way towards explaining the decentralization of cities, the polarization which appears to have occurred and the existence of ghetto areas. Also the way finance is raised for development and dominance of large property companies helps to reinforce the concentration of high-profit activities in city centres, the 'invasion' of capital into the (often old) heart of urban areas.

The other, non-residential activities, including the big manufacturing establishments supplying the national market, the major office activities, exporting services from the central area, and main consumer-serving activities, all reinforce the pressure on land in key locations, especially in city centres, and make the land-use planning problem more difficult. (Some specific urban case studies are cited.) As the scale of these activities grows and the differences in the returns generated by different activities increase, the choice between types of development at key locations becomes an increasingly severe *economic* choice.

2 The UK: an urban economy

Introduction

Up to now we have given the city only a shadowy existence within the national and regional economy of the UK, which is itself struggling to survive and compete in the face of increasingly concentrated forces of international economic power. We have occasionally surmised what we might expect when we do reach the city. Almost the whole of the economy is fully incorporated into the market economy. The overwhelming majority of the population are *employed* for a wage or salary, and most economic activity is concerned with producing products or services for sale on the market, or with the provision of public services for collective consumption. But, as we shall explain below, the use of land is crucial to a satisfactory definition of urban structure, and we have already excluded the agricultural sector, which is the largest land user, from our definition of the 'urban' part of the economy. Agriculture is certainly part of the industrial, market economy, but its use of land is different in kind from that of other activities. It is the major land user, but it is the least labour-intensive land-using activity, so it does not compete for labour with other activities in urban centres. The relationship between the investment of capital, especially in buildings and the land itself, is obviously unique in agriculture. The land itself is part of the productive capital, and buildings only occupy a tiny fraction of the land used for agriculture. Urban activities, in contrast, generally take place within buildings or other immobile physical structures.

If we take as given for the moment the fact that economic activity and the population is distributed in a number of discrete cities or urban areas (Section 4 shows how and why such a clustering of activities occurs), we already have from Chapters 1 and 2 some idea about their structure and development.

Almost 95 per cent of the working population participate directly in the industrial urban economy, earning wages and salaries. Approximately 75 per cent of the population live in the 'coffin' megalopolis (Hall *et al.* 1973) stretching from London through the Midlands to Lancashire and Yorkshire with outgrowths in the North-East, Glasgow and South Wales. We do not expect each of the 100 city regions (or 'standard metropolitan labour areas', SMLAs), which can be discerned within England and Wales, to have a similar structure of activities to that of the nation as a whole. Apart from the legacy of early industrialization in certain areas (which means that the traditional, resource-based industries are still heavily represented in a few traditional areas), the whole mechanism of competition, the increasing concentration of production and the increasing separation and differentiation of the different stages in the production process would lead us to expect a massive degree of *specialization* on the one hand, with certain cities concentrating on certain activities, and *interdependence* on the other, with flows of goods and services between urban areas. With the trend to larger firms and plants noted in Chapter 1, urban areas will tend to have a very 'lumpy' pattern of industrial production. One or two plants and one or two national or multinational companies may effectively own the whole town, or at least its economic base. There will also be large differences between different urban areas, if only because of the growing difference in profitability and in the labour-to-capital ratios in the industries and firms on which they depend.

We should be unsurprised to find that the big firms which dominate a local economy are controlled from headquarters in the metropolis where they can easily communicate with each other and with the government, and where specialist services are available. We know that in general the physical fabric of many cities will have changed only slowly, although over one-third of the national housing stock has been constructed since the war. In the 'traditional' areas, the UK economy will have a massive legacy of outworn physical stock, decaying factories and plant, pre-First-World-War houses, hospitals, schools, etc., many of them inadequate to the social and economic needs of the day.

Because the UK economy is a highly cohesive and integrated structure, we would expect only limited differences in the wages and salaries, and standard of living, of different urban areas even though, as we have already seen at the regional level, such differences will persist and be strengthened by differences in the industries upon which different cities depend. We shall be looking at the effects, at the local level, of the decline in manufacturing, the role of national finance and property sectors in providing capital for building development and the highly uneven distribution of income. We shall be looking for the local effects of the short-term fluctuations in the national economy which so preoccupy British national planners.

But how is the UK economy organized in space, where are these cities, how big are they, and indeed why do they exist at all?

The conventional way of defining the boundary of a city or urban area is the limit of the journey to work — the commuting limit. (There are other definitions, which will be clarified later.) Although the definition by journey to work is so common (Hall *et al.* 1973), the real social and economic significance of this definition often seems to be overlooked. We shall show in this section how the general outline of economic structure of the UK economy outlined in Chapter 1 helps to define the role of the urban area within the national system. We shall argue that urban areas usually defined as self-contained labour markets are, indeed, first and foremost *pools of labour* when they are viewed from the perspective of the national economy and especially from the point of view of the big firms which so dominate the economy.

Superficial trends: urban growth and decentralization

Overall pattern

In the definitive Hall *et al.* (1973) study, and the follow-up by Drewett *et al.* (1975), it was established that 75 per cent of the population of England and Wales lived inside 100 of these standard metropolitan labour areas (SMLAs). Each such areas contained at least 70,000 population (60,000 in the extended study of Drewett *et al.* (1975) which identified 126 of them in 1971; including Scotland). The studies investigated changes occurring between 1951 and 1971. The areas are formed by grouping together those administrative areas (local authorities) which send workers to a defined employment core. The

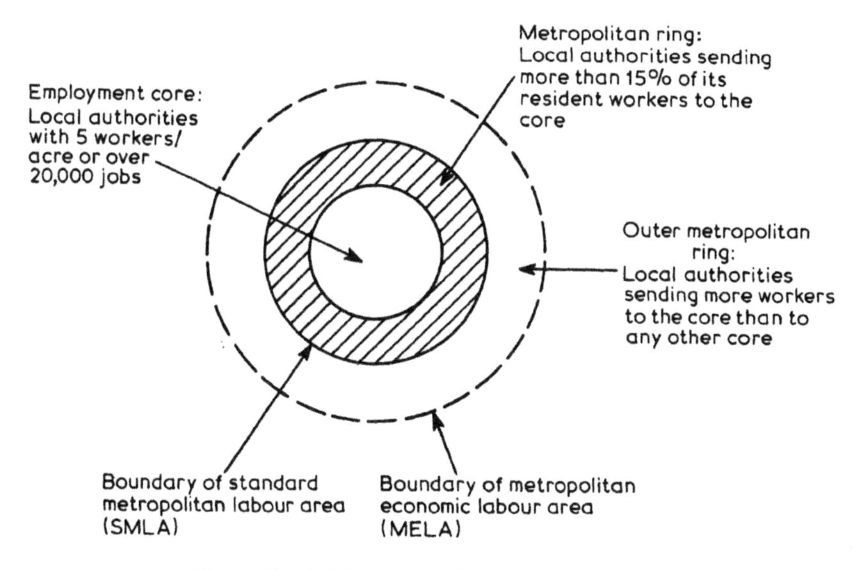

Figure 18 A labour area. Source: Hall *et al.* (1973).

complete definition and structure of an individual labour area is shown in Figure 18.

This type of analysis is derived from original work first undertaken in the United States. The division of an urban area into a core and a strongly connected ring surrounded by a weakly connected outer ring is clearly an arbitrary definition. But these studies, even though they refer only to very aggregate variables ('population' and 'employment', with some socio-economic breakdown), do produce a comprehensive, descriptive picture of the urban system of the UK and the way it is changing.

The overall conclusions emphasize the theme of urban growth plus 'decentralization', with first population, and later employment, beginning to move from urban core to the metropolitan ring and finally to the outer ring. Urban economists often emphasize the advantages of urban concentration and there is plenty of it in Britain. Not only is the main axial-belt of population the 'coffin', one of the most concentrated 'megalopoli' in the world (see the map on page **00**), but there is an even greater degree of concentration within it. The London labour area alone takes one-quarter of the population of England, and, with its neighbouring urban areas, over one-third. The top eight labour areas in the whole of the UK take 40 per cent of the whole population; so once again we are immediately faced with a strongly hierarchical structure, a concentration, which as demonstrated in Chapter 1 is such a feature of the economy as a whole. These major 'conurbations' are now covered by new large local government authorities, called metropolitan counties (see Chapter 4).

This immediately points to a limitation of the Hall study, since a very large proportion of urban change is concentrated *within* these very few big centres. Although having 126 labour pools gives an impression of the rich variety of areas which exist the smallest 60 of them (almost 50 per cent) only account for about 14 per cent of the total population. Any generalization of the pattern of urban change based on this disparate collection of cities has to be treated with caution. It would appear that a 'labour area' like London must be very different from a labour area 100 times smaller. What happens *within* London is at least as important as what happens within and between fifty of these smallest 'cities'. Nevertheless, the study does help to pose the question: are there identifiable 'urban processes' which are common to *all* these cities? Sections 4 and 5 argue that there are, even though in a very large city these same processes operating on a grander scale can produce very different effects.

Land has traditionally been the urban planners' basic resource, although they have more recently been concerned with large-scale strategic issues of population and employment change and the spatial implications. Deciding how to use this basic physical resource efficiently might reasonably be expected to be one of the specialist skills of the planner. In planning journals land-use standards are given relatively little attention. The book by Stone (1973) contains an interesting and valuable discussion of physical standards for urban areas.

Given that even a highly urbanized area has plenty of open space, planners have adopted three rule-of-thumb definitions to express more clearly the relationship between the use of the commodity 'land', the buildings occupying the land and the activities using the buildings. They define 'total', 'gross' or 'net' residential density for a given area, depending upon whether the population is divided by the total area, the total area actually built on, or only the area occupied by houses and local service streets — excluding land occupied by other activities. Champion (1974) shows that the residential activity was far and away the most extensive land user occupying over 40 per cent of the urban area, followed by open space (20 per cent), industry (3 per cent) and education (3 per cent). Roads and railways probably occupy about 15 per cent. Retailing and commercial activities occupy a minute proportion of the total area. A typical low-density residential development is 12 houses per acre net, and residential net densities can go to over 50 dwellings per net acre in multistorey building. The average urban density of all activities increases with increasing urban size.

Hall's work tells much about the physical growth of urban areas in the UK. The overall picture is one of large-scale change, much more than is generally recognized. The largest cities had increased by 5 million in the thirty-five years up to 1961. This growth has concentrated in the South and Midlands, although towns in the North have also grown, at least in physical terms. The consistent pattern depicted is growth coupled with persistent *decentralization*, first of population and later jobs, within the labour areas. Although no doubt the broad conclusions of the study are correct, there is a danger that some of the results might be dependent on the way the SMLA is defined. In any area which is growing extensively (i.e. where the boundary of the development area or the commuting area is extending) almost *any growth* is going to appear either as 'decentralization' or as 'centralization' in Hall's terms, at least relatively, depending where the boundary of the core is drawn. The studies do try to limit this problem as far as possible, but the Drewett study does seem in some doubt as to how to allow for zone boundary change. The problem becomes most severe because the definition of the core is based on 'absolute' measures like 'total population' and 'density'. (Ideally it should be based on a relative definition, e.g. the percentage commuting inwards, just as the metropolitan ring is defined in relative terms. However, *density* is related to relative location within the urban area and so this partially solves the problem.)

The majority of the population live in the *cores* of these labour pools; even though decentralization has been going on for twenty years in many cases the *absolute* decline in population of the urban cores is still fairly small — 3 per cent in ten years. In relative terms the share of total urban population living in urban cores has declined from 52 per cent to 47 per cent of the country's population since 1951. Part of this is due to the growth of population outside the cores, rather than to a migration from the cores. Even in 1971, only 32 per cent of the UK population lived in the metropolitan rings and 16 per cent in the outer rings. The growth in the rings (17 per cent in the decade 1961-71) was faster than in the outer ring (10 per cent), consistent with the idea of a 'frontier' of decentralization, pushing out from the core but not yet always reaching the outer ring.

In contrast to population, urban employment is much more concentrated, even at this very aggregate level of study. The cores contain some 60 per cent of employment; jobs in the cores increased by 7 per cent between 1951 and 1961, but declined by 3 per cent from 1961 to 1971. But there are indications that there is local-scale employment (service employment, foodshops, etc.?) which is fairly ubiquitous, because the spread of employment between the two rings is more even. The inner ring has 23 per cent and the outer ring 14 per cent. Again the overall conclusion is that employment also is now decentralizing, the inner rings having grown by 15 per cent in the decade to 1971. Overall there has been an increasing imbalance between jobs and population in all three zones, with the cores concentrating more on employment, and a consequent increase in commuting distances.

Looking at the urban system as a whole, the studies concentrated on two main aspects of growth and decline: first, the effect of urban size and, second, the regional pattern.

Effect of urban size on growth and decline

The overall rank-size distribution of cities has not changed, that is, the number of cities of a given size remains the same — a large number of small 'cities' and a small number of large ones. Most of the *absolute* growth in population in the 1950s occurred in the very large urban areas — Birmingham, Liverpool and London all figured in the top ten for absolute growth. By the 1960s most of the growth was concentrated in large urban centres outside the conurbations, in towns such as Coventry, Reading, Southampton and Leicester. Not surprisingly, the highest *relative* growth occurred in much smaller towns, especially new towns, largely in the South. Urban 'decline' where it occurs has tended to do so at a much slower rate — only one area has had a reduction in population of more than 5 per cent over ten years. This contrasts with population increases of up to 50 per cent for newer towns. Large absolute decline usually means large relative decline and is mainly concentrated in the North. But whereas in the 1950s most of the decline was concentrated in medium-sized cities based on older industries, by the 1960s it was the very largest cities that were declining most rapidly; London (a 'dramatic reversal of fortune'), Liverpool, Glasgow, Newcastle and Manchester were all in the top ten.

Employment growth and decline broadly follows the population trend but with a time lag and with many local exceptions. Large towns in the Midlands and North, for instance, did not gain as much in population as in employment during the 1950s. Most of the declining urban areas are in regions qualifying for regional assistance (see Chapter 2).

The main overall pattern of urban growth, in relation to the average, can be sketched out broadly as follows:

Slower than average growth over the twenty years:	cities in Northern UK, London and other very large cities
Slower than average growth followed by faster than average growth:	cities around conurbations, the cities in the very outer parts of the South-East
Higher than average growth over the twenty years:	small cities, especially in the South

The few largest cities account for most of the absolute total population and employment decline. London, for instance, accounts for nearly 30 per cent of all population decline and 44 per cent of *all* employment decline which has occurred in the 1960s. Summarizing the growth and decline pattern:

Accelerating decline:	Largest cities — London, Liverpool, Newcastle
Decelerating decline:	A 'broad band' of cities in Yorkshire and East-Central Scotland
Decelerating growth:	New towns, other towns close to London; Midlands — Birmingham, Coventry; outer regional growth centres

Accelerating growth: Outer South-East, towns near large conurbations, free-standing towns

The largest cities had the largest absolute gains in population and employment in the 1950s and generally the smallest gains in the 1960s. In both periods, the percentage change in employment and population is inversely related to city size. This shows up the difference between effects at different spatial scales, because, looking only at the urban *cores*, it was the smaller not the larger cores which were gaining population in the 1950s. For employment, on the other hand, both the cores and the whole urban area grew proportionately with size in the 1950s and declined proportionately with size in the 1960s.

In regional terms the South-East had the most absolute population and employment growth in the 1950s, followed by the West Midlands and the East Midlands. The largest declines were recorded in the North-West, Scotland and Yorkshire. In the 1960s the South-East had almost zero total urban growth, although it did experience a massive internal shift.

There is often a sizeable difference between total population and employment change of a given region. In the 1960s the South-East gained more in employment than in population, as did the Northern region, whereas the West Midlands gained more in population than in employment. Regional employment growth was more variable than that of population.

Migration and natural increase

The growth and decline of these urban areas is a function of two factors — births and deaths on the one hand and migration on the other. The percentage changes in population were very closely related to the rate of migration, except for retirement centres where deaths exceeded births. Since only youngish age groups in the child-bearing stages tend to migrate, high migration tends also to be correlated with high natural increase and a youngish population. For the large declining cities (see Figure 19), like London, Liverpool, Manchester and Newcastle, out-migration exceeds natural increase but in many cities natural increase is more important, e.g. in the West Midlands and Yorkshire.

	Natural increase	*Net migration*
London	524 000	–920 000
Birmingham	228 000	–105 000
Glasgow	141 000	–181 000
Manchester	112 000	–139 000
Liverpool	111 000	–163 000

Source: Drewett, Goddard and Spence (1975).

Figure 19 Components of population change in large cities 1961–1971.

The large cities are thus almost population 'factories' (see below) *producing* over one million new people in a decade and *exporting* 16 million. In regional terms, the migration pattern is very much 'the South-East versus the rest', since, even allowing for massive exports from London, cities in the South-East had by far the largest proportion of net in-migration, followed by East Anglia, the South-West and the West Midlands. The relative changes due to natural increase vary much less than the changes due to migration. A long migration move -- such as a move *between* cities -- implies both a job change and a housing change. Most migration studies show that long-distance moves are primarily for job reasons. The effects of short moves within a given labour-market area (or within the wider metropolitan system of neighbouring labour areas) is discussed in Section 2. Inter-city migration has been studied by Cordey-Hayes and Gleave (1973) and they have shown that in-migration to a city might be influenced by factors such as income and unemployment in a rather different way than out-migration. Cities with higher than average in-migration may also have high out-migration.

Throughout their discussions, Hall *et al.* (1973) and Drewett *et al.* (1975) constantly come up against the problem that many of these labour areas are not really independent systems but are often part of a system of cities surrounding a large conurbation centre. The growth and decline of the individual cities is then best explained in terms of outward movements from the conurbation centre.

Centralization and decentralization

Hall outlines a typical four-stage history for a typical UK city:

1 Centralization of people and jobs in the centre.

2 Population begins to move to suburbs (e.g. 1900-50 in many UK cities).

3 Redevelopment of the inner city for commercial uses and slum clearance.

4 For the very largest cities (e.g. London and Manchester) a total decrease in population accompanied by the growth of smaller areas surrounding it.

In the process of decentralization as Hall defines it, population decentralization occurs before employment decentralization, although this is not a straightforward lead-lag causal relationship followed exactly in every city. Sometimes population decentralizes while employment is still centralizing. The authors distinguish between 'absolute' and 'relative' decentralization, the latter referring to the changing proportions of population or employment between ring and core, and the former to an absolute loss of population or jobs by the core and an absolute gain by the ring.

In the 1950s many labour areas were still centralizing population (46 cities) and jobs (73 cities). By the sixties 20 of these had moved from centralization to decentralization and 20 had moved from relative to absolute decentralization. For employment 24 cities had moved from relative or absolute decentralization and 13 from relative to absolute decentralization. The authors distinguish the following main trends in the two decades:

Relative centralization in both periods: Towns around London, new towns, free-standing towns
Relative centralization to relative decentralization, together with rapid growth: medium-sized, free-standing towns -- Bedford, Cambridge, Ipswich
Relative centralization to absolute decentralization: 'Backbone' of England -- Birmingham, Sheffield, Leeds
Absolute decentralization followed by relative decentralization
or
Decentralization with its decline followed by absolute decentralization : Towns in the North-West
Absolute decentralization and overspill to neighbouring areas: Very largest cities, e.g. London, Manchester

In general, it is in the largest cities that the decentralization process has gone furthest, a picture consistent with a wave of population spreading out from the core into the ring and the outer ring and, finally, spilling over into the surrounding centres. This wave is sometimes, but not always, followed by a wave of employment decentralization. But in showing that size is associated with decentralization the authors do acknowledge the limitations of the approach. To an extent 'growth' is synonymous with 'decentralization'. Land is limited in supply and unless the density of activity at the centre continues to increase at a fast enough rate, any additional population is bound to take up more land at the periphery and, therefore, to create 'decentralization'.

The decentralization of larger cities is far from being a neutral phenomenon. The 'inner city' is the problem on which most commentators seem to agree. It is the richer, 'more dynamic', 'more independent', 'self-sufficient' sections of the population who have left the inner areas. Hall *et al.* (1973) show how the proportion of upper socio-economic groups in the urban cores have declined as decentralization proceeded. The population left in the urban cores therefore tends to have lower than average income and to be older on average. These areas are less and less able to generate income as firms close and employment moves to the suburbs. Inequality between small areas *within* an individual urban area has always been greater than it has been between the regions. Now that these intra-urban differences are becoming more apparent, the spatial pattern of wealth and poverty is becoming more clearly revealed. Poor areas become fewer and larger. This progressive 'rationalization' of the internal geography of cities we can call 'zoning'. The inner-city problem cuts across recent discussion of both regional policy reform (Chapter 2) and local authority finance (Chapter 4). The problem is further discussed in Section 5.

Broad conclusions

The following conclusions emerge from these studies of global urban change:

1 An overall increase in urbanization: 93 per cent in 1951 and 96 per cent in 1971 lived inside urban areas.
2 New urban centres have been formed in rural areas and in former dormitory suburbs around London.
3 A generally accelerating pattern of decentralization with, first, population and, later, employment, spilling out from the urban core into the surrounding strongly connected ring, later into the outer ring and, finally (in the largest cities) spilling into the neighbouring towns. The overall process has led to an increasing imbalance or polarization between home and work and also to an increase in segregation between social groups.
4 The whole process is most rapid in the urban heart of Britain and especially in the largest cities.
5 Migration differences are the chief cause of decentralization, but natural increase is also important and is likely to remain high in the 'rings' of the cities because of the young population structure.
6 Cities which were growing or declining in the 1950s continued to do so in the 1960s, but the largest cities which in the earlier period already had a lower relative growth rate began to lose population in the 1960s. Declining towns were mainly in the North, growth cities in the South, in both decades. Exceptions to these trends are usually due to overspill from conurbations into surrounding towns, or else to movement into rural centres from the countryside.
7 Although population change generally leads to job change, this is not always the case. Major towns in the West Midlands and Yorkshire gained more people than jobs in the 1960s in contrast to their marked increase in employment during the 1950s.
8 There is evidence of reduced job losses in some towns in the North even though the population is still declining (regional policy?).

These aggregate descriptive studies give a comprehensive account of urban change which need explaining (although, as we have said, the conclusions beg some important questions, since *any* urban growth is bound to mean 'decentralization' unless density increases). We first of all have to explain concentration of population and jobs and why concentration differs between the two, and then go on to explain why and how decentralization occurs. We should perhaps give more weight to the decentralization of large metropolises. Such a process might even be interpreted as increasing centralization rather than decentralization since surrounding labour areas which previously were relatively self-contained are becoming more dependent on the metropolis. This would be a geographic parallel with the increasing centralization of the economy noted in Chapter 1.

We can learn much more about the implications of decentralization by looking in more detail at the individual city, and this we do in the following sections, as we try to identify the main urban processes which are behind these trends.

3 Towards an explanation of the urban hierarchy

Introduction

The preceding description of urban trends considered the classic geographic aggregates of population and employment. We inspected at a distance the aggregate concentration of population and jobs across the country. But where is the tie-up with the structure of the economy described in Chapter 1? If there is an explanation of urban development, it must be found as a specific geographic expression of the development processes in the national economy. We should be looking, therefore, at the process of *competition*, operating through reductions in the cost of production and the drive to extend markets. We should also be looking at the specific geographical effects of the structure of production and especially at the effects of concentration of economic power, through large firms and through the state.

In technical terms we must 'map' the national structure of production on to the urban structure. There has been very little research which gives anything like a satisfactory account of the way the urban system has adapted itself or has been influenced by the modern economy. The subsections below illustrate some of the ideas that have been put forward to shed light on the relationship between the urban system and the national economic system. These ideas have never really been brought together in a coherent overall explanation of the role of the city in the economy as a whole. The subsection below puts forward an outline of such an explanation.

In this section we are mainly concentrating on the relationship between the individual urban centre and the outside world, rather than on its internal structure. In economic terms this means focusing on its importing and exporting activities -- its 'economic base'.

Size and function

We turn first of all to a well-known theory of urban structure. This theory is at least implicitly related to the concept of perfect-market competition first met in Chapter 2 and further elaborated in Chapter 5. Nevertheless, it is a useful starting point as a possible explanation of the concentration in the urban system and the fact that some cities are larger than others.

This body of theory associates differences in city *size* with differences in city *function*. On this theory London as the 'first-order' urban centre has a unique function. In London are a unique but finite set of national-serving activities, undertaking functions which London performs for the nation as a whole. In paying for these goods and services, the whole nation thus contributes to the payment of a specialist labour force in London which undertakes these activities. This also partly explains why London is the largest city. The other part of the explanation is that only a tiny proportion of any given workforce can perform these specialist functions. Therefore, London has to be large in order to provide enough specialists to work in these national-serving activities.

The theory goes on to suggest that there is another set of goods and services which have a smaller 'spatial range' and which require less specialized skills — and hence a smaller total population to perform them. They can be provided by the next lower order in the city hierarchy — Merseyside, Manchester or Birmingham. These serve their respective regions, and London also serves its own region with these second-order functions. As we go down the hierarchy, we come to smaller and smaller urban areas, themselves producing goods and services for smaller and smaller surrounding regions. The smallest areas are receiving their specialist goods and services from all the many levels in the hierarchy above them. This elegant 'central-place theory' corresponds in some respects with the ideas of division of labour outlined in Chapter 1. The idea of specialist functions being concentrated in certain places corresponds with the general economic process of development, whereby, with increasing division of labour, the different parts of the production process become separated from each other, and become concentrated in distinct, large-scale production processes of their own.

In the Hall *et al.* (1973) study described in Section 2, in the explanation of the growth, decline and decentralization of cities, there was some reference to urban function — 'rural service centres' or 'outer metropolitan dormitories'. There have been remarkably few attempts to classify British cities according to function. The classic study of Moser and Scott (1961) did not attempt to define cities according to commuting hinterlands or labour pools, but used the existing administrative areas. They used a standard statistical technique known as component analysis and found that what most distinguished one town from another was a general combination of census variables said to measure 'social class'. Much of the social polarization found by Moser and Scott lies *within* Hall's labour areas.

One of our own PRAG studies by Webber and Craig (1976) also used a development of component analysis, using a smaller number of areas — the new (1974) larger local authority districts. Webber classifies these areas into six main groups:

1 High-status suburban growth areas
2 Rural and retirement centres
3 Traditional manufacturing and mining areas
4 Service centres
5 Public housing areas
6 Central and inner London

Here again some of these differences occurring between these districts are occurring *within* the labour areas, defined by Hall.

This general notion of specialization of function in different cities must clearly be of some relevance in a modern urban economy, where a significant proportion of local manufacturing and service production is for export from the urban area. However, as we have seen, even at descriptive level many manufacturing activities, wherever they are located, do tend to serve the *whole* national market, not just a local region. The 'lower-order' urban centres are all 'first-order' centres in this sense.

As a caution against making an oversimplified equation between city size and city function, some American studies suggest there is no such simple relation. Nelson (1955) divided economic activities into various categories showing how each varied from one city to another. Activities directly serving the local population tended to decline with increasing urban size, and manufacturing tended to be the most variable activity of all -- some towns have no manufacturing.

Organization structure

Since an individual large firm may comprise a significant entity within the national economy, then individual cities may well be said to be contained 'within' the firm rather than the other way around. It is, therefore, appropriate to go beneath the bland surface description of 'employment' change and look more closely at the production process itself. This is a continuing theme throughout this book and is discussed further in Chapter 5. Two different types of work have recently focused on the role of the large corporation and its effect on the urban system. They do not explicitly consider the full economic implications of the concentration of production into large firms for the whole national economy, but they tend to concentrate on the single large firm, whose production may be located in several different places. Keeble (1968) notes that as long ago as 1926 geographers recognized that within one enterprise there exist 'packets' of manufacturing functions which may be located at different places. He considers such functions to be end-product manufacture, component manufacture, sales and distribution and research and development. Keeble stresses that the whole concept becomes more and more relevant the larger the firm.

The more recent development of 'organization theory' does not look specifically at the flow of goods and materials within the firm, but rather at information and other flows specifically related to *dependence* or *subordination* of one level in the organization to another. Goddard (1975) cites Warneryd (1968) and Westaway (1974) in a discussion of the way the long-term, 'high-level' goal functions of a national corporation (e.g. profit maximization) will be cited in the national capital, the medium-term planning of production in a regional centre, and the day-to-day management of production will take place at a peripheral factory located in a peripheral town. Some of this work has also been given a 'post-industrial' slant (i.e. by implication 'post-economic' or 'post-market-economy' slant), but it does help to show how the large corporation tends to reinforce the existing hierarchy of urban centres, and in particular the dominance of the capital.

The influence of a single firm can be very widespread; the ICI company has some 100 establishments scattered widely across the country (Rees 1972), concentrating in the main centres of population in the North-West and the North-East. Parsons (1972) showed that, even within manufacturing, the South-East, although having less than 30 per cent of the production plants, had nearly 75 per cent of head offices, mostly in London. The South-East also has some 80 per cent of all research and development establishments in the whole country (Boswell and Lewis 1970). Westaway has shown that the larger the firm the more likely is its head office to be in London -- 86 out of the 100 largest are in the capital, and this concentration is still increasing.

All the evidence reinforces the impression that the hierarchy of the cities reflects a hierarchy of economic control emanating from London and stretching out through the conurbation centres into the peripheral areas. This could mean that, even if a peripheral area's economic base is in a 'growth industry', it may still have only a relatively peripheral part of the total production process within that industry. However, all major urban areas produce some goods and/or services, for the whole national economy, although this is usually restricted to a very narrow range of industries.

The economic bases of Teesside (Cleveland) (Barras, Booth and England 1975: North 1975), Merseyside and Coventry are dominated by big firms (Morrison 1976; Palmer 1975). On Teesside, ICI and British Steel (nationalized) employ at least 50 per cent of the labour force in manufacturing. Teesside therefore has an extremely simple economic base -- a chemical firm, a steel firm and an engineering sector. In the more complex economy of Merseyside a few big firms account for a large proportion of the manufacturing base — namely, Pilkington (glass), Plessey (telecommunications) and Unilever (oils and fats) — each employing above 10,000 workers. The Ford Company, the Harbour Company, Cammell Laird Shipbuilding, British Rail, the Post Office, British Insulated Cables and the National Coal Board provide more than 5000 each. Together these firms probably employ a good half of the local manufacturing employment. (Remember, however, that there are more than 20,000 firms on Merseyside in all.) In Coventry British Leyland (vehicles) employs upwards of 15,000. British Leyland and Chrysler car firms employ upwards of 20,000 or 50 per cent of the manufacturing base. The Teesside petrochemicals complex represents the top of the world profit and capital/labour table even though it is in the 'declining' manufacturing sector. ICI is the largest British company. In this sense the Teesside economy is therefore buoyant — over £500 million has been invested in petrochemical plant in the past few years. (Investment in North Sea oil installations rises into thousands of millions.) But the local impact of this investment on the local economy (which as we have seen is largely transmitted through the wages of its thousands of employees) has not been anywhere near proportional to this investment. The public money going in investment grants may have increased profits, but its impact on job creation has been considerably less. Teesside has two valuable scarce resources needed by modern, large-scale, capital-intensive industry — a deep water port and a lot of flat land (ICI bought 2000 acres just after the war). The large-scale production plants are clustered together along the river. But where are the local benefits of having a highly profitable, multinational, capital-intensive firm? The wage payments injected into the local economy through the labour force are only a relatively small part of the outlay of the firms in these types of industries. This is clearly reflected on Teesside, which at first sight is a stark illustration of the problems of a latter-day overdeveloped county — masses of capital in shining new industrial plant in the midst of considerable urban decay, poor social capital and lack of services. Clearly the other economic input by industry into the local economy — the rates paid on the land, plant and buildings — has been insufficient to regenerate the area.

One peculiarity of the integration of the local economy with the national is that the UK is far more dependent on Merseyside for certain industrial products (e.g. oils and fats and glass) than Merseyside is on these particular industries for employment. In both Coventry and Merseyside basic manufacturing employment is also concentrated. Again quite naturally, in all these towns the big firms are concentrated in one or two places, often away from central areas; they are not locationally sensitive to the local labour force.

Much of the discussion in Chapters 1 and 2 of the structure of industry and the economic fluctuations is reflected in a local study of Coventry — the Coventry Community Development Project (1975). We saw previously how consumer durables have been particularly affected by national tax and credit restrictions, and this applies especially to cars. In order to be able to cut back production quickly, British car firms traditionally buy more components from outside suppliers (50–65 per cent of the total car value). This compares with 25–40 per cent in other countries. Even a prosperous area which has benefited from the boom in car production since the war has recently suffered severely: the 1975 recession has reached far into the heartlands of British productive industry. In Coventry there are a group of component firms linked to the major vehicle builders, and these suffer first in times of cutback — laying off first the unskilled manual workers, their unemployment rate was over 9 per cent in Coventry in 1970. The Coventry study shows how these workers tend to live together in the 'marginal' or twilight zones of the city, showing a clear link between the national economy and its local aspect, and the social areas so often recognized through census studies (see Section 5 below).

In looking at the control exercised by big firms, we are not saying that Unilever, British Leyland or ICI holds the whole of Merseyside, Coventry or Teesside in the palm of its hand. Many large firms have *small* plants dotted over the country, none of which crucially affects its local city. But, more often than not, a handful of firms which produce for the national market usually have a crucial influence on the economy of even a large city like Liverpool. Each of these firms is likely to dominate only a small number of cities. On the distribution and service side (activities serving local populations, often from city centres), there is also a national concentration of big firms, and these are likely to appear in a very large number of city centres (e.g. big retailers, banks, insurance companies). So, on the 'consumption' as well as the production side, the concentration of economic power reinforces the dominance of metropolitan centres over the periphery.

Office space and office rent

If we are interested in the geographical distribution of economic control, the distribution of office space is a useful indicator. Offices are places where information is received, processed and dispatched (Goddard 1973), and the demand for office space tends to outstrip the demand for many other types of building, owing to the growth of services generally. The increasing division of functions in the economy requires more communication, and the increasing complexity of capital intensive manufacturing also increases the need for office-type activities.

If we regard office space as an indicator of the distribution of 'ownership and control' activities, the concentration is striking. London has half the total office space in the UK. The largest regional centres, Manchester and Liverpool, have only a fraction of London's total, Manchester itself having about 7 per cent.

Even more startling is the distribution of office rents. If the rents were taken at face value as a measure of the true economic value to the country of the office activities, then we would find that the 'value' of a unit of ownership and control activity in London -- that is, a unit of office space -- is up to ten times the value of a unit of office space outside London.

Although the difference between London and the rest of the UK has decreased in the recent past, the variation in rents between *all* other centres in the UK is tiny when compared to London rents. In terms of control, then, we have a simple 'dual' economy: a large central mass (London) containing 80–90 per cent of the total office value in the country, surrounded by a small peripheral region -- the rest of the UK. (For a more detailed account, see Location of Offices Bureau 1968; Gower Economic Publications 1974).

This distribution of office rent throws an interesting sidelight on the dependence of the city on the national economy for the provision of capital for development and renewal. We have already seen that most of the national capital stock is buildings, and how financial institutions have a grip on much of this capital. Local government spends a large proportion of its capital on building houses (over £700 million out of £2000 million in 1971). Most of the rest goes on schools, roads and sewage (now transferred to water authorities). Most of this capital has to be borrowed, half from central government and half from private sources -- mainly financial institutions. At the same time, even larger loans are made by financial institutions to individuals buying private houses (£1600 million in 1971).

In commercial development, which is the most profitable type of building activity, we saw how within ten to fifteen years a small group of specialist property companies and financial institutions came to develop a significant proportion of all new office buildings. One or two such institutions may well have a dominant position in a city through a major town-centre redevelopment scheme. The city is dependent on them for raising the capital, organizing the construction and letting the buildings. The local authority might help to assemble the land (by compulsory purchase if necessary). The six largest property companies have a nationwide network of properties covering all major cities and towns (Broadbent and Catalano 1975). These properties are a source of rent, and rent payments from local commercial and retail activity (and even from local authorities who often themselves occupy these properties) are channelled back to the national financial centre.

These companies have assets stretching into hundreds of millions of pounds, sometimes up to £1000 million. This is comparable with the value of all the commercial property in even the second largest office centre in the country (Manchester). So this is a stark illustration of how dependent even a large city is, not only on outside capital for developing the physical stock, but also even on individual firms.

It is argued below that the concentration of office space and office rent is due on the one hand to the processes operating within the city, and on the other hand to the fact that financial capital is concentrated in a very few institutional hands. Both these factors lead to a build-up of office space and office rent in a few key locations.

Summary: the city and the economy

We have now taken the large firms which so dominate the economy, and spread out the various parts of their production and control processes over geographic space, producing a pattern of dependent and highly specialized urban centres. This pattern reflects first and foremost the centralization and concentration in the economy as a whole, with control functions concentrated in London. The urban system also reflects the integration of the economy whereby some of the surplus generated in the economy is channelled and circulated through large financial institutions. Investment in commercial property in city centres by large financial and property companies operating from London provides another illustration of the *economic* dependence of even large cities on these large national resource-allocating institutions.

The competition between economic activities and the consequent reductions in the cost of production, provide the elements of an overall explanation for the concentration of people and jobs in cities and of the decentralization process within cities (see below). Conventional urban economists point out three main reasons why activities and population cluster together into urban centres (e.g. Mills 1972). These reasons are broadly 'efficiency' arguments, the cost of production being reduced when urban activities are close together:

1 *Economies of scale.* The overall cost of production may be lower if production takes place in one large plant *at a single location* instead of being scattered in several smaller plants. This, then, tends to lead to the concentration of production.
2 *Comparative advantage.* A scarce natural resource (e.g. coal, flat land or a harbour) will tend to lead to a concentration of activity at that point.
3 *Agglomeration economies.* Production costs are reduced on account of the existence of other activities and large populations nearby.

These economies of production costs resulting from the concentration of economic activity do still operate in the new, large-firm economy where the division of function *within* the firm now goes on *between* different urban areas and transcends the individual city. So, although the pattern of urban centres is still very much as it was fifty years ago (albeit with fairly large-scale changes in the balance and size of these centres), now cities are much more interdependent than they were. Individual cities are dependent on cities higher in the hierarchy and especially on London. Nevertheless, individual cities may have a very significant proportion of the total national production of a single commodity. The individual city appears less important because many processes which used to occur *within* the area, such as the different stages of production within a firm, or the circulation of funds from sales and profits back into investment in new building plant or equipment (i.e. circulation of surplus), now go on *outside* the area through the centralized industrial/financial and state concentration of power in London. The individual city may appear to be nothing but a small dependent satellite, but it nevertheless does retain some bargaining power, *vis-à-vis* the rest of the economy, in respect of the very specialized product or function it undertakes.

As these economic processes and the size of firms have outgrown the individual city, so the area in question is reduced more and more to the status of a *labour pool.* The great majority of internal linkages in the city are those which pass through the household (see Figure 20). The other linkages involving the sale and purchase of products between firms now occurs across the urban boundary.

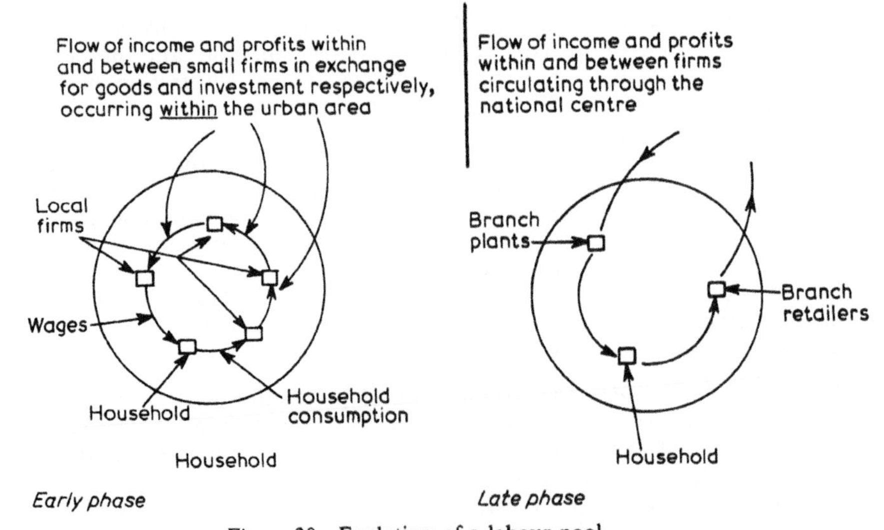

Figure 20 Evolution of a labour pool.

4 The city as a pool of labour: the first urban process

Introduction

This section argues that even large urban centres are first and foremost 'pools of labour' and shows how this influences the development of the area. This labour pool idea focuses the attention directly on the role of *people* as producers and as consumers. Both these activities begin and end *within* the labour pool. Viewed from the household, the urban area is a relatively closed system, even though in economic terms it is a very open system. This view of the 'labour activity' of the population has philosophical and social overtones; it represents a concrete way of measuring the contribution to the *economy* made by individuals, namely the exercise of skill and effort over a period of time. We therefore leave behind the 'machine-like' aggregate notions used to describe the national economy and focus specifically on what people actually *do* on a day-to-day basis. This way of measuring human activity can also be applied to the non-work activity within the household. The household can be viewed as a production activity -- a 'small factory' -- in its own right, producing a commodity (labour) for the market by employing other inputs (such as consumer goods) like any other production activity. We can, therefore define an urban area in terms of this generalized home-work activity, and the precise description of the household production process becomes possible, for instance by measuring the *costs* of the various inputs. Once we begin measuring the cost of production, we can see how we might explain the development of the urban system in terms of the reductions in the production costs of households just as the development of the economy at large can be explained in terms of reductions in the costs of production of other activities. It also highlights the fact that some labour activity (mainly performed by women) which contributes to the economy takes place *within* the household and is not paid a wage. It therefore does not appear in the national accounts as making a contribution to the economy as a whole. This production-view of households (and the people in them as producers) emphasizes their active contribution to society rather than their passive role as consumers or recipients of social welfare. Once applied to households, this analysis of production can be extended to cover all the other commercial, industrial and state activities in the urban area; the urban area can thus be regarded as a system of production. A preliminary explanation of the decentralizing process in the city would then look for reductions in the cost of the inputs to the production processes at various locations. If a city is decentralizing, it might be because the production costs at the edge of the area are becoming cheaper in relation to those in the centre.

It is worth recapitulating why urban centres can be regarded first and foremost as pools of labour, and why all the main economic flows go through households rather than between firms.

1 Most production in the national economy takes place in large firms, firms whose economic size is much larger than even the larger cities.
2 The whole national economy (and, to a considerable extent, even the world economy) is becoming integrated and centralized into a single unit. The circulation of income and profit from production and investment from plants, factories and offices in an individual city is drained off from the city and circulates through national institutions -- either within large firms or through financial institutions or the state.

3 The studies of individual urban economies show the lack of linkages between firms, even within quite large cities like Liverpool (e.g. Morrison 1976; Merseyside Metropolitan County Council 1975a, 1975b). This causes the so-called non-local multiplier effects (Pred 1973), which means that an increase in production in a factory outpost of a large firm does not generate increases in production in other firms in the *same* urban area.

The activity of people in cities

So far our economy looks like an impersonal machine. Aggregates like 'manufacturing', 'the state', 'finance' and 'growth' are all anathema to the many urban research workers in the USA and the UK who are overwhelmingly concerned with 'the people'. This includes both the pluralists and the ultra-radicals whose theories are outlined in Chapter 5. To many writers, the word urban is synonymous with 'social problem'. The city is essentially where the marginal people live, the old, the poor, the young and the delinquent.

Now, while this is undoubtedly true, it is so because nearly *the whole* population lives in 'urban' areas. Many of the so-called 'urban' problems are simply pre-existing *social* problems which are merely revealed and laid bare by the urban economic processes. It may well be that social problems appear more amenable to solution if they are called 'urban' problems, but the fact that low-income groups tend to live together in multiply deprived areas ('marginal' areas) does not indicate that the urban system itself created the problem of deprivation.

This is not the place to get involved in philosophy but the definition of the city as a pool of labour has wider connotations. According to some philosophers it is this activity of 'labour' which distinguishes 'man' from other living forms. 'Man' makes him/her self through his or her activity, an activity which, unlike the automatic behavioural patterns of animals, is exercised according to a preconsidered purpose. It is not only purposive, it is cooperative, operating on the natural environment (using and transforming natural resources). The exercise of this purposive, cooperative activity is the means by which people gain their living and also by which human experience is developed and enlarged.

It is the way this activity is organized, and the scale, place and time over which it takes place, which helps us to define the city. Human activity is organized in a very particular way under the modern market economy, a way that is fundamentally different from that prevailing in previous eras. First, the 'means of production' -- that is, the materials, machines, tools and implements needed by the individual to produce the goods and services -- are no longer under the direct ownership and control of the individual. Under the feudal system the individual or small groups of individuals often lived and worked in the same place, using their own personal tools. A considerable proportion of the means of subsistence was produced *and* consumed by the single individual or small group, within the local village community. There were, of course, higher levels of cooperation and organization than the village. Trade did occur, cities did develop, especially as market places and financial centres. (It was trade and the development of merchants' capital in the UK which led to the development of modern industrial production.)

Under a fully developed market economy industrial production becomes a 'socialized activity'. It requires large numbers of people to cooperate simultaneously, using means of production which involve massive amounts of complex machinery and input materials, within a network of interlinked organizations -- firms, factories, finance houses and the government. Most of the population do not own any of these means of production -- the productive wealth of the country. It is probable that 5 per cent of the population still own 70 per cent of the wealth and 95 per cent of the shares in companies.

People have not, therefore, got immediate access to the machines and materials necessary for them to work; they do not own them, either as individuals or collectively. The way labour is brought to work is the same as the way any other commodity is brought into use in the market economy, that is, it has to be sold. Labour has to sell itself on the market to the industrial or commercial firm before it can gain access to the means of production and work can begin. Under modern conditions, therefore, where production is socialized (and not only under capitalism), there is a separation between work and the work place where the goods are produced, and the home where the worker subsists, reproduces and carries on other social activities. In fact, the human purposive activity is divided into two parts. The first part -- 'work' -- is concerned with producing the goods and services used by the society at large; the second part -- 'home' -- is where the worker subsists, reproduces and undertakes social activities. (This fundamental division between home and work has been given some attention recently, especially in literature on the women's movement. The radical strand of this argument points out that the industrial market economy, and the requirements of mass production, has made the worker a 'free' individual (i.e. free to sell his or her labour on the market), no longer tied to a feudal lord. Work is an 'impersonal' activity, and so the emotional, social, personal side of life has to be fulfilled in the home. This separation between home and work also made a separation between males who were 'producers' and females whose prime concern was the rearing of children.)

The following aspects of the home–work separation need mentioning

1 The purposive exercise of skill and effort is common to *all* activity, whether at home or at work. Activity in the houshold is also 'work', it takes time, skill and effort.
2 A corollary of this is that much of the activity within the household contributes directly to the economy.
3 'Consumption', which is a traditional function of the household and which is usually regarded as an end-product of social production (i.e. all production is aimed at satisfying the needs of the consumer -- see Chapter 5) is actually only a means to an end. It is the essential input needed by the household in order to produce its output product -- labour.

The city, then, can be defined as the place that circumscribes *both* the home and the work activities. Although the home–work separation is a fundamental division in society, nevertheless the individual urban area, viewed from the household, is a closed system. It is suggested, therefore, that the urban area should be looked at first of all in terms of the day-to-day activity of the people who live there, and in terms of *both* their roles — as producers and as consumers. It is not so much an interlocking economy of producing and consuming firms, but a *community of people* who produce and who also consume in order to produce.

This separation between home and work, together with the large scale of the various production processes (factories, offices), defines the first and most immediate characteristic of the city — its finite extent. The other reason why the city is limited in size is the 'economy of time' and the daily-life-cycle pattern. Individuals need a considerable time at home and at work within the space of a single day. The large numbers of workers needed in the production process, therefore, need to live in relatively close proximity. Everyone knows that the growth of cities in the nineteenth and twentieth centuries went hand in hand with the expansion of industrial production, the drift from the land and the population explosion. Subsequent improvement of urban transport again allowed cities to grow spatially while still allowing the population to be 'in touch' on a daily basis.

The idea that people economize on time in the same way as they economize on money has received some attention (Becker 1965), especially through 'time-budget' studies (e.g. Cullen and Godson 1975). These exercises usually involve individuals keeping a personal diary over a given period. But the limitation on time available for both home and work has not been put forward very strongly as a reason for the existence of urban centres. What this amounts to is really a more precise elaboration of the 'scale' and 'central place' arguments for the existence of cities. That is, if production is on a large scale or if a large population is needed to supply specialist skills (and if transport is not instantaneous — see below), then the need to be at home and at work simultaneously in one day will lead to the existence of large populations living in close proximity. Labour, above all other commodities, is produced for the *local* scale, the *local* market.

We have defined an urban area as a place limited in space within which the population can and do interact with each other on a daily basis to produce goods and consume their daily needs. But there is no hard-and-fast boundary around the city, and much research is directed to defining the correct city limits. Even if the geographical boundary is drawn very far from the centre there will nearly always be some people who cross it every day to work. Researchers are thus often concerned to decide what is the appropriate cut-off point (e.g. Senior 1969; Hall *et al.* 1973). Hall, as we saw in Section 2, includes a given area within the city region (MELA) if it sends workers into the region to work, provided it does not send more workers to some other region. Hall points out that there is usually a choice of two alternative definitions of an urban area — a 'physical' one and a 'functional' one. These two categories correspond to the distinction between the specifically human aspect of an economic system, namely the *activities* of groups and individuals on the one hand (e.g. work, travel) and the inanimate objects and products of that activity on the other, that is, (the commodities produced and consumed, the buildings, vehicles and the roads used, etc. (Broadbent 1971a). The first are 'functional', the second 'physical'. The need to separate *people* and *things* when looking at cities is now being stressed by several authors (e.g. Harvey 1973). It is clearly important if we are to understand the reasons for social and economic problems, and especially the relationships between groups or classes of people. In our own research on 'activity analysis' we have tried to maintain a clear separation between human activities and their products -- 'commodities' (Broadbent 1971a; Broadbent 1973; Barras and Broadbent 1975a).

Of these commodities, the actual buildings, roads and other structures used by production and transport activities can be said to define an 'urban' area, since these physical objects are the commodities used by urban activities (see Section 2). Hall notes that the immediate physical appearance of the buildings may be imisleading; what looks like a village may actually be part of an industrial town inhabited by commuters. In the definition used above, however, this should cause no problems since we shall define the city as including the spaces occupied and used by the population in both its home and work roles and in the interaction between them. In a sense, therefore, we are defining the geographical limits to the city in terms of the single combined home-work activity of the population.

We are now in a position to make a preliminary distinction between the *city* or urban area on the one hand and the *region* on the other. A city is defined first and foremost in terms of the daily home-work system which delimits a clearly defined local labour market and a working population that can all participate together in production on a daily basis. The region, which is a larger area, can be defined first of all not in terms of the generalized 'labour' activity of a resident population but in terms of the range of demand and supply for certain goods and services. The urban economists' scale and agglomeration economics and resource-advantage concepts come into their own. So, for the supply of water, the building of motorways, the distribution of wholesale goods and the development of natural resources such as coal and ports, one must consider populations and distances which are much larger than the local labour market. As we have seen, as the economy develops, activities supplying goods and services become split into different stages with some of the stages operating on progressively larger and larger scales. We should expect, therefore, that goods and services previously provided within the urban area are now provided on a regional scale. Regional water authorities which have taken over what was previously a local, urban function provide a good recent example.

This picture of the national and regional urban system being composed of relatively independent labour pools is clearly only a first approximation. The separate labour pools are linked together as we saw in Section 2 in the case of the expanding conurbations, and also we have so far only considered the situation at a fixed moment in time. At this point in time all population and other activities have fixed locations within these urban centres. But over time these activities themselves move; they do not just continue 'trading' with each other from fixed locations (commuting is 'trade' in this respect). Activities 'migrate', and this process still needs explaining. Migration occurs both within and between these labour pools. (A concrete way of describing these various production activities would be through the scheme described in Chapter 2, which calculates precisely the costs of input to each production activity, i.e. the 'activity analysis' approach. The pattern of production costs in different places determines whether or not an activity will obtain a cost reduction by migrating.)

The city as a system of production

So far we have considered only the daily home–work system centred on the household. We now briefly consider all the other activities in the urban area which provide the jobs and hence the income of the area, on the one hand, and the various consumer goods and other consumer needs, on the other. Just because we have defined the city as a labour pool this does not mean that the labour pool is the first cause of the city's being where it is. It is an entirely dependent entity as we have seen, and is entirely dependent on there being *jobs* available to produce the wage income for the household. Indeed, although the really significant *exchange* relations inside the city involve the household, most of the other activities especially the large firms exchange with the outside world; inside the city they are purely *producing activities.* Similarly with the retail and other consumption-providing activities: they import goods from the outside world and can be looked upon as 'producing' activities — they 'produce' retail goods.

The best way to understand the system of activities in the city is to follow economic logic, that is, we should begin with the most important characteristics of the national economy and follow their effects through as they are worked out at city level. Although this will not provide a historical explanation of the existing size and location of an urban area, it may provide some insights. We have said the UK economy is an integrated, competitive structure, dominated by big firms and characterized by state intervention. In looking at a particular city, therefore, it is quite natural to look for those production activities including manufacturing, commercial, financial and other services which are most fully integrated into the national economy, that is, which produce for the entire national or international market. The more the UK economy becomes an integrated, cohesive unit, the more will the role of the city be first and foremost to supply this integrated economy with a narrow range of goods and services.

As production becomes more concentrated, a relatively large proportion of the nation's requirements may be produced in a single city.

Next, it would be logical to look at the local input requirements of these national-serving activities -- especially land and labour, labour being by far the most important of these. The combined home–work activity of the resident population provides us, as we have seen, with our best preliminary definition of the size and geographical extent of the city. (We shall see below how the home end of the home–work system can also be regarded as a production activity similar to the usual production activity — industry, etc. — although clearly it is a rather special activity. (Logically, the next group of activities would be those providing the transport linking the two parts of the home–work system. We would then go on to look at those activities that supply goods and services directly to support the population, especially the retail system, noting that some of them will be supplied on a larger regional scale. These, together with the publicly owned activities such as education, hospitals and other government services which provide social support directly to the population, will often use the bulk of the rest of the workforce. From this point it would then be logical to look at the whole range of often smaller-scale activities which provide goods and services to local firms and to local government.

What is especially important is the use of inputs and their costs (especially land, labour and capital) used by each of the different groups of activities. In spite of all the complexities, and the special local factors in what often appears to be an immensely complex, interacting system of activities and households, looking at the city as a system of production does indeed help us to understand more clearly the forces at work and the forces with which urban planning is called upon to deal, and to perform the exact calculations needed to understand the relationship between the various activities and the effects of planning upon them.

Households: a production activity?

We now return to the central concern of this section — the household. We have said that the city can be looked at as a system of production. But can what takes place within households be described as a production process? With certain provisos, households can indeed be looked on as 'small factories'. We have shown how both home and work can be counted as part of a single combined home–work activity. Modern economics, on the other hand, begin by separating these two activities as much as possible. At one end of the spectrum people appear in the guise of 'labour' — a basic resource for industry, a factor of production. At the other end they appear as 'consumers' whose sovereignty finally determines overall patterns of production. We shall try to show in Chapters 5 and 6 what effect this 'schizophrenic' approach (Cairncross 1970) has on ideas about planning and on planning techniques. But if what is carried out in the home constitutes 'labour' in our general sense (i.e. the purposive exercise of skill and effort), in what other sense is the home a small factory?

Well, in the first place it produces a commodity for the market, a worker or workers. The commodity produced — the worker — is sold on the market to industry for a wage or a salary. The income from this sale is used by our small factory, just as it would be by an industrial firm, to buy the materials (the inputs, that is) necessary for producing the output commodity. The inputs include capital goods, such as the house and the land on which it stands, and other durable equipment such as domestic appliances, vehicle, etc. Current inputs (i.e. those which are more or less immediately consumed) include first food, followed by transport, clothing and various 'non-essentials' like pleasure pursuits. This ignores the debate about which part of total household consumption is 'productive' and contributes directly to the efficiency of the worker and to the economy, and which part of household consumption does not so contribute (Becker 1965).

The vital element needed to convert all these household inputs into an output product is the same as it is in industry — human labour. The expenditure of human time, skill and effort is required within the household just as it is in the factory, and this reinforces our initial reasons for considering home and work as aspects of a single activity. The residential activity — the households and the people in them — perform a function which is vital for the functioning of the whole economy.

Treating households as production units in the same way as other sectors of activity allows us to describe more precisely what happens when some part of its function is taken out of the household and 'socialized', that is, becomes the 'social consumption' of Chapter 2 rather than private consumption (a recent example would be the care and education of pre-school children). The rearing of children or care of the old *in the home,* can be given a cost in terms of the input of various commodities required — food, clothing, space, heat, etc. — and above all in the time expended by whoever in the home (usually female) is doing the work. It can also be costed in monetary terms: it may cost society (i.e. the market economy) less to have the work undertaken free of charge in the household. However, it may well be more efficient from the market point of view to have the care undertaken by the state and to release the responsible person from full-time household work for paid work in industry. If the economy wants to bring more women into the paid workforce it has to make arrangements for the work currently undertaken in households to be undertaken in some other way, e.g. crêches, nursery schools. There is thus no 'female reserve army' to be simply tapped as a new resource without cost and this kind of 'production cost' approach to the household provides a unified and logical framework for analysing the implications of the increasing socialization of activities which were previously household activities.

But an even more important reason for looking at households as active producers relates back to the earlier comments about the 'poverty' view of social and urban problems. People and households are now seen not in the role of passive consumers, nor as passive recipients of social provision, but as actively engaged in producing products which are essential to the economy. In other words, people *are* the society and their activity underlies everthing that is produced and consumed; 'industry is nothing more than the British people at work', to quote a recent political catchphrase. This dispenses with the Victorian hangover whereby people are 'provided' with consumer goods, or with social welfare by the society, by an economy which appears to be an assembly of institutions, e.g. 'firms' or 'the state'.

The notion of conscious 'purposive activity' implies a conscious will, something which might be exercised by choice. This means that the hitherto passive 'consumers' or recipients of 'social welfare' are capable of organizing themselves into groups outside the household and applying leverage on society — through trade unions, for instance. It is a failing of both the traditional British Victorian view of social provision and the ultraradical view which puts people under a system of total oppression that both see no hope for or ability of, people to exercise any leverage on the society, even if it is acknowledged they are a part of it; this is needlessly pessimistic.

But surely these small household factories are also very different from factories in industry? They tend to be uniformly small, with between zero and six individuals; there are no giant households as there are giant firms (other than those which own the bulk of the nation's wealth and these are deliberately excluded). Secondly, they do not as a rule make profits, accumulate capital, or invest in more and larger equipment in order to produce more and more output over the years. Capital accumulation does occur on a small scale but by and large all the current income, wage or salary tends to be spent *in its entirety* on current inputs (goods and services), on interest and rent and, of course, on taxes. Households do not generally buy labour on the market as firms do, but rather perform their own. They are also more mobile than firms, with complex household fission and fusion, and migration mechanisms throughout the life-cycle. Finally, households do not buy and sell goods from each other as firms do.

But if households can be described using a standard description of a production process — that is, as a series of inputs producing certain outputs — what about the massive differences and inequalities between different households? It is certainly true that different social groups have very different lifestyles and consumption patterns, but this production view allows many of these differences to be described in a precise way. The consumption of different goods varies with household size, age and income.

The author (Broadbent 1974) in a primitive attempt to classify the households into a series of different production activities, each with its own particular combination of inputs and outputs, came up with twenty-nine types of household. The subsequent analysis of household inputs showed how the different combinations of inputs varied with income and number of persons in households. Food is generally by far the largest item, the proportion decreasing from 50 per cent of total expenditure for lower income groups to less than 20 per cent for higher groups. Some items are a relatively stable proportion of expenditure, regardless of income. Expenditure on transport remains a constant proportion of total expenditure over a wide range of income, apart from the very lowest.

Census data and other social surveys have often been used to study what appears to be the immense variety and complexity in the social structure of cities so beloved of the 'pluralists' (see Chapter 5). Meaningful patterns can be identified within this complexity, although again the interpretation is largely a matter of choice. In one of our own PRAG studies six social areas can be identified in Liverpool — 'high status', 'lower middle class', 'Victorian terrace housing', 'outer council estates', 'inner council estates' and a 'rooming house' area (Webber 1976; Cullingford, Flynn and Webber 1975). Even within these areas there is a complex of interrelationships between the size of households, socio-economic status, life-cycle, and the size, age and condition of housing. For instance, even with public housing, the inner council estates showed a far higher concentration of 'problems' — a greater proportion of unskilled workers, much more overcrowding, unemployment, etc. Many of these census studies try to test the validity of the 'ring' or 'sector' theories of urban development outlined in Chapter 5, usually concluding that elements of both types of development are present, but neither provides a complete explanation — the Liverpool social areas seem to exhibit a pattern of rings.

So, what we have here is on the one hand a fairly simplistic account of urban centres as labour pools, within a larger simplified picture of the national economy based on the three 'estates' — the private (profit) sector, the working population and the state. On the other hand, we have set up a general way of looking at the city as a production system which is again fairly simple: each activity produces and consumes outputs and inputs. Even though households can be regarded as producers, a wide range of income differences and social differences must be recognized, and this shows up on the ground as distinct social areas within the city. To understand how, we have to look at the other urban process which determines the development of the city.

5 Competition for space and location: the second urban process

We now have to look to the process of competition between our production units (households) to provide an explanation for city development. We have shown that activities compete on a national scale, and that there are few direct transactions between commercial and industrial firms inside the urban areas. The second urban process, on the contrary, does involve *all* the activities in the city. A typical selfcontained urban area will have most of its employment clustered near the centre but the centre itself will be mainly commercial office and retail activity, with large manufacturing plants sited in flat places away from the centre itself. The sea of population will be located around these centres, with high-income groups often towards the fringes, but as we have seen there will be a whole complex of social areas with, as in Liverpool, public and private housing located both near the centre and on the fringe. The whole structure will be decentralizing as we saw in Section 3. To show how this second urban process works, we begin by looking at a single production activity — households — and then extend the analysis to cover all the other commercial and industrial activities in the system. We have seen that competition operates through activities reducing the production costs of their product. The product of households is labour, and, in so far as there is a labour market, households try to reduce the costs of producing labour. They are in competition with each other, and, therefore, in the local urban labour pool households try to reduce the cost of transport, and the cost of housing and space, two input commodities to the production process. This competition explains to some extent the existence of social areas and also the process of urban development and decentralization described by Hall.

Our small household factories are trying to reduce their production costs by locating close to their markets (the main job centres of the city) and also within easy reach of their input commodities — retail centres, schools and other services. These numerous 'small factories' are jostling each other in competition for the fixed and limited amount of space (land) in favourable locations near the sources of supply of their inputs and the sources of demand for their output. We have already noted that the residential activity uses more land than any other; it is the extensive activity *par excellence.* Competition in the land market exists because housing needs a lot of land (and therefore occupies a large area) and because the demand for the labour output and suppliers of commodity inputs to the household are located at only a few places. No matter how good the transport system, some residential locations are bound to be worse than others, being further from employment and services than others. The amount of space available at good locations is fixed: land is a 'God-given' natural resource and cannot be manufactured. (Nevertheless, the investment of capital in a given piece of land can obviously increase the housing space on it.)

We can now get an inkling of why the price of land and housing at different locations varies. We are all familiar with the idea that housing often costs less in areas and cities far away from London. We also know that in large cities housing that is not accessible to jobs and services often costs less (even though we also know that high-price, high-class housing areas may also be located far from jobs and services — see below). First consider a much more simple situation where everyone in the city gets paid the same wages and the employment centres are all in the centre of the city. Housing has to be spread out over a large area in order to accommodate the whole population. Consider those who live at the edge of the city; their expenditure on essential inputs, like food and housing, would also include the cost of travelling to work in the centre. Everyone else lives closer to the centre, and *if they are paid the same wages* they make a profit in relation to those on the edge of the city because they need to spend less on the journey to work. If the inhabitants do not own the land under their houses, the landlords could charge rent for the land nearer to the centre. The householders could pay this rent out of the surplus profit that they make by living nearer the centre and still be no worse off than those living on the periphery. Those on the edge of the city pay no rent but they do pay a large transport cost to get to work; those in the centre pay little or nothing to get to work but pay a high rent. Figure 21 shows this elementary relationship between increasing transport costs and declining rent as we go outwards from the city centre. This elementary abstract example, although highly artificial, does serve to illustrate the essential characteristics of competition for land and the nature of the rent. Rent is a 'residual payment' which is extracted by the owner of a piece of land from the profitable activities using it, even after paying rent the land user gains no advantage by moving elsewhere. The amount of rent depends on the relative profitability of the land in question for the using activity *vis-à-vis* other land in other locations, that is, it is determined by the costs of the various inputs to the activity in question at one place *vis-à-vis* the same costs at another place.

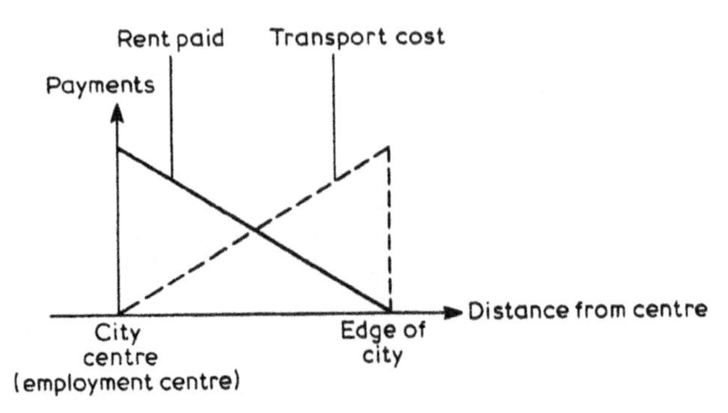

Figure 21 Increasing rent and decreasing transport costs for a single extensive activity (households)

So far we have considered the situation where only one production activity -- residential -- is using the land in question. But a more typical case in a city is where many production activities which go to make up the urban area are *competing* for the limited amount of land available. We give some indication in the next section of how the rent actually paid by an activity is determined not only by the favourable location of land for that activity but also by the level of rent which could be paid by some other possibly more profitable activity. If offices and big retailers could use land near the city centre this will force up the rent for any other activities -- households or small shops -- which wished to remain there.

Activities with a high transport cost to the centre will tend to locate near the centre, since their costs of production (i.e. cost of transport) would be increased very much by a move even a short distance from the centre. These activities would, therefore, be prepared to pay more for city-centre land than would activities with lower transport costs to the centre; they will therefore *push out* other activities with lower transport costs. This type of rent operates when there are *different* activities competing for the *same* land. The presence of one activity increases the cost of production of another activity because land is limited and therefore one activity is displaced by another. The activity displaced has to pay an increased transport cost to the centre. This is called an 'opportunity-cost' type of transfer payment in conventional economic theory -- see Chapter 5.

The main point here is not to explain how the competition between activities for space and for location actually works in detail but rather to show the basic point that this competition for urban space is something that involves *all* the industrial, residential and commercial activities in simultaneous competition. The first result of this as seen in Figure 22 is to *separate* activities into distinct zones, depending on the structure of their input costs and on the cost of transport in relation to their other input costs.

This process explains why different social groups even within the single residential activity live in separate areas. This could be because they are, in

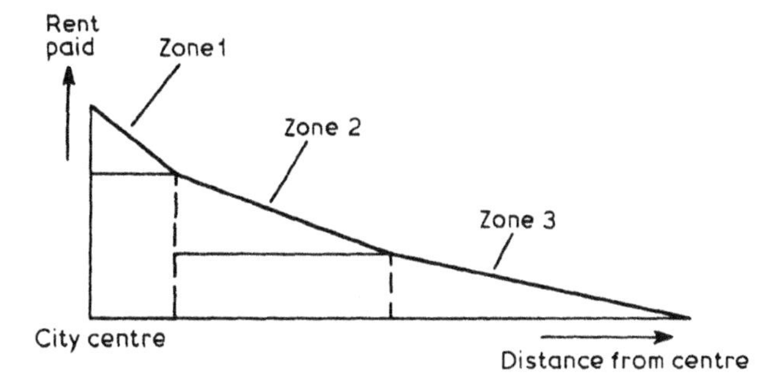

Figure 22 Zoning competition for space and location between several activities

effect, a set of *different* activities each with a different input cost structure. Transport costs are much higher *relative to income* for lower-income groups, and this might result in their living near the centre of the city (but their income may be only high enough to outbid other groups if they live at higher density). But we cannot at this stage produce any specific ordering of social areas within a city, nor a standard rent profile. The only general result possible at this stage is that *all* activities compete for land and that this competition leads to polarization, that is, to zoning.

There are many factors complicating this basic picture of competition in the local land market. Transport costs in smaller cities are relatively small; people are not only trying to live near the centre but also want space and a good environment. The rent payments themselves are not all determined by the market; much urban land and urban housing is publicly owned with rents held down below market level. Although many wages are fixed nationally transport costs may vary locally and therefore this affects the ability to pay rents. The existing housing stock and the size and distribution of houses are a historical legacy which cannot be changed in the short term. All this tends to result in a complex pattern of social areas such as that in Liverpool (described above). Despite this variety and complexity, the key feature remains *zoning* and *polarization* between activities. Thus the six social areas in Liverpool are very well-defined zonal bands, forming an approximate ring pattern around the centre.

We can now make a second crucial distinction between city or urban areas on the one hand and the region on the other. In the city, it is competition for the limited land available between a multitude of production activities that largely determines how the area develops. The total size of the city, its population, and the number and type of different activities using land and buildings determine the space needed and the rent payable for sites at various distances from the centre. High rents force out non-profitable activities and there is then a progressive decentralization as the scale of many activities increases. As the transport system develops, activities become more separated and the city decentralizes. Competition for land inside cities is fierce because it takes place within the spatial range of the labour area where many households and commercial activities are competing for good locations. In the larger area of the region, competition for land is not so fierce and any given piece of land will not be competed for by a large number of activities. We have, therefore, supplemented our elementary definition of the 'city as a pool of labour' with the 'city as a land market'.

The process of competition in in the urban land market does not simply end with the displacement of activities and zoning due to competition within one urban area. We still have not considered the effect of the rest of the national economy on the city itself. We have already described how the investment capital for new building comes from *outside* the urban area. Similarly, the activities which *use* the office floor space in the centre of the city are often branches of national enterprises, and their ability to pay rent in a given city is also determined by the position of the firm in the national economy — in the national productivity and profitability league.

When we realize that commercial 'land users' are often in reality the large national organizations — big firms, financial institutions and property companies with assets running into thousands of millions of pounds (Chapter 1) — we begin to understand the nature of the land-use planning problem. The forces of competition are now so extreme that a planning decision between a low-profit use such as housing and a high-profit activity such as offices can mean the difference between a building of a few thousand pounds and one of several million pounds — factors of over a hundred. The total rent paid in the economy has recently been increasing (although much of this is the 'imputed' rent of owner-occupiers). We would expect this in a general sense, since, as the economy and its capital stock grow in relation to a fixed land area which doesn't and which cannot be increased, land becomes relatively more valuable.

The rents and values of commercial properties fluctuate less than other values (e.g. share values) and property assets are a good hedge against inflation. We can now begin to see the local effects and mechanisms by which the property assets in the economy have recently increased. We see large blocks of financial capital seeking to undertake large-scale developments as a source of rental income and increasing asset values. The greatest opportunity for this lies in prime sites, especially in city centres where much of the total consumption of the area takes place and where it is still often the most accessible point to the labour force — a labour force which, as we have seen, is increasingly employed in non-manufacturing, in services and other office-type activities. The picture, therefore, is one where the forces of competition are invading the cities, where the activities competing for space and locations are often small components of large national firms, financial institutions, big retailers, etc., who are essentially engaged in competition with each other at the national level.

This massive effort in town-centre redevelopment has been documented in a passionate and emotional way by Amery and Cruickshank in *The Rape of Britain* (1975) but there has been little cool, hard description and analysis of the process. What we see in city centres are both the cause and effects of steeply rising rents and land values, which in turn result from intense competition for space and location between large organizations disposing of large quantities of development capital. These large-scale redevelopment schemes are a concrete (sic) manifestation of the growth of financial capital and the financial sector, noted in Chapter 1. This capital circulates in the economy and must find a profitable outlet; competition in the land market generates rents and values and thereby provides this outlet. One of the incipient contradictions in this situation lies in the fact that competition for land is a local, urban-scale phenomenon whereas the capital propelling this local competition is circulating through the national economy — through national organizations. The scale of new central-area development is often an order of magnitude larger than the scale of many of the historic centres being replaced. Ambrose and Colenutt (1975) document some precise case studies of large-scale commercial developments in London and Brighton and the pressures to maximize returns by making developments as profitable as possible by limiting social components like housing.

Once the urban labour pool is described as a set of production activities, each with a specific cost of production defined by the cost of its inputs, and all these activities are in competition for space and locaton, it becomes possible to explain in broad terms why urban areas might be decentralizing. As the scale of the various activities increases, and as the real income of the population increases, more land will be absorbed by the 'household' activity; this itself leads to an expansion of the area, a decentralization. Second, the relative cost of transport might decrease, again reducing the cost of locating far from the urban centre, or in another nearby urban centre. However, if transport costs are decreasing, if this explains why population might decentralize, why does commercial activity become so concentrated in the centre of cities? Reduced transport costs put the centre in touch with a larger and larger total labour and consumption area so that more and more income can be generated at the centre. The limitations of transport technology mean that fast commuter links and bus routes and main roads cannot be provided for *every* location, and this tends to reinforce the advantage of the city centre. New small establishments will generally find cost advantages by locating near to the existing centre. However, once transport costs become reduced still further, for example through suburban ring roads, and the *scale* of developments increases, the provision of a new shopping centre is no longer dependent on a nearby conglomeration of shops in an existing city centre, but it becomes an 'independent' locator; a large new hypermarket, for instance, can be located away from the city centre, and decentralization of employment has set in.

The decentralization process then depends on the costs of production associated with different locations, the difference between so-called private costs (the costs as seen by whoever is undertaking a development) and the public costs (the total social costs). It could be cheaper for a major property company to put a large development into a town centre, because many of the other costs are borne by the community at large, or by the local authority. Housing is demolished, journeys to work become longer, and there is no need

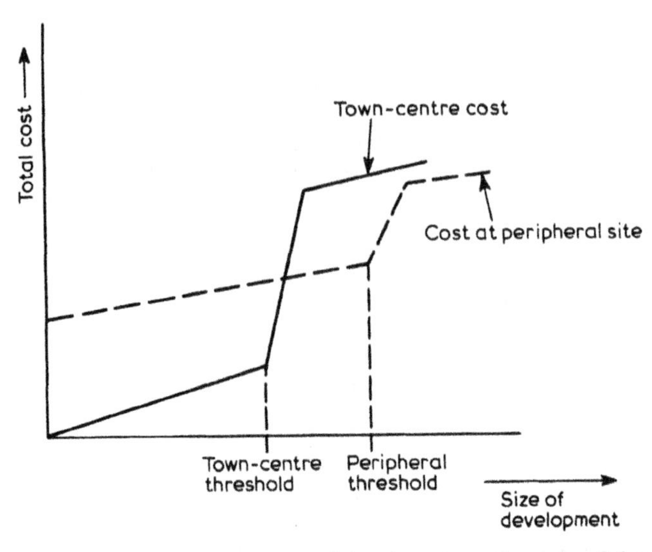

Figure 23 Town-centre costs versus peripheral costs as the scale of development increases

to build new infrastructure and services (roads, sewage pipes, bus stations, etc.).

The overall pattern of costs will vary with the total size of development. A small development will not need much new infrastructure, but as larger developments are considered a whole range of extra costs will be incurred (demolition, car parking etc.). This is an example of the well-known cost 'threshold' (Kozlowski 1972). More thresholds may be reached as the scale of the development increases. The costs of development at a location away from the centre may be initially very high — say because a new road or car park is required — but the next cost threshold may not be reached before the town-centre threshold — say because the cost of land is so much less (see Figure 23). But the crucial question for the whole urban area (and for urban planning) lies outside these two partial cost curves — with the reasons why the thresholds exist. When the costs to the developer in the town centre are low, other costs are being borne by other sectors of the community. What is needed then is a total calculation of costs for all sectors in the urban area.

This whole phenomenon of urban decentralization simplifies the pattern of land uses. This is an important point, since it is often said that the pattern of urban activities and their interactions is becoming *more* complex, not less. In the Appendix (Section 9) to Chapter 2 we saw how a single organizing principle can be imposed progressively across a whole system. The 'market principle' is an example; more and more goods are being produced for the market *as a whole* — either for a whole urban area or the whole national economy.

The progressive simplification of the city into mono-use zones is then a result of competition between land-using activities, and this is a reflection of competition in the national economy. Any attempt to suggest that it is a result of bureaucratic planning (see Chapter 4) must be treated with suspicion.

Competition in the land market in the first place creates no *new* problem of poverty and unemployment — it simply *reveals* the problem which already exists. Inequality and poverty among lower household groups already existed,[1] high-profit office uses already existed, and small shops and small firms were already in a poor competitive position *vis-à-vis* large shops and multinational enterprises. The geography of the city itself has done nothing to create these initial economic differences. The geography of the city can reinforce inequality, but first and foremost the 'inner-city problem' should be seen as merely a euphemism for 'national inequality'.

Planners therefore need to use techniques that clearly distinguish the *total poverty* or problems in the urban area — not only the relative concentrations of problems in the inner city itself.

Summary

Chapter 3 has shown how the pattern of urban centres in the UK has been changing, how geographically concentrated the population is and how the centralization and decentralization of cities and the relative decline of the inner areas of cities have been proceeding. It has tried to show briefly how this national urban system is affected by the structure of the national economy — the big firms, the financial institutions which have invested in urban development, especially in the commercial profit-making central areas. The individual city is first and foremost a pool of labour, and the city itself is not an economic system in its own right; there are few links within the city economy other than those which flow from the employment of labour, and from the consumer spending of the labour force. These flows pass through households: households should be regarded as 'producers', as small factories. The second key process in the development of urban areas is a reflection of the competition between firms and other agencies in the economy within limited resources and where there is continual displacement of one activity by another. This process is the competition for space — for land. Competition (not planning) tends to create zones of specialized land uses, especially in key locations like city centres where profits can be generated. Some real urban areas within the UK are used as examples. Channelling these forces of competition and coping with their effects is the essence of the urban planning problem.

1 Differences in regional unemployment are a factor of two; differences between inter-city and outer-city unemployment rates for skilled workers can be a factor of three (Webber 1976), but for Merseyside as a whole unskilled workers have an unemployment rate up to six times that for professional workers.

4 The local state: urban government and urban planning

1 Introduction

We saw in Chapter 2 the potential threat to the continued viability of the mixed economy in its present form posed by the fact that the national economy is still a *relatively closed* system, so that certain types of growth in the public sector threatened to undermine the profits of the private sector. The resources needed by the state are at least in part drawn from the potential profits of the private sector itself. Now it is immediately clear that the local economy is a much more *open* system than the national economy, and by the same token we shall see in Section 2 that the link between the local state and local urban economy is much less direct than that between the national state and the national economy. The local state does not only gain its income from the local economy but also from central government; it is thus more 'relatively autonomous' with respect to the local economy than is the national state *vis-à-vis* the national economy. The local state does not pose such an immediate threat to the urban market processes described in Chapter 3.

As the scale of production increases, it does so in the public as well as the private sector. Over the years the local state has been deprived of various productive activities, especially those that generate income through payment for services — the so-called trading activities. There is a nostalgia in some schools of thought for the idea of the 'unitary' authority which supplies all relevant services for the local population. This is largely an outdated concept: there is no reason why the *supply* authority should be the same as the *demand* authority, since the scale of supply is becoming so much larger than the scale of demand in the individual urban communities or labour pools. The prime function of the local authority is becoming more and more to represent the local population, to be its corporate, community expression as workers (producers) and consumers, rather than to supply it with water.

As the complexity and scale of urban development processes increase, local government itself becomes 'big business' and adopts longer-term, more rational planning and management techniques. There has been a growth in strategic planning for transport, health and social services besides the strategic, longer-term structure plans.

The whole system of local states (which takes an increasing share of public spending, even though local discretion and autonomy are being reduced) is nevertheless part of the national state. As such its operations are equally proscribed. It keeps to non-profitable activities as far as possible, responding and programming its actions sector by sector to the requirements of the local economy for social consumption and social investment. The legal restraint of *ultra vires* ensures that 'public enterprise' is kept to a minimum. Local government cannot act as a business, manufacturing its own products or selling profitable services to keep down its costs.

Urban planning itself is a peculiar and special activity. It has few resources and therefore has no *direct* economic impact. It is geared to the *second* urban process — competition for space and location. Both the wider local state activity (the provision of infrastructure and services) as well as the urban economy at large need urban planning. It has a very real task to perform — a task that continues to demand a better knowledge of urban processes. The history of UK urban planning shows a continued increase in its scope and scale, beginning with individual sectors such as public health and housing, and ending with broad considerations of economic and land-use competition within and between whole urban areas.

The present planning system, essentially unchanged since 1947, is a negative-response system, which originally assumed that most development would be undertaken by public authorities. It suffers from the separation between plans and powers described in Chapter 2. There is a separation between 'execution' (principally development control) and 'planning'. An attempt is made to describe what the planning machinery actually does. Much of it is an *ad hoc* response to market forces; planning has no specific socio-economic brief. This system does not work very well, even in its own terms, and many plans are out of date.

Section 5 discusses the economic role of land-use planning in more detail. Planners tend merely to *move* land values from one place to another; they do not create overall shortages of land in the medium term, except in exceptional cases. The history of various attempts to take land into public ownership provides some support for the 'barrier' theory of the encroachment of the public sector on to the private, outlined in Chapter 2. Land cannot be taken into public ownership by stealth; this has been tried and failed. There is a starker choice: either enable the market to work, or replace it completely with a universal system of public ownership.

2 Urban government: increasing size, decreasing power

Growth

While total public spending has risen as a proportion of GNP, local government spending has done so even faster. While public spending in money terms rose by 120 per cent during the 1960s, local authority spending in the UK increased by 170 per cent — twice the growth of GNP. Local spending amounted to about one-third of total public spending or nearly 15 per cent of GNP in 1973. We might therefore see in local government an exaggerated version of what is happening to central government — with all the contradictory implications of the growth of the public sector magnified.

Note however that this total spending includes some transfer payments and, when these are excluded, local government took 12 per cent of GNP in 1973 (8.9 per cent in 1963).

30 per cent of local government spending is capital expenditure (5 per cent for central government), and it spends the greater proportion of all public capital spending (£3100 million in 1973–4; compare £834 million for central government). We have seen that most capital is buildings, and therefore local government and the construction industry are intimately linked, sometimes too intimately, as recent corruption cases show.

Of three broad classes of state spending, local government is overwhelmingly concerned with 'social consumption'. Looking at Figure 24, education dominates local spending and has grown massively since the war. The major item is teachers' salaries, and one recent growth element has been the expansion of further education through the new polytechnics.

Housing expenditure has also increased. Costs in the building industry increase faster than other industrial costs; it is an inefficient, backward industry and the price of land and interest payments on past loans add still more to costs.

Note how the so-called 'trading' functions have declined. These are services which sell commodities and services. They have been progressively taken from local authorities as the scale of their supply-and-demand technology has increased. Electricity and gas were removed in the late 1940s and more recently regional water authorities have taken over water and drainage.

In capital spending, housing takes over from education as the dominant sector, and often takes well over 50 per cent of the total. If we are looking for the role and significance of public-sector debts to the private sector, housing provides it: the outstanding debt was £6636 in 1967 and cost £412 million to service. The total local government debt had risen to over £20,000 million in 1974–5.

In the two-tier system of local government which emerged in England and Wales in 1974 (and in Scotland in 1975), the way spending is apportioned between the two tiers in the major English conurbations is different from that in the rest of the country. The so-called non-metropolitan areas outside the conurbations are top-heavy; each of the forty or so English and Welsh counties takes up about 90 per cent of all spending, with only housing of the major services undertaken as a lower-tier district function. (There are some 400 of these districts which include major cities like Bristol and Cardiff.) In the conurbations the 'metropolitan counties' are the upper tier, and they spend only 20 per cent of the total. Highways and transport are their major function. It is a weird system indeed, with a huge variation in size between the lower-tier, non-metropolitan districts (24,000 to 422,000) and also among the metropolitan districts (173,000 to over 1 million). The non-metropolitan, upper-tier counties also vary from 99,000 to 1.4 million.

Clearly there is some correspondence between these local states and the system of city labour markets or labour pools described in Chapter 3. The larger counties cover more than one such area, but they were not generally designed to fit in with function areas. The metropolitan counties' boundaries

	1947–8				*1972–3*			
	Rev. £m	*%*	*Capital £m*	*%*	*Rev. £m*	*%*	*Capital £m*	*%*
Education	209	(21.8)	10	(3.3)	2 903	(36)	385	(14)
Housing	59	(6.1)	213	(70.1)	1 400	(18)	838	(30)
Trading	279	(29.0)	54	(17.8)	408*	(5)	205	(7)
Police and fire	44	(4.6)	1	(0.3)	573	(7)	47	(2)
Highways	70	(7.3)	5	(1.6)	476	(6)	267	(9)
Public health	62	(6.5)	7	(2.3)	447	(5)	289	(10)
Industrial health	19	(2.0)	1	(0.3)	164	(2)	16	(1)
Welfare, aged and children	–	–	–	–	391	(5)	51	(2)
Other	218	(22.7)	13	(4.3)	1 242	(16)	697	(25)
TOTAL	960	100%	304	100%	8 004	100%	2795	100%

From: Hepworth (1971); Local Government Financial Statistics (1972–3).

*Water supply has since been removed to water authorities

Figure 24 Revenue and capital expenditure of local authorities in England and Wales

are drawn very tightly around the built-up area, often leaving important residential areas and industrial plants outside the county. In spite of this widespread 'mis-match' between the properly identified labour pools and the system of local states, and despite the added confusion and complexity resulting from the generally criticized two-tier system, there is sufficient correspondence to warrant a discussion of the relation between the local state and the urban economic system.

Finance

Although local government appears to be of increasing importance *vis-à-vis* central government, this is to some extent an illusion, since its income has been provided more and more by central government. This is in keeping, of course, with the increasing centralization of the economy at large, discussed in Chapter 1. Nevertheless local authorities do have a degree of autonomy, as the Figure 25 shows.

	1973–4	*1973–4*	*1947–8*
Local rates	£2414m	28%	29%
Central government grants	£3897m	45%	28%
Rents, interest, trade, etc.	£2349*m	27%	43%
Total	£8660	100%	100%

Source: Local Government Financial Statistics (1973–4); Layfield Report (1976)

Figure 25 Source of local government current finance (England and Wales).

Rates are a tax on the value of buildings; the rateable value is supposed to relate to the market rent which could be derived from the property. Although local income has declined, mainly owing to loss of trading and rent income, at least there is now more local discretion over the use of central government grants than there used to be. A decreasing proportion of the central grants are earmarked for specific purposes.

By 1974–5, the total general grant (called the 'rate support grant') had risen to £3488 million or 90 per cent of the total central government grant. The size of the general grant is based on a fairly complex calculation involving three main elements. The first relates to the so-called 'needs' of the area. This is intended to measure the characteristics of the area, including its particular socio-economic and physical structure. Weight is given to population, the number of children, a high density of population, dispersed rural population, declining population, the presence of high costs associated with large urban areas, and so on. This 'needs element' makes up by far the largest part of the central government grant — over £2000 million in 1974–5. The 'resources' part of the grant is meant to give added income to those areas whose total rateable value and rate income is less than average. If the shortfall is 20 per cent, then the authority gets a grant of 20 per cent of its total net expenditure. Note that this isn't a *redistribution* of rates from rich to poor authorities but a grant to those with low rateable values to bring them up to the average. This amounted to £820 million in 1974-5. The third item is meant to reduce the burden of *domestic* rates on householders, and amounted to £471 million in 1974-5.

About 50 per cent of total rateable value in the country is in houses, 12 per cent in industry, 9 per cent in shops, 9 per cent in offices and 12 per cent 'other' (hospitals, education, entertainment, power, etc.). Since the domestic burden is reduced by the central government 'domestic' grant, more than 50 per cent of the actual rate income comes form non-domestic property.

Criticisms of the rating system are obvious and often stated. Rates bear unfairly on lower incomes — Nevitt (1966) has shown that they amount to 14 per cent of lowest incomes and 4 per cent of the highest groups. They act as a disincentive to house improvement, raise revenue unrelated to the services provided, do not rise atomatically with income as other taxes do, and they are charged in a large lump sum — all this making them very unpopular. On the other hand, rates have practical advantages over other local taxes, whether on income or local sales. They are clearly based on the local geographic area, they are thus very independent of central government, rebates can reduce the burden on the poor, they are easy and cheap to collect, and they don't fluctuate as much as the economy generally.

On the capital side, Figure 26 shows that by the late 1960s the majority of local capital had come from private sources.

	1973–4	*1973–4*	*1947–8*
Loans from central government	7 956m	41%	24%
Other loans (mainly from private sector)	11 450m	59%	76%
	£19 406m	100%	100%

Source: Local Government Financial Statistics (1973–74); Layfield Report (1976).

Figure 26 Outstanding local authority debt (England and Wales).

Throughout the 1960s annual borrowing from central government increased in relation to other borrowing (£760 million compared to £659 million by 1971). The proportion has varied over the years, and at some times local authorities have only been able to borrow from central government (the so-called public works loans board) as a last resort. At the present time they are allowed to borrow up to 30 per cent of their total capital payments in a current year, plus 3.5 per cent of the outstanding debt (40 per cent and 4 per cent respectively for development areas). These loans generally operate at 1 per cent below market interest rates. Interestingly, the private money market reckons that local authorities are not as good a risk as central government.

Since 1973 local authorities have also borrowed in Europe. Over the period 1968-73 there was a marked trend towards shorter-term loans, going hand in hand with an increase in interest rates.

It is not surprising that the public-sector debt problem emerges here in an acute form, since local government is the big capital spender in the public sector. From 1965 to 1975 the debt increased from £10,000 million to £24,000 million. The annual payment of interest on these amounts is now a large element in current spending (£1403 million in 1973 or nearly a quarter of current spending on services). The local authority debt has remained at roughly 18 per cent of GNP from 1963 to 1973 — in contrast to the total public-sector debt which has taken up a smaller and smaller proportion of GNP.

We have seen that inflation is one way of repudiating debts and thus largely accounts for the apparently leisurely growth of the total debt. However, the annual national deficit has increased massively in the last two or three years, in contrast to the budget surpluses of the late 1960s, and this necessitates large-scale borrowing at high interest. Budget deficits plus inflation ensure local authorities have a vested interest in further inflation in the future, to keep their total accumulated debt and interest payments within bounds. Interest rates and inflation tend to move together, although with the recent very high inflation the rate of interest has fallen behind the rate of inflation, the 'real' rate of interest being negative at present.

Reforms in borrowing are often suggested. Authorities are often in competition for funds and this tends to force up interest rates; it seems iniquitous to many commentators that housing rents should be used largely to pay interest to the private sector. Nevertheless arguments in favour of subsidizing local authority borrowing are usually countered by the fear of giving an artificial incentive to capital spending. The Labour Party (1974) says we must spend on 'people not buildings'.

Central controls

Current spending of local authorities is only under partial control of central government, and again illustrates the general difficulty of public expenditure control as shown in Chapter 2. Forecasts of local government spending are included in the new three- and five-year forecasts of public spending, but because they are once removed from immediate government control forecasting is even more difficult. Obviously the main central government sanction is through its own rate support grant, but when a short-term cut in local spending for 1974-5 was imposed in December 1973 and the RSG fixed accordingly, local authorities were unable to comply, and local rates rose to compensate. Other central controls include approval of bylaws, auditing of accounts, inspection of services and approval of schemes for the development of services. Longer-term plans are now required for specific subject areas, for example, ten-year social service plans by the DHSS, transport plans and programmes (TPPs) by the DoE, etc.

Central government pushes in the other direction by laying down minimum standards for the provision of services by statute. Central control on *capital spending* is much more direct. Local authorities have to get government approval ('loan sanction') for all borrowing, usually going through a specific ministry to the Treasury. There are some 'block' loans which do allow some discretion, but the sanction generally applies quite strictly and determines also the time period for the loan; for example:

ten years for vehicles, office machinery and library books
twenty years for furniture, machinery and plant
sixty years for land and housing

It has been argued that this system doesn't allow enough local discretion, even given the need to control overall capital spending in the economy. It leads to a sector-by-sector approach, rather than allowing the local planning of a bundle of projects specifically geared to local needs. This 'sectoral' approach can be used to project an insensitive 'central government' pattern of capital spending ill suited to the needs of the local area (e.g. road building — see Chapter 6). The Layfield Report (1976) feels that the system of central control is confused, that it leaves local authorities uncertain about the future availability of resources, and that the short-term changes in policy have lead to disruption.

Several authors have suggested that local authorities are largely 'agents' of central government (e.g. Maud Committee 1967; Griffith 1966), and this seems to be supported by most of the evidence available, even though it is difficult in practice to define the concepts accurately. The 'agent' theory would imply that local government action is to implement objectives set nationally; but of course such objectives will not necessarily imply uniform levels of service or levels of spending between all authorities. Indeed, the existence of a wide range of standards across authorities has led some to infer that this implies much local discretion, but it more likely implies a uniform relationship between the public and private sector across the country, so that prosperous areas have good local services and poor areas have poor services.

The Layfield Report (1976) urges that a more definite choice should be made between a strengthened system of central government control and one of greater local autonomy. On balance it favours the latter course, sustained by a local income tax. This seems on the face of it to fly in the face of the overall trend to a highly cohesive centralized economy noted in Chapter 1. With central government all the time attempting to get a firmer grip on the economy, such decentralization would only make its task more difficult.

As local government get larger and becomes 'big business' the annual budgeting procedure is turning much more into a forward-planning exercise. Just as central government is forecasting expenditure further ahead, so local government is going even further in adopting business planning techniques — 'corporate planning' — where the decision making is more centralized: individual departments, given less authority and made responsible to an overall 'central policy' committee, and planning by objectives — 'programme planning and budgeting systems' — are being adopted. The idea here is to get away from planning by input to planning by outputs. For example, the care of old people may be advanced by several different inputs — providing special residential institutions, building individual houses, providing domestic help, and so on; a proper output plan would devise the best combination of these three services rather than planning them separately. This could pose a threat to the various professional vested interests (architects, planners, clerks, treasurers and engineers) who are often rivals for status and power within the authority; so far they seem to have accommodated the new management techniques without losing any of their privileges. Such moves towards more rational, more predictable planning is also reflected in the urban-planning function itself (see below).

The ultra-radical state-as-oppressor theories would see the growth of business management techniques in local government as further evidence that the state is becoming indistinguishable from the market sector of the economy. This is to mistake an *appearance* for a *cause*. The growing sophistication of local state activity and the increasing use of formal techniques in the first instance simply reflect the increasing scale and complexity of urban development. Second, they reflect the increasing level of state intervention. But this of itself does not prove that the state is becoming more of an oppressor or that sophisticated techniques would not be required even under a system of 'popular control' rather than 'bureaucratic control'. Many of the methods and techniques used (see Chapter 6) reveal the *impotence* of the local authority rather than its ability to 'control' the people. They use a whole battery of methods in an attempt to 'forecast' all kinds of economic and social changes; they are largely responders to events in a world of uncertainty.

The local state and the urban system

We shall now try to examine the relationship between the 'three estates' at local level. We shall concentrate particularly on the local state as part of the local economy. The local state does not 'threaten' the motor-generator of local development — the local market processes — in the same way as the national state threatens the national economy. The main point is to contrast the local state with the way national government operates within the national economy. We have looked at the two main conventional justifications for public-sector intervention — the micro-view (to overcome market imperfections) and the Keynesian macro-view (to sustain growth and employment through deficit financing if necessary). We have also looked at more radical views to see if the role played by the state threatens the market economy in the longer term.

Since the local urban economy is so open, and most of the economic activity is owned and determined from outside, there can be no question of the local state beginning immediately to threaten the survival of the local economy. Local government income is so dependent on the channelling of central government funds that it is not simply 'feeding off' the local economy. The rate burden is a quite small proportion of total taxes and national insurance paid by the general population and by the private sector of the economy.

The main input to the local economy by the local government is the employment of labour. Wages and salaries are by far the largest component of local authority current spending, and therefore provide a large 'Keynesian' consumption input to the local economy. As far as the local economy is concerned, this is a much less ambiguous injection of income than is central government spending in the national economy. At least 50 per cent of it is a *deus ex machina* 'gift' from central government. So at first sight this 'agent' of central government is a beneficent one. Of course, when looking at the system of local authorities as a whole, this gift has been financed in part by taxes raised from residents of the local authority in question, But this is now cycled through the national economy, reflecting the increasing centralization observed in Chapter 1.

The other major increase in local consumption generated by local government is not through the purchase of goods and services but on capital construction; again much of this will leak outside the city in question through the large construction companies, but some building firms are still small and local, and of course local employment will also be generated.

We saw how the relationship between the numbers employed directly by the public sector in relation to its total spending provides some indication of how far government is acting as a producer, pre-empting real resources, and how far it is acting as a channel to redistribute the surplus back into the economy, either by financing contracts (not-for-profit production) or else simply redistributing profits from one sector to another by grants and tax relief. Using this indicator, local government acts far less as merely a channel for funds (transfer payments) than central government, and far more as the *direct provider* of services — social consumption. It employs far more people than central government in proportion to its local current spending. It takes money out of the local economy — from the surplus generated in the private sector and from wages and salaries — and uses it not to buy goods and services but to provide services directly for the local economy which reduces its production costs. It does this rather more 'fairly' than central government since rates bear more heavily on business than on households.

The locally generated income of local government is directly related to the two local economic mechanisms discussed earlier — the labour pool (the labour market area) and the rent-generated competition in the land market. Although there is a far-from-perfect match between local authority areas and labour-market commuting areas, despite the overall irrationality of existing local boundaries, the increased size of local government areas at reorganization did represent an attempt (albeit grossly inadequate) to encompass the increasing scale of activities, the expanding commuter areas and city decentralization, within local boundaries. It is also significant that the local tax should be related to rent, the major mechanism promoting city development.

Eversley (1975) shows how in the main conurbations the 'fiscal gap' tends to become more extreme through the decentralization process sketched in Chapter 3. Rate-paying activities move outside the local boundaries, and, because the leading sectors and the most prosperous income groups leave first, and even though some local needs appear to have declined (e.g. numbers of children), in fact many needs are increased; there are relatively more non-economically active old people, for instance. The costs of providing services do not decline in proportion to the decline in needs, partly because they are public goods and partly because of 'indivisibilities' (see Chapter 5). A school involves certain fixed costs even if there are fewer pupils.

On the other side of the coin are the suburban areas, whose rate base is expanding and who have more than enough to provide the fewer services needed by a younger, more dynamic population. They have every incentive to continue attracting more rateable value, since there is no 'equalization' of rate income between authorities, whereas the poor areas have no such incentive because of the asymmetry of the rate support grant, the resource part of which they lose in proportion to whatever new rates they attract. This disincentive for poorer areas parallels the disincentive of lower-income groups to gain higher wages now that the tax threshold has decreased.

Debt burdens are also increasing faster in the larger cities — the increase in debt per head for England and Wales was £79.4 between 1966 and 1974, £103 for Birmingham, £135 for Manchester and Liverpool, £86 for the GLC and £147.8 for the London boroughs. The big contrast with the national case is that this is a purely *external* debt: it is not owed directly to the *local* private sector, but to national financial institutions in London. In this sense the debt is worse for the local economy than the national debt is for the national economy. To repay it implies that an absolute deduction will have to be made from the local surplus. Paying off the debt does not feed anything back into the local private or domestic sector. Of course, if all other authorities pay off debts, then there will be a feedback from the national economy to all local urban areas — just as with a payment of the total public-sector debt. Unfortunately, as we have just seen, large urban areas have more debts. Because most of local authority income comes through central government, the local state does appear to be more independent — more 'relatively autonomous' *vis-à-vis* the local economy, and more of a *deus ex machina* than is the national government *vis-à-vis* the national economy. The total picture does seem to support the idea of the local state as an agent of central government.

Does the local state undertake Keynesian-type short-term management to maintain the local economy in a crisis-free condition? Directly, no — only as a backwash to the national attempt to curtail public spending. In general local authorities do not act independently to do this, but are of course involved when central government increases or cuts public spending to stimulate or restrain the national economy. The local labour market is affected when public-sector salaries get out of line with other wages and salaries — when, for instance, incomes policies are applied to the public sector, as happened in the early 1970s.

The stabilizing effect of the local state on the local economy stems from the continuous, relatively stable injection of central government money which does not fluctuate as quickly as the local economy. The local tax (the rates) is related to rent which is also a relatively stable income — but it may not be *stabilizing* because in times of local economic difficulty the rate burden will appear higher and bear more heavily on costs. It will not decrease to compensate an economic down-turn and may therefore help to exacerbate the restructuring.

Most serious commentators recognize that central government will increasingly call the tune; many suggest that this could improve and simplify the redistribution of resources between authorities. One idea is that essentially national services (like education) could be paid for directly by central government (e.g. Eversley 1975; Ilersic 1974). Eversley would also transfer housing to the national level, leaving local authorities with only parks, libraries, museums, arts and administration, but would allow higher standards to be financed locally if desired. Not surprisingly this doesn't find much support among the professional vested interests, the town hall treasurers like Hepworth (1971).

There is little scope for local authorities to act as local Keynesians on their own behalf — to specifically spend to create local employment — although, interestingly, the increase in the 'mixture of powers' and the increasing integration of public and private sector is also beginning to be felt at local level. Some of the new structure plans (see below) have advocated local public works programmes to counteract local employment crises.

In other cases, authorities have set up joint companies with private firms as a way of gaining profits from new development schemes (Minns and Thornley 1976). They have even on occasion rescued or partly rescued ailing local firms. But here we come up against our other big limit on the public sector — the ban on profitable 'commercial' enterprise.

The so-called 'trading' functions of local authorities include all the usual types of public enterprise — the natural monopolies, basic services, etc. — but they also include a few more directly commercial activities. Trading functions have gradually been whittled away from local authorities over the years, to become supervised by non-elected mixture-of-power bodies — often at regional level. Unlike other services, trading is broadly expected to cover its costs from sales; activities include passenger transport, cemeteries, restaurants, markets and slaughter houses, airports, industrial estates, other general estates and odd quirks like Birmingham's bank and Doncaster's racecourse.

In general, though, even when selling such goods and services local authorities *do not compete* with local private enterprise, nor do they specifically undertake activities which would make a profit, even though private enterprise isn't providing them. The most explicit reason for this is the law of *ultra vires.* It is illegal to undertake activities beyond the specific functions of local government laid down by law. Also 'public money' cannot generally be put 'at risk'; and since by conventional definition there is a risk in all commercial ventures this imposes another limit on profitable enterprise. Even where the law could permit more trading, this often doesn't happen.

A crucial passage in Hepworth (1971) is a fascinating *cri de coeur* — 'why should they not compete?' He cannot understand because he does not recognize how the economy works, and how the local level of government relates to the economy as a whole. It may well seem logical to Hepworth, as a local government official concerned with the public benefit, for a local authority to run a life assurance scheme, to compete with private builders, or to provide a computer service for local business; but what would happen to the motor-generator of the national economy — the profits of the private sector — if authorities throughout the country started running profitable trading enterprises? They would be diminished, they would be encroached upon, and the whole motive force driving the economy would be threatened. It would threaten the existing conventions governing state intervention (Chapter 2) whereby in general the public sector bears certain basic costs of production but the private sector alone derives income from goods and services which can be sold.

The one area of trading which might increase under current legislation is the buying and selling of land. Again this links in with the thread of our discussion. That local authorities should own their land can be justified within the conventional arguments for public intervention. First, private landowners do not contribute anything to the production of goods and services as capitalists do; they merely deduct part of potential profits as rent. Therefore it might well be more efficient for the market as a whole if business paid no rent or else if the public authority channelled the rent payment back into social investment which helps to reduce production costs. We have seen that the urban land market abounds with 'externalities', where a single transaction affects many other parties either by limiting the total land available or by increasing land prices around a new development. An official report (Uthwatt 1942) saw the public ownership of land as the only fair and practical way of collecting and distributing the rise in rents and land values that results from the total economic development of an area. In extreme cases, the local authority itself creates rent by investing in new streets and sewers, or by building new civic centres. The authority may have to buy this land at higher prices for its own housing development and thus pays twice over. But detailed discussion of the land market is further developed in the section on urban planning below.

An essential fact about all these local states is that they are all in competition for central funds and also for private capital. We saw in Chapter 2 that regional policy was essentially market-led, and that, although there are incentives, there is no system of direct planning of national and regional resources which ensures that urban areas get their fair share — of employment or development capital.

The local authority often acts as an agent or advocate to attract private enterprise to its industrial estates, using glossy advertisements and other methods. Councils go on trips abroad to help promote local business. Here there appears to be no incipient conflict between the local state, local business and the local population — 'everyone benefits' from increased employment and prosperity. But the fortunes of the area, as we have seen, depend on the particular narrow economic base situated there, and nearly all authorities (in vain) try to broaden their own base. Authorities can try to make things as attractive as possible for outside capital. Again this points to land-use planning. We have seen how market forces, through finance capital and property companies, are invading cities and redeveloping town centres and other key sites. The authorities' legal powers over land use and their ability to buy land are a crucial element in enabling maximum exploitation of the rent-generating capacity of the local economy by outside capital. The authority can make development more profitable by allowing high-density, high-profit development, less open space, cultural space and social space, and fewer low-profit activities. There is a much more immediate conflict here with local interests — as useful activities are forced out of town centres, housing is demolished and more of the local income is drained off as rent.

But this kind of local authority intervention in the local economy is not so directly dependent on the actual money resources available to it. This is true of land-use planning as a whole, which involves little expenditure by the authority. It is through land-use planning that the local state intervenes in the major internal market of the city — the land market — and helps to smooth the way for market forces to operate. This role becomes even more crucial as these forces become stronger.

3 The growth of urban planning

Early history

Town planners must be just about the most unpopular of all the professional groups in Britain today. The growth of 'community' pressure-group politics in the United States and UK since the late 1960s has often been fuelled by opposition to some local planning scheme. Most of this reaction is aimed at preservation or conservation of local communities or existing buildings. Most community action tends to be defensive. Planners are seen to be basically destructive — uprooting the local community, dispersing them into soullness tower blocks or antiseptic new towns, replacing homes with roads, local shops with offices, and generally being the agents of oppression at the local level. They are seen to work hand in hand with developers, property companies and other private interests rather than being responsible to the local community who pay their salaries and in whose name they operate (e.g. Dennis 1972). In this section we shall try to sketch the history of town planning to illustrate some of the general points put forward in Chapter 2 on state intervention in the UK economy. This will help us to understand why town planning developed as it did, and how it operates at present. At the same time it provides a practical case study which helps to make the arguments in Chapter 2 more concrete.

Most of the standard works on urban and regional planning do, of course, show that planning has developed as a response to the particular problems thrown up by a developing, industrializing urban economy. But most of these discussions stop short at the point where they are becoming most interesting.

Peter Hall's book (1974) on urban and regional planning is a brilliant synthesis, giving the reader a highly coherent and convincing picture of the growth and development of ideas and the parallel development of planning legislation in the last hundred years. But while this approach shows how planning grew up in response to the explosive, unsanitary growth of the Rochdales or Hartlepools of the nineteenth century or to the transport-led sprawl around London in the first half of the twentieth century, the economic analysis seems to lose its sharpness, and becomes more vague and elusive as it approaches our own time. Planning is still often portrayed as an adjunct to society — necessary, perhaps, but not really cogently explained as part and parcel of the economic role of the modern state. Urban planning was not mentioned in our previous list as a major resource-allocating function of local government. This is the most important thing about urban planning as it is practised in the UK — it preempts no or few resources directly, as compared to other functions of local government. It belongs to that part of government concerned with statutes, legal powers and regulations. Since even the spending departments of the local state operate largely in response to the private market, holding the ring or providing the underpinning, this is even more true of urban planning; since the driving force for urban development lies in the market economy, planning can only respond. Planning is a function of the local state, although it is nationally organized, and it still exhibits the main features of all state activity in the mixed economy:

1 It becomes more necessary, more sophisticated and more visible as the economy develops, as the size and scale of urban development activity increases, and as competition for location and space becomes more severe.
2 As its influence becomes more pervasive, it still attempts to avoid competing directly with the private sector at local level.
3 Thus, in superficial terms, it cooperates with business, smoothing the path of private development and regulating competition, by providing some degree of certainty for developers.
4 But it aspires (at least by implication) to achieve a socially optimum allocation of land uses, which often conflicts with the present system of allocation by the market.
5 There are strict limits on its ability to intervene and perform its allotted role effectively, limits which are determined by the private sector itself. Therefore, it can change the distribution of land uses and hence the development of the city only to a limited extent.
6 This is thus the contradictory situation, a contradiction which keeps cropping up in the debates within the planning profession.

The gap between aspiration and the reality of the very limited public-control powers available is often much wider in urban planning than it is in national economic regulation or in other parts of the local authority machine. Hence there are wild swings of fashion between one utopian solution and another. These fashions are also used on occasion to disguise the real nature of the urban-development process and planners' ability to control it.

A planning decision to allow or forbid a particular development can mean differences of millions of pounds to the private sector; the differences in profits (and hence the rent-generation possibilities) between one type of activity and another are extreme. Since rent is a pure deduction from the economy, paid to landowners, and since rent grows because the city grows, then it is often said that land should be publicly owned, the rent collected for the community. Attempts have been made to do this, but so far all have foundered on the barrier discussed in Chapter 2. If the private-market mechanism (the motor-generator of urban development) is undermined, it has to be replaced with something else, otherwise development will not take place. Just as with national and regional economic planning, neither urban planning nor any of the several attempts at land reform has so far challenged the role of the market in urban development.

It must be made quite clear that planning in cities predates the modern, capitalist, industrial society. As Hall (1974) points out, in ancient Rome, in medieval London, in medieval Winchelsea or Caernarvon, in seventeenth- and eighteenth-century baroque Rome, Paris or Karlsruhe, or in merchant-capitalist St James's, Mayfair or Bloomsbury, planning abounded (Hoskins 1970; Ashworth 1954; Bell 1972; Cherry 1972). Hall shows graphically how the industrial revolution spread from scattered, rural textile mills and iron foundries to become concentrated in towns as coal replaced water as a basic fuel. The canals and railways connected up the existing market towns to the new industrial ones, and a 'third-world' type of growth occurred in towns such as Leeds, Huddersfield, Rochdale and Hartlepool in the nineteenth century. Tiny villages became huge towns within twenty years. We look back 130 years to an unbelievably different world. Life expectancy was a little over forty years and as low as twenty-five years in places like Manchester, and a quarter of all births were registered as deaths within a year. There were several cholera epidemics in mid-century. It is quite clear from the series of reports on the conditions of the population in cities, and from the various public health acts, housing acts and local government acts between 1830 and the end of the century, that there was a slow but growing awareness of the need for the public authority to assume major responsibility for control and enforcement of sanitary standards, housing provision, water supply, and so on. It is also clear that this was not only seen in terms of improving the conditions under which people lived, but also of benefiting industry and commerce and their representatives who at that time effectively controlled local authorities directly.

The main pieces of public-health-type legislation in the nineteenth century are shown in Figure 27.

There was considerable local resistance to public health measures in the name of the freedom of localities and individuals. Housing improvements interfered with an even more fundamental right, that of property. No one should be 'dictated to as to the way his property was to be managed'. Standards laid down were difficult to enforce. Acts specifically concerned with housing are shown in Figure 28.

It is interesting to see how the scale of concern progressively increased

1835	Municipal Corporations Act	Boroughs not exclusively responsible for services and sanitation.
1844–5	Two reports by the Royal Commission on the state of large towns	Proposals for a single health authority in each area, regulating drainage, paving, cleansing and water supply. Also powers governing the standards of new buildings.
1848	Health Act	Set up a general board of health and local boards of health (primitive local governments). Brought in local surveyors, local officers of health, and drainage, sewage and cleansing requirements; restrictions on dwelling in cellars.
1849	Report of London Ministry of Health	'. . . no sanitary system . . . can cure those radical evils which infest the underframework of society unless . . . improve the social condition of the poor.'
1866	Sanitary Act	Imposed a duty on local authorities to improve provisions for sanitation. Powers to regulate the use and condition of individual houses and to prevent overcrowding.
1871	Royal Sanitary Commission report	
1872	Sanitary Act	Established two sets of sanitary authorities, 'councils' and 'guardians', to appoint medical staff.
1875	Public Health Act	Recognized public health as aspect in its own right, not merely part of the poor law. Set up urban and rural sanitary districts supervised by a local government board (i.e. a reform of local government).

Figure 27 Major public health legislation in the nineteenth century.

throughout the century, as the scale of the problems and processes became understood. The early acts laid down regulations for individual houses, but later it was realized that the spacing of houses and the layout of streets was also important if decent health standards were to prevail, so that whole areas had to be improved or demolished (Ashworth 1954).

Both public health and housing legislation therefore reflected a growing awareness of the need for local and national state intervention for the specific purposes which were later justified by neoclassical welfare-economic theory (Chapter 2). Monopoly services were firstly to be regulated and then later run directly by local authorities (water, drainage, sewage and, later still, public transport); externalities and spillovers were controlled by imposing sanitation and overcrowding standards. We can also see the beginnings of 'social consumption' noted in Chapter 2 through cheap housing for the working class.

1851	Ashley Act	Powers for local authorities to build houses for the 'labouring classes'.
1866	Public Works Commissioners	They were authorized to lend money to local authorities for building purposes.
1868	Torren's Act	Powers to clear insanitary buildings given to local authorities.
1875	Cross Act (artisans and labourers' dwelling improvement act)	Powers to clear whole areas.
1884	Royal Commission on overcrowding	Proposed no effective remedy.
1884	Manchester Police Act	Imposed conditions on privys and pollution.
1890	Housing Act	Permitted the building of single houses and gardens, consolidated previous acts, but main emphasis still on clearing insanitary areas.

Figure 28 Nineteenth-century housing acts.

The nineteenth-century debate on these issues was dominated by the classic Victorian concerns with improving the moral behaviour of the poor, rather than their physical wellbeing (Palmer 1972). Provision for the poor was often seen to encourage shiftlessness, and for both these reasons the poor had to show by improved behaviour that they deserved the charity bestowed on them. Echoes of these attitudes still persist in discussions of public-sector provision of services, as we have seen in Chapter 2. The most effective argument for reducing poverty and ill health was that this reduced the cost-burden on the ratepayer. A healthy workforce would improve overall productivity, increase total output in the economy and reduce the need for and cost of public transport for the poor in the longer term. There was also a fear of disease spreading from poorly serviced working-class districts to the rest of the population.

As the effect of the later nineteenth-century legislation spread outwards from the individual house to cover the neighbouring streets, open space provision, lines of building frontages and the heights of buildings (Ashworth 1954), as the separate concerns of housing and public health began to overlap and as the need for 'zoning' to preserve the character of different residential areas (separating working-class zones near the factories from the prosperous merchants suburbs) become more cogent, there was a growing realization that the overall use of land in a given city had to be considered as a single problem. This led to the first town planning legislation in 1909.

From 1870 onwards, many local authorities adopted model bylaws enforcing minimum standards for two-storey houses, a minimum width of street, and separate external lavatories with access to a back alley, via which earth closets were emptied. This housing is still quite ubiquitous and at an average density of 50 houses per acre is even now capable of being upgraded to modern standards.

Three acts consolidated the reforms of local government instituted in the large public health acts and set up a system of counties, country boroughs (large towns) and country districts which survived until the major reform of 1972, mentioned above.

1882 Municipal Corporations Act
1888 Local Government Act
1894 Local Government Act

Hall points out that between 1870 and 1914 most British cities acquired cheap and efficient public transport systems, first in the form of horse trams and bus services and then, from 1900, electric trams and finally motor buses. In London there were also electric tube railways from 1890. From the arguments of Chapter 3 we should expect this reduction in transport costs to lead to the physical spread of the city. London only expanded from a radius of two miles to three miles between 1801 and 1851, during which time the population doubled from 1 million to 2 million. But the city stretched up to fifteen miles in the north-east and the south directions by 1914. According to Hall this represents the apogee of the 'early transport city.'

This process speeded up even more after the Second World War. The growing number of white-collar workers in non-manufacturing employment could borrow money on credit to buy houses, which in any event were cheap because of the depression in the price of building materials and labour. These so-called inter-war suburbs made London into a roughly circular city by 1939 with a radius of twelve to fifteen miles. This — which Hall calls the 'later transport city' — was produced by a system of transport which combined motor buses and electric tube trains. In other cities similar low-density, sprawling development occurred. Local authorities themselves developed large new estates of single houses built with relatively generous standards of internal and external space. These developments were encouraged by several housing acts (see Figure 29).

During this period, town planning became recognized and officially institutionalized as a reaction to the problems posed by the increasing urban sprawl and longer journeys to work. Preservation of the countryside was also an important factor. Planning was brought in through the reports listed in Figure 30.

So much for the formal mechanisms of early town planning, but what of the content? What kind of cities were planners supposed to be trying to create? Well before the development of formal town planning, from the early nineteenth century, visionary private-enterprise schemes for 'new towns' and been constructed by several enlightened industrialists. Robert Owen built New Lanark around 1810, Titus Salt Saltaire in 1860, George Cadbury Bourneville in 1890 and W. H. Lever Port Sunlight in 1890.

These schemes showed that pleasant, low-density residential areas planned together with factories which were decentralized from traditional towns made good business practice. Private enterprise thus led the way in planning for decentralization, a path followed later by the state, which now bears many

1900	Act	Enabled local authorities to develop outside their boundaries.
1900	Act	Extended powers of local authorities and banned back-to-back housing.
1918	Tudor Walters Report	Recommended model housing standard of single-family homes at twelve per acre (i.e. one-quarter the density of bylaw housing).
1919	Housing and Town Planning	Began the direct involvement of central government in housing and accepted the principle of local authority building – 'Homes for Heroes'.
1919	Housing (Additional Powers) Act	Provided grants for private building.
1923	Conservative Act	Treasury grants for private building, administered by local authorities. Ended local authority building.
1924	Wheatley Act	Guaranteed a fifteen-year building programme for public and private housing. Restored local authority building with rate support and Treasury backing.
1935	Slum Clearance Measures	Five-year programme which, to 1939, cleared only 50 per cent of its target.

Figure 29 Early twentieth-century housing acts.

1909	Housing and Town Planning Act	Powers to prepare schemes for controlling the development of new housing areas. Emphasis on *new development*, very few schemes actually completed.
1914	Town Planning Institute formed	
1919	Housing and Town Planning Act	Obliged towns over 20,000 to prepare 'development schemes'.
1923	Housing Act	Extended the time for the preparation of schemes.
1925	Act	Town planning schemes extended.
1925	Council for Preservation of Rural England formed	
1932	Town and Country Planning Act	Extended planning powers to cover nearly all types of land, developed or undeveloped. This scheme was in fact not really a planning system, but merely ratified existing trends. There were no real powers over land use.
1935	Restriction of Ribbon Development Act	Partially controlled development on radial arteries but was generally inadequate.

Figure 30 Town planning measures 1900–1935.

of the costs of decentralization of cities through housing provision, new towns, roads, railways, planning legislation, industrial estates, grants to industry and the rest.

Private consumption of housing and services direct from the individual industrialist has been replaced by social consumption. The provision of finance for urban development and the construction and delivery of services has been divided into separate specialist stages. No longer can they be provided by the firm which provides the employment. At the same time the scale on which they are organized has continued to increase until they are often organized in institutions operating at a national scale, within an increasingly integrated national market.

But these early experiments were important in showing that economic forces, although they could not be ignored, nevertheless could be *channelled* to provide much more satisfactory urban environments — satisfactory for people and profitable for industrialists.

These practical schemes preceded a whole line of visionaries, some utopian, some more practical, who put forward schemes for the future of cities. Howard (1902) was probably the most important. He absorbed some of the basic economic theories of his day and confronted the urgent problems of urban squalor by devising essentially practical schemes for building a system of linked garden cities. As a champion and propagandist of the 'suburb', combining the benefits of both town and country, he was not entirely utopian, as his 'garden city' ideas seem to suggest. He was influenced by Marshall's view that there were many footloose industries free to locate new places away from existing towns. The scheme was also based upon the fact that the value (rent) of land increases after it has been developed, so that private enterprise could provide the initial capital to develop garden cities on green-field sites and reap a return from the rise in values. The full system of cities would have a population of 250,000, each garden city having 30,000 and the central city having 60,000. Housing densities were relatively high at eighteen per acre. Howard helped to begin the first two practical experiments, at Letchworth and Welwyn Garden City; however, both showed the entirely private-enterprise new town to be untenable.

Howard's garden city is well known, but his ideas about the public ownership of land much less so. He believed that the landlord class were oppressors and that rent was an unjust burden on the working class. The aim was for the workers themselves to run the garden cities — with all land owned in common and the rents used to run the social services. He understood, therefore, that rent is a pure deduction from the market economy by landowners. Where he was utopian was in believing that the transition to a national system of public ownership would come about gradually because garden cities would possess inherent advantages over other cities

Howard's ideas were later implemented in three 'garden cities' — Letchworth and Hampstead Garden Suburb around London, and Wythenshawe near Manchester. The ideas were further developed with the notion of segregation of precinct neighbourhoods, and separation of traffic in hierarchical road systems. Abercrombie put some of these notions into practice in a plan for London in 1943. Patrick Geddes (1915), another non-planner (a biologist and ecologist), helped to make planning more scientific, with a logical step-by-step procedure — 'survey–analysis–plan'. He stressed the importance of the local economy in identifying the 'relevant' region for study. Abercrombie (1945) combined the ideas of Howard and Geddes in a massive plan for decentralizing population and jobs for the London region. Since Howard, several other 'solutions' to the planning problem have been proposed, including self-sufficient, low-density, roadside civilizations, and Le Corbusier's high-rise/open-space cities — a concept blamed for the British penchant for high-rise blocks in the post-war era.

It is often said that these early town planners were utopian visionaries concerned to impose beneficient blueprints which ignored the conflicting complexities and aspirations of individuals and groups in a mixed economy. It is also said that they tried to solve social problems with bricks and mortar, feeling that the moral wellbeing of the population would be improved by building a new environment. These two fashionable criticisms can themselves be questioned. Both Howard and Geddes were very concerned to use the existing trends and forces in the market economy in the implementation of their ideas. Howard used the idea of footloose industries which were already decentralizing; he saw that overcrowded urban areas involved real costs and tried to use the generation of rent and the rise in land values to finance his garden cities.

Where Howard was utopian was in the notion that public ownership of urban land would bestow its benefits so equably to the population and industry alike that the superiority of publicly owned garden cities would become obvious, and that this form of urban development would gradually become universal, spreading 'by example'. Nor did he foresee that the entirely privately financed city would not be viable and that neither the free market nor the planning institutions could arrange for land to be available for urban development at pre-urban prices, except within a few special cases — such as the post-second-World-War new towns which absorbed only a small proportion of new urban development.

Much of modern planning thought tends to obscure the real economic and social forces operating in cities under an obscuring mist of social and technical jargon or radical rhetoric. The idea we shall develop in Chapter 7 is that we should be looking again at the economic forces operating in the UK economy in its present overdeveloped state. As to the so-called 'physical determinism' of the early planners, two points can be made. First, basic physical necessities such as decent housing and a pleasant environment may not be a *sufficient* condition for social progress, but they are quite likely to be a *necessary* condition. In any event, bricks and mortar in houses and local amenities represent real quantifiable resources which benefit the people who use them. Much of the current vogue for so-called societal planning serves to avoid the real issue of how real resources are controlled and who benefits from them. Resources are resources are resources, and they cannot be replaced by vague notions about 'linkages', 'communities', 'cooperative approaches' or 'cybernetics' (see Chapter 5).

The inter-war period

In 1921 the UK economy was still dominated by the three 'pillars of the industrial revolution' — mining, metal manufacturing and textiles. These industries were based on coalfields, around the Clyde, in Lancashire, the West Riding of Yorkshire and also in South Wales. The Clyde specialized in coal for export and steel. To return to our 'competition' theme, Hall points out that by the 1920s, with the impetus of the industrial revolution gone, population growth, and therefore demand, was slowing, and oil was beginning to replace coal. As other countries began to industrialize and the prices of some primary products increased, the three basic industries were no longer so important in the world economy and the UK was no longer pre-eminent in them. The newer industries, such as electrical engineering, motors, aircraft, precision engineering, pharmaceuticals, processed foods, rubber and cement tended to grow up around London, Birmingham and the East Midlands, Leicester, Nottingham and Derby, rather than in the older industrial areas. Many of the newer industries were geared more to the home market than to exports. During the 1920s and 1930s, when there was a worldwide recession with mass unemployment, the older industrial areas were by far the worst hit. In 1934 unemployment was 60 per cent in Glamorgan and less than 10 per cent in London. Migration was one response: in the 1930s 160,000 people left South Wales and 130,000 left the North-East.

To most writers on planning, Barlow is the most important single name in the evolution of the distinctive British planning policy after 1945. A Royal Commission was set up (1937-40) to investigate the regional industrial problem in the light of suggestions being made that government should aid distressed areas by giving grants to industry to encourage development in these areas. To Hall, the Barlow Report (1940) for the first time presented the physical growth of the conurbations and the problems of regional decline of the basic industries as two faces of the same problem. The commission pointed out that although the London area had less than one-quarter of the total employment at the beginning of the thirties, it was receiving over two-fifths of the employment growth. They examined the factors influencing industrial location, and argued that nineteenth-century industry was located near fuel, raw materials and navigable water, whereas new industry was less geared to these basic supplies and was therefore gravitating towards its main markets. In other words, industry was becoming demand-orientated rather than supply-orientated. The result was an increasing concentration in large cities for the reasons outlined in Chapter 3 — the availability of labour skills and specialist services. The commission saw an increasing tendency towards further growth in large cities, but concluded that the disadvantages of large urban concentrations far outweighed the advantages. They concluded that housing and health were worse in larger cities, that journeys to work were longer, traffic congestion was worse, and land and property values were too high.

The commission was split on remedies. They all argued against moving population and, in what then appeared an extremely radical suggestion, stated that controls on the freedom of industry were essential. The majority of the commission felt that controls on industry in and around London were sufficient, whereas the minority, under Professor Abercrombie, recommended controls on industrial location throughout the whole country. They also discussed how to control urban sprawl, how to preserve agricultural land by bringing in a better planning system, and how to ensure that the community as a whole benefits from the general rise in land values. The report spawned more detailed studies on these ideas, and many of the recommendations have been implemented and continue to influence policy to the present day. In the Barlow Report we therefore have a clear and unambiguous statement of the urgent need for government intervention at national level in the vital area of industrial development and physical planning. The inter-war history of planning, therefore, seems to reflect the wider changes taking place in economic and social thought, largely in response to the depression and the inability of the unfettered market economy to overcome it. State intervention became respectable, the Keynesian theory was born, and more sophisticated analyses began to link the problems of the national economy with issues in urban growth and urban planning.

Wartime reports and the birth of post-war urban planning

Some might see the great burgeoning of official report writing during the war — which advocated massive social reforms including education for all, universally available social services (Beveridge 1940), a free health service and land nationalization — as an establishment conspiracy to bribe the population to make the war effort. It might have been a farsighted attempt to head off demands for more revolutionary changes which would be forthcoming immediately the war ended. An atmosphere of optimism and the breaking down of social divisions was essential for the mobilization effort. The population should be fighting not just for the existing way of life but for a better one. Thus, according to Titmuss (1974), the increase of social discipline necessary in wartime would only be tolerable if social inequalities were not intolerable. These reports are listed in Figure 31.

Cullingworth (1972) shows how the depressed areas virtually disappeared during the war when a national industry policy helped to provide 13 million square feet of arms factory space in the depressed areas.

The burst of report writing during the war was immediately followed by a flurry of legislation between 1945 and 1952.

1945 Distribution of Industry Act
1946 New Towns Act
1947 Town and Country Planning Act
1949 National Parks Act and Access to the Countryside Act
1952 New Development Act

1942	Scott Report on Land Utilization in Rural Areas	Stated that agricultural land was a priceless asset and argued for a planning control system to operate in the rural areas as well as in towns. This report still influences planning through the green belt idea.
1942	Uthwatt Report on Compensation and Betterment	They asked who benefits and who gains from local authority actions such as road building and housing construction. They proposed that all non-developed land should be nationalized and all property owners should pay a regular tax on the value of the site. It was never implemented, but there have since been several attempts to enable local authorities to buy land cheaply.
1944	Abercrombie Greater London Plan	An attempt to end the sprawl of London, it suggested a 'green belt', created eight new towns (400,000 people), and proposed that 600,000 should go to expanded country towns.
1945	Dower Report on National Parks	
1946	Reith Committee on New Towns	Proposed new towns of 40,000–50,000 organized under a special development corporation responsible to parliament.
1947		Hobhouse Committee on National Parks

Figure 31 Wartime reports

This period saw the start of the regional policy described in Chapter 2. Proposals for new industrial plant or factory extensions of 10,000 feet (or 10 per cent) needed special government permission. There were also special grants for industrialists prepared to expand development in Merseyside, North-East England, Central Scotland and South Wales. Government-built factories were provided at low rents. These measures did not cover the main growth sectors of the economy (i.e. non-manufacturing employment), and incentives were given to capital-intensive firms rather than to those using a high quantity of labour. The controls could be avoided in several ways — by piecemeal expansions of a factory, by reorganizing warehouses and offices within a firm or by purchasing secondhand factories.

The 1947 Planning Act essentially created the whole planning system of the UK as it exists today. It instituted very effective control on development in some ways, e.g. the green belts (Hall 1973). It can be argued, and indeed it is essential to the arguments put forward in this book, that the key to its success in these areas is to be found in the way it made a comprehensive, once and for all change in property rights. It effectively nationalized the right to decide whether land should be developed. It is often said that it nationalized the 'right to develop land' (Hall 1974). We shall argue in Chapter 7 that another fundamental revision in property rights in land is now required. From 1947 onwards, all landowners only owned rights to the *existing use* and the *existing value* of their land (although the value provision was later rescinded). Any development required first of all 'planning permission'; second, developers had to pay 100 per cent of the increase in the value of the land resulting from development. At the same time the pre-war machinery of planning which was still largely optional for local authorities, entirely regulatory and restrictive, purely local and still bedevilled with compensation problems, was entirely reformed and given a national dimension. Before the war, hardly any of the land in England and Wales was subject to development schemes although most land was under 'interim development control'. The 1947 act also gave a 'positive' dimension to planning in that all planning authorities were required to produce a quinquennially revised development plan for their area. This was meant to provide a framework for development control and to guide developers to suitable areas within the city. The number of planning authorities was reduced from 1400 to 145.

4 The post-war planning system: plans without powers

Unforeseen change

The financial provisions of the 1947 act (in addition to the 100 per cent development charge) included a once and for all payment of compensation for lost development rights amounting to £300 million, to fall due in 1954. The supply of land for development had all but dried up by the time the Conservatives were elected in 1951 and the government abolished both the charge and the compensation in 1953. This still left local authorities free to buy land cheaply for their own use by compulsory purchase, but from 1959 local authorities again had to pay the full market value for all land.

The 1947 system has been working for almost thirty years. According to Hall, the system was initially logical, under the assumption that most development would be carried out by local authorities. In practice, of course, not only the majority of housing has been constructed by private developers, but market forces in the form of rents and property values have invaded the city in a massive way. So the system has turned out to be one of negative controls and response to developers' initiatives. It has worked well where such control has been possible and market forces containable, namely in the green belts around built-up areas. Inside the urban areas themselves, we can argue that it has been seriously, or indeed in many cases catastrophically, inadequate. Many of the existing local authorities were too small, and there were conflicts between those wishing to export population and those wishing to restrict it. There was little regional or central coordination between authorities.

This negative planning system was faced with a larger population growth than expected, hence with larger-scale changes in local areas. The dominance of short-term considerations in forecasting methods was not restricted solely to Treasury officials. A relatively short-term boom in population growth was translated by planners into a mid-1960s forecast that the UK would need to build a whole Bristol every year until the end of the century. The population forecast has been reduced in recent times (but now, equally falsely, some leading planners are saying 'growth' is now out at all times in all places). The size of the average UK household dropped from 3.7 in 1931 to 2.19 in 1971.

This urban growth created conflicts between large cities, such as London and Birmingham, and the surrounding areas. Between 1930 and 1966, the car ownership rate rose from one household in ten to one in two. This imposed massive pressures on the centre of major cities for urban motorways and multistorey car parks, and went hand in hand with the increased suburbanization of the largest urban areas. In the fifties central government encouraged all local authorities to make green belts, and in the absence of effective regional coordination several of the larger cities lost major planning inquiries in their attempts to expand into neighbouring rural counties. Large slum-clearance programmes were started again in 1955. Around 1960 it was stated that many big cities had simply 'run out of land' (Cullingworth 1972). In the metropolitan ring around London the net growth of population in the 1950s was 800,000, one-third of the total in Britain. As some of these problems began to be recognized, a large number of regional planning studies and *ad hoc* subregional studies were undertaken during the 1960s (e.g. Leicester 1969; Notts-Derby 1969). Hall shows that most of these reports continued the Howard–Barlow–Abercrombie formula of planned decentralization of conurbations, green belts, and new towns, with an overall emphasis on housing policy.

In 1968 a new Town Planning Act brought in a two-tier system of plan making. The aim was to reduce the amount of detail in the plans which were submitted for central government approval. 'Structure plans' would be submitted to the ministry containing the main, broad policy proposals in outline, for a large area. Detailed land-use maps would then be solely the responsibility of the local authority.

In the reorganization of local government, planning was one of the functions that was split between the two tiers. Everyone in the country is now living simultaneously in two planning authorities — with the upper-tier counties preparing the strategic structure plans and (where possible) the lower-tier district councils preparing local plans. A rather bizarre aspect of the reform was that, while the total number of authorities was reduced, the number of planning authorities (and hence the number of planners) was massively increased (to over 400).

At the time of writing, the planning system remains essentially as it was laid down in the 1947 act, with its financial provisions removed by the Conservative government of the 1950s. As we have seen, the control exercised by planning authorities is essentially one of negative response to private initiative. The state owns the rights on whether a specific development will be allowed on a piece of land, but the individual landowner or other person wishing to develop has the initiative, and nothing can happen until he or she makes the planning application to develop. On the other hand, all local authorities have to prepare a development plan which is essentially a combination of the existing pattern of development and forecasts of trends, together with the aspirations of the planning authority. The plan is not implemented by the planning authority itself. To be practical, therefore, it can only attempt to channel the forces already operating in the land market, in the demand for space and in the demand for locations, into what it perceives to be desirable directions. As with all legislation of this type in Britain, there are, of course, a large number of qualifications and exceptions to this general picture. The local authority itself does carry out a large proportion of development. Housing uses by far the most urban land and in some areas the local authority may own most of the housing. But even here, as we have already seen, the market comes into play: the authority has to buy land at market prices and often has to raise capital from private sources, on which interest has to be paid. The authority has every incentive to reduce its costs and maximize its profits.

In Chapter 2 we discussed the separation between planning and executive powers. It is important to understand that the major modification to the 1947 planning system, embodied in the 1968 act, did not in any way alter the essential characteristics of the 1947 system, nor the powers available to the local authority to change the course of city development. It remains essentially a negative-response system which has not been substantially modified, despite the mistaken assumption of 1947 that most development would be undertaken by local authorities.

There is, however, an additional power available to local authorities under this system. Local authorities can designate land for compulsory purchase. But, under the present system, this power, although necessary in order to ensure that land is available for essential complementary services and basic infrastructure to service private development schemes, nevertheless is such an arbitrary and selective instrument that in many ways it makes planning more rather than less difficult. The powers are most often applied in 'comprehensive redevelopment areas' to assemble large blocks of land from existing small parcels. The existence of the powers naturally creates uncertainty among landowners and it also puts the dead hand of 'planning blight' on to an area which is to be compulsorily purchased and redeveloped, but has not yet been so. Compulsory-purchase powers have been used only sparingly by most local authorities for reasons we shall outline below.

The power to declare a comprehensive development area has, according to some, been the main instrument for the 'Rape of Britain' and it is true that many city-centre redevelopments are covered by these designations (Amery and Cruikshank 1975).

The 1947 system: negative control

Under the pre-1974 system of local authorities, there were 150 or so planning authorities in the UK. These were essentially counties and county boroughs.

Under the 1947 system, the plans were largely in the form of maps showing land uses in some detail. But there were supporting written statements, and all these plans had to be submitted in detail to central government for approval. All plans had essentially the same form but differed in the degree of detail. The three main types of plan were:

1 The county map plus written statement.
2 A 'town map' plus written statement for the urban areas within a county.
3 Comprehensive development-area maps and written statements for areas within a county.

The county map was at a scale of 1 inch to the mile, the town map 6 inches to the mile and the development map at 1 to 2500. The maps and written statements generally tried to show the main land uses, both existing and proposed, within the area, using a system of symbols and colours. But they did not very clearly distinguish between what already exists and what is proposed.

The most fundamental concept in the act is the definition of development itself (Heap 1973). This is the 'carrying out of building, engineering, mining or other operations in, on, over or under the land, or the making of any material change in the use of any buildings or land'. There are thus two parts to the definition of development and these correspond to the distinctions we made earlier between 'human activity' on the one hand, and the objects and products of that activity on the other. Clearly the use of the land for building is essentially the activity of the people and/or organization using it. The building and the land itself is the 'material object' which is used and/or produced. As with most of the UK government practice, the planning powers and the definition of development permit a very wide area of discretion and interpretation for the public authority itself. The system therefore depends on the local administration being sensitive to changes in public opinion, having sufficient expertise and being completely impartial and above influence by interested parties. Heap's (1959-75) *Encyclopaedia of Planning* gives a selected compilation of acts, circulars and planning appeals and decisions.

Under the first part of the definition of development, any extension to a building, any change in its appearance or any new building is 'development'. But to define a 'material' change of use is much more complex. A change in the kind of use — from a house to a shop, for example — is material, but a change in the *degree* of use may not be. To avoid the system's being bogged down in endless legal precedents and arguments over what is development, the system has two provisions by which certain types of operation are specifically declared *not* to be development. These are, first, the 'General Development Order', which specifies classes of permitted development, and, second, the 'Use Classes Order' which specifies groups of uses within which interchanges are permissible. The Use Classes Order has changed very little during the years (Broadbent 1975d). There are four broad categories of use: residential, office, shops, 'general industry and 'light' industry. Of the eighteen use classes, the rest are much more detailed categories which, perhaps curiously, reflect the public health and nuisance concerns of the early town planners. They include land uses such as industrial processes which produce smoke or noxious fumes, and places of entertainment which are intensively used by the public. It is quite clear that many socially and economically significant changes of use within the four major categories are not subject to development control at all. The character of urban areas can change because different types of industry move into existing factories employing fewer people, because local food shopping is replaced with specialized shops serving wider areas, or because the households originally occupying the area as tenants are displaced by higher-income owner-occupation in the so-called 'gentrification' process.

In spite of the increasing pressure of market forces (rents for shops and the price of housing) which make these changes, the only recent change in the Use Classes Order has been to expand the special category of takeaway fish and chip shops — to cover Chinese takeaways — well within the traditional nuisance concerns of the earlier planners. It is as though the development control powers, which are the only precise direct power that planning wields, have been left dormant and untouched by all the fashionable concern for economic and social planning which has influenced the plan-making process itself.

Nevertheless, of the changes of use that do constitute development, the local authority has very wide discretion to attach all kinds of conditions on a development to restrict the use to a much narrower category than the Use Classes Order specifies, and to impose conditions on the height, size, standard and appearance of the buildings themselves. Again, there are powers to revoke or modify planning permissions where a development has not yet gone ahead. A 'discontinuance order' can be served, but this is extremely rare and requires ministerial permission. If an existing use is discontinued, compensation is payable, and in general the planning acts basically permit the continuance of existing uses even though they would not have received planning permission in the first place. This means that, even in the broad use classes of residential, shops, offices, industry, much of the *existing pattern* of use is 'nonconforming'. Thus in London in 1956 it was estimated that there were 2000 acres occupied by nonconforming industry. This problem has scarcely been tackled, although small adjustments have been made, either by purchase, by agreement or by serving discontinuance orders.

Perhaps one of the largest gaps in planning control (in its own terms) has been the so-called statutory undertakers — gas, electricity, railways, and so on — which have all been allowed to develop without planning permission, under the General Development Order, but are progressively becoming more and more subject to planning control.

The day-to-day practice of development control shows quite clearly the precise role planners are undertaking in the city's development process. Even after the land-use maps have designated areas for given land uses, the specific decision has to be made on the individual planning application. The development control 'advice notes' coming from central government stress the importance of measuring precisely the demand (e.g. for shopping space), and only allowing development sufficient to meet the demand. Similarly, retailers themselves, although they may urge local authorities to give more planning permission for large developments like hypermarkets, do not want too many permissions. Most of planners' rule-of-thumb techniques are based on notional levels of service (so many libraries, shops, etc., per 1000 people) — elementary methods for ensuring a balance of supply and demand in the competition for space and location. It would do private capital no good at all to have too many shops or offices (although this is a special case) or other massive developments competing for too little income. But here we have a dilemma — the supply and demand system operates on a larger and larger scale, activities are more and more interrelated, and therefore understanding the process becomes more difficult. But the market requires that planners should have the understanding — and it is therefore the developments of the market economy itself which promotes urban planning and makes it more sophisticated.

The three maps submitted to the ministry under the 1947 system showed, first of all, zones for primary land uses and their density, for major roads, and for social, educational and welfare centres at the county level. The county map also showed which areas were to be covered by town maps, comprehensive development-area maps and new towns. The development-area map defined specific sites for given uses in some considerable detail, allocated other zones for more general uses and also designated land for compulsory purchase. All these plans could be modified from time to time by policy resolutions the local authority. These policy resolutions could clarify general policy, and define specific policies, on land uses, building heights, mineral exploitation, new buildings, and so on. They thus constitute an important aspect of the overall control powers over development (Solesbury 1974).

The 1968 and 1971 acts: revision but no change

The report of the Planning Advisory Group (PAG), called *The Future of Development Plans* (Department of Environment 1965), was published in 1965, It did not aim to change the basic characteristics of planning control but it did try to improve the *planning process*, which was just not working as originally intended. Originally the plans were intended to show land uses and locations in broad terms, but in the event, because of the need to give a firm underpinning to development control and also because the plans were largely based on maps, they became more and more detailed and essentially had become no more than local land-use maps. There was a clear contradiction here. First of all, such plans took a very long time to prepare, so they could not respond to rapid changes in car ownership, income or the business climate. It also took the ministry a very long time to approve them because of their great detail. Indeed to specify land uses in detail for extremely small areas over a long period is not possible. It is difficult enough to predict or plan for the broader-scale strategic changes in city structure.

The 1968 act implemented the main proposals of the PAG report. These were to separate out the broad *strategic* issues in a new type of plan called a structure plan, which would be submitted to the central government for approval. Within this, detailed land-use planning would be restricted to the local level. The structure plan was essentially a written statement with diagrammatic illustrations to be used as supporting material. The primary emphasis on mapping was thus removed. The structure plan would give a much clearer statement of the local authority's policy: where development should occur, in what places and of what kind. It should take into account a much wider range of social and economic issues both within and outside the local authority itself. The structure plan was also intended to take into account regional economic planning policies, and the resources likely to be available in the private and public sectors.

The structure plan itself, therefore, consists of a written statement and also a report of survey; the 'survey' now becomes a major part of the planning process. This survey is supposed to include all the principal physical and economic characteristics of the area and also neighbouring areas where relevant. The survey, and hence the structure plan itself, is intended to be continually updated, and continually under review. There is a statutory requirement for the local authority to involve the public in the consideration of the major alternatives in the structure plan. All the other plans under the 1968 system are termed 'local plans' and do not require specific approval by a minister, although reserve powers are available. Flexibility and discretion available to local authorities, on which local plans should be prepared and how, is massively increased in comparison with the 1947 system. Local plans are essentially a written statement and a map on ordnance-survey basis. The local plan is intended to implement the structure plan in detail. There are three types of local plan (DoE 1970):

District plans are comprehensive policy statements for relatively large areas in which change is expected to take place slowly and in a piecemeal fashion.

Action area plans are comprehensive policy statements for areas where rapid change is expected, through redevelopment or improvement by public or private agencies (these take over from the comprehensive development areas).

Subject plans can deal with one matter which is not confined to one specific area (recreation, for instance).

At the time of writing, the new system, although undoubtedly an advance in concept, is having considerable teething troubles. First of all, although many plans have been prepared, few have yet managed to pass through all the stages and to receive ministerial approval. The aim, however, is that all the counties should be covered by structure plans by 1978.

Second, there are problems with the *content* of the new structure plans. Because they are supposed to account for a wide range of social and economic change, there is a tendency for planners to stray into areas in which they have even less influence or control than they do over land use. Anxious government circulars have urged planners to concentrate only on key issues — especially employment, population and housing (DoE 1974).

Third, although we saw earlier that local government is now being asked to produce other long-term plans (e.g. social service plans), the structure plan is the only explicit, wide-ranging statement of strategy which is statutorily imposed on the authority. Consequently, there is a temptation for other local authority departments to try to use the structure plan as a vehicle for their own policies. For this reason, it has been suggested (Drake *et al.* 1975) that there should be another broad statement of local authority strategy, the 'county report', which would leave the structure plan much more clearly limited to spatial arrangements and land use. In Scotland a version of this idea is already operational; the upper-tier authorities (called 'regions') produce their 'regional report'.

Fourth, the new, integrated 'corporate-planning' system in local government and the new techniques of planning by objectives have imposed strains on the planning process in many cases. This means that there are now in many local authorities dual systems of responsibility, with the traditional departments still in existence alongside new structures of cross-departmental 'programme areas' operating under a powerful central committee for 'strategic planning' or 'policy and resources'.

Fifth, the new two-tier system of local government has split planning responsibilities and involves massive problems of coordination and political bargaining between the counties preparing structure plans and the lower-tier district authorities who are supposed to prepare local plans were possible, and to deal with most of the planning applications.

Sixth, the definition of 'strategic' policy matters is extremely difficult. It has to be concrete enough and detailed enough to have real meaning and to involve real political choice, and yet at the same time try to avoid getting into too much local detail.

Seventh, the training of planners as non-specialists with few technical skills is inadequate to many structure planning tasks. Economic analysis, statistical and mathematical expertise, and a generally scientific and quantitative approach are essential if any sense is to be made out of the large and complex social and economic system covered by a structure plan.

Finally, of course, an adequate basis of power and control over socially and economically significant changes of land use is still not available to local authorities.

As a result of these shortcomings, many of the early structure plans are very unsatisfactory documents indeed, although some are much better than others and good practice might begin to prevail as more and more structure plans are produced. As an example of the shortcomings inherent in early structure plans it is worth looking at the South Hampshire plan (1973). This is not because it is worse than the others, but rather that since it is fairly ambitious it illustrates a wide range of possible inadequacies. The mere fact that a document exists to be criticized illustrates the strength of the new system, but it seems fairly clear already that these plans will aptly illustrate the contradictions inherent in public-sector planning which we are trying to spell out throughout this book.

First of all, the sheer weight of documentation — a Written Statement of over 300 pages and a Report Survey of a similar length — itself immediately makes the key issues relatively inaccessible to the general public. The issues tend to be obscured and buried beneath a mass of detail, and the logic of the plan is split between the two volumes. The whole thing is thus by no means easy reading. There is a tendency throughout to retreat into generalities and abstractions upon which either everyone can agree (e.g. objective: 'to see that financial resources and investment are used wisely and efficiently, especially where these are the responsibility of public and local authorities') or into jargon which is not well defined.

The concrete issues and choices do emerge from time to time but, whereas the existing concrete problems of the area are spelt out in only a few lines, much more space is given to what turns out to be in the end a largely wasted effort in examining the wide range of 'alternative strategies' up to thirty years ahead. To take one of the key problems in the existing situation, namely, that while employment has been concentrated in a very few major centres, especially in the town centres of Portsmouth and Southampton, residential development has concentrated in new housing areas well outside these cities or in new developments in smaller towns. There are thus imbalances in local communities, lack of services and local employment in residential areas, massive competition for land in the centres of the city with congestion and redevelopment, and a very overloaded transport network, partly caused by the peculiar geographical characteristics of the area. Yet in reading through the whole plan, at the end of the day, having considered a whole plethora of aims, objectives, policies, alternatives and strategies, the plan in effect throws up its hands in the face of these trends over the next ten years.

In this period the concentration of jobs, and particularly offices, will continue in the centre of the cities, the surrounding estates, towns and villages will continue to be under-served in employment and services, and extra strain will be put on the transport system. Now it could well be in the light of commitments already made and the powers and resources available that this is inevitable, although perhaps not desirable. But instead of admitting this, the plan makes tremendous efforts to show that this is indeed the best of all possible worlds. There is an impression of inevitability, and an irresistible honing down, from the initial wide range of alternatives considered, down to the single preferred strategy. This is a 'corporate state' indeed, where a whole panolpy of evidence, often including vague references to computer 'allocation models', is claimed to point in only one direction. Not only is there no *political* choice (which might give alternative emphasis to provisions by the public sector as against the private sector), but also there is no feasible *technical* choice between differing economic or spatial patterns. The other point worth emphasizing is that, although strategic issues may not be spelt out very clearly, there is still a remarkable amount of local detail on certain subjects. A whole welter of technical points could be made on this plan and some will be discussed in Chapter 6. It is clear that planners are still far happier dealing with individual subjects such as population, employment, recreation, and so on, than they are with analysing and devising strategies for the urban system as a whole. For this latter exercise, the South Hampshire plan retreats back into earlier traditions and looks for imaginative spatial arrangements, star-shaped cities, linear cities and other spatial forms, rather than looking first and foremost for basic economic and social alternatives.

These negative aspects of the South Hampshire Plan have very little to do with the fact that 'systems analysis' and formal urban planning techniques were employed. There is a wealth of useful and detailed analysis on individual subjects buried in various parts of the plan (on retailing, housing, etc.). This analysis has been proved useful in other structure plans, and there is every reason that many of the techniques used could be harnessed to a more explicit and concrete analysis of options for South Hampshire.

There are many other indications of how the new planning system, with its essentially passive role, acts to accommodate the inevitable — the market forces which really shape the city. In one county a rather strict policy to restrict development was adopted following a subregional study in the mid-sixties. This was to encourage development in older mining areas of the county rather than in the more prosperous main city — a regional centre. Developers were none too keen to build in mining areas, and six months *before* the first structure plan report was to be published a huge quantity of land — enough for many years' residential development — was provisionally released all round the city. At the time of writing there is still a controversy over this and the policy may be rescinded, but three points emerge. First, significant policy changes can, are and will be made irrespective of the plan — again we have the classic separation between plan and powers and a lack of executive power attached to the plan itself. Second, it shows that the negative power to deny or restrain development of large tracts of new land at the edge of the city does actually exist — that is, that there are strategic decisions to be made about land use. Finally it shows that if the private market will not develop land, planning cannot do so either and has to cut its cloth to accommodate the market.

Effects and operation of the system

Hall *et al.*'s (1973) *Containment of Urban England*, Volume 2, is probably the most ambitious attempt to make an overall assessment of the impact of the UK urban planning system. It is not a rigorous econometric or statistical analysis and its conclusions are at least open to question, especially as to how much of the pattern of change in urban development has been due to planning as such, rather than to the market. Hall sees the major success of the system as the 'containment' of urban sprawl through the green belts. It is seen to be less successful over the planned expansion of large towns (other than London). The system was designed with a 'no-change' situation in mind and it has therefore slowed the rate of change, especially the decentralization of employment and services from the town centres. It has helped to promote 'apartheid' through increased geographical separation between home and work. It may also have helped to increase land prices overall because of its 'drip-feed' attitude to land release for development. The gainers from planning are seen to be 'rural residents'. Commuting owner-occupiers have made questionable gains, with their long journeys to work and poor housing-space standards.

However, more recently this result has been challenged. The 'Second Land Use Survey' is a mammoth and detailed exercise to examine the local, microscopic details of land-use change which has occurred in specific local areas. This survey seems to indicate that planning has not been at all successful in stopping piecemeal, urban-fringe development which is wasteful of land. In a hard-hitting, forceful and strikingly specific argument, Coleman (1976) suggests that, far from tightening up on land wastage, there are large areas of wasted and unused space in new developments in nearly every part of the country. Can this be blamed on planners' lack of power to implement development directly? Surely there is nothing to hinder planners from devising standards of land utilization for specific types of development? Even within their present limited powers they could reject an application for development which does not conform to required standards. Nevertheless there is a catch here. Such standards would have to be imposed consistently in all areas. This would imply some form of national land-use planning. It is quite likely that the failure to impose requisite local standards on developers results from the anarchic, competitive situation between local areas. Large developers can play off one area against the other — they can choose to build in the area which imposes the least stringent conditions.

Coleman argues cogently that the failure to control land use stems from recent changes in planners' techniques, and specifically from the change from 'maps' to 'computers'. This will be discussed in Chapter 6.

Coleman's results at the micro-level may not entirely contradict Hall's findings of a limited success for planning in 'containment'. The containment of large cities may well have occurred but it might have occurred inefficiently from a land-utilization point of view. To anticipate the discussion on planning techniques, if planners ever had any kind of specialist technical skills, one of them would surely be that of organizing an optimal and efficient pattern of local land uses, an 'engineering' skill. In their own eyes, many modern planners are no longer land-use engineers, they have become 'social' engineers. The key question is why?

Free-market critics of the planning system (e.g. IEA 1974) suggest that what is good about the system is largely due to the market, and that planning has essentially reduced the rich interplay of ('micro') market forces which produced natural, organic, 'untidy' cities and towns by imposing a rigid, bureaucratic, simplistic planning image with its 'zones', 'neighbourhoods', 'precincts', etc.

Ambrose and Colenutt (1975) forcibly point out the failures of urban planning to curb the power of developers, owing to the fact that the whole system of development control is *ad hoc*. The 1918 act has failed as yet to achieve a coverage of up-to-date, legally approved plans across the country. This must surely be a central failure of the system within its own terms, since the development plan is the only real 'positive' aspect to the system. In the next two years more authorities will have approved structure plans, but this will still mean that many areas will have to rely on plans that are several years out of date.

The only success (one which must be severely qualified as regards the efficiency of land use) — 'containment' — has been at the 'margin', not only at the geographical margin of cities but also (as will be argued in later chapters) at the margin of the land market: its geographical effect reflects its economic role. The major decisions about land use in city centres are still determined by market competition.

The added scope of the new structure plans highlights a further paradox in the planning system. Although these plans are supposed to take into account wider socio-economic issues, in fact, planning itself has, in legal terms, no terms of social or economic reference. No specific objectives are laid down for the Town and Country Planning System. McLoughlin (1973) gives examples of how economic and social considerations have been specifically ruled out of order in certain 'appeal' cases. Each case is regarded as unique; there are few general terms of reference to which decisions can be related. In some cases it has been explicitly stated that it is for 'oil companies' or 'betting shops', or whoever is undertaking the development, to decide whether or not there is an oversupply of developments in the local area. In other words it is not for planning to take economic decisions. The pressure of demand for planning permissions is increasing — up to approximately 10 per 1000 population per annum. Most 'operational' plans are in fact non-legal and are still 'bottom-drawer' plans which have not been approved by central government. Again McLoughlin notes that the development-control policy of the Department of the Environment emphasizes that only 'planning' considerations are supposed to affect the decision — green belts, use zones, safety, disturbance, etc. — and the effect on public services, drainage, and so on.

But what is the *de facto* economic function of planning? The Uthwatt Committee was quite clear that the effect of planning was to move land values from one place to another, not to create or destroy values. Others contest this; some micro-economists say it increases prices because it limits the supply of land. 'Ultra-radicals' say it also restricts the supply of land and that this merely benefits landowners *vis-à-vis* the rest of the population. Some say that the object of planning and urban public investment generally should be to remove the differences between one piece of land and another — that is, to reduce transport costs — in order to reduce differences in profits made by activities which use land nearer the centres of cities or the national market centre. This in turn would tend to reduce the rents which can be extracted by landowners from land users in good locations. This would suggest that planning should aim to reduce land values.

Others maintain that planning should aim to *increase* land values, so reflecting the overall efficiency of the allocation of land to its various uses. This reflects a wider view of the development process, recognizing that, as transport costs get less, larger and larger market areas may be created, and new linkages between activities may be formed which did not previously exist. Lean (1969) notes that the best planning works 'in harness' with economic forces rather than against them. Evans ?(1973, 1974) discusses specific planning controls on density and plot ratios from the point of view of conventional micro-economics (see Chapter 5).

So how can the role of planning be summarized? The preparation of *ad hoc* plans, the statutory plans and large-scale planning decisions, and the other short-term control functions all require an increasing ability to understand the interrelationship between systems of interlocking land-using activities and the pattern of supply and demand for goods and services, and space over large areas. The larger the scale of investment, the more certainty the developer requires about the likely return, and the whole panoply of development controls, statutory plans, comprehensive development areas and compulsory-purchase schemes can be brought to bear for this purpose. Planning redistributes values, and it may also increase them in the short term, but there is no overall mechanism whereby the total land supply for a given activity in the country can be restricted, except perhaps temporarily in really unique locations such as central London. Planning reduces the anarchy in the local land market, but it imposes no *overall* limit on the amount of land allocated to specific uses nationally; therefore it is best to regard planning as first and foremost a redistributor of values rather than as a creator or destroyer of values.

5 Land: the planners' resource?

In Chapter 3 we tried to show how the private market increasingly dominates the city and how market forces increasingly determine the major pattern of development — what gets built, when, where and for whom. We tried to show that when land is privately owned its owners benefit from rent income and that the level of this rent is determined by the ability of the activities using the land to pay rent from their potential profits. As the urban economy develops, more and more activities compete for given pieces of land in favourable locations, either in city centres or in accessible points near road networks, and so on. Access to markets for products and supplies, especially of labour, are factors in industry and commerce's decisions where to locate. In addition to this, the booming sectors, the service sectors, the office sectors, can afford to outbid most other possible land users for prime sites. All this competition generates the rent which attracts private capital from the large national institutions which, on favourable terms in cooperation with planning authorities, then undertake development or redevelopment of the city.

In the face of this, all planners can do under the existing 1947 system of control is to use their negative powers to refuse planning permissions to channel development where, in their view, it is most needed and where it fits either into the development plan or the new structure plan and its ancillary local plans. But, as we have seen, planners in an individual local authority are in no position really to impose their will on property developers, industrialists, insurance companies and large construction companies. The most important of these operate on a nationwide basis, and it is they who are not only able to outbid existing landowners and existing land users in particular areas, but also to impose conditions on the local authority to gain its cooperation to make the necessary compulsory purchase orders, designate the comprehensive development areas and give the necessary planning permissions. Not the least of reasons why this is possible is because these developers can go elsewhere, and play off one local authority against another. And so, even within the present limited powers of planning control, there is a large gap in the planning system itself. There is no regional or national coordination which could limit the abilities of these entrepreneurs to 'rape' particular towns.

The Uthwatt Committee in 1942 found that the only way to compensate those who lose from planning decisions — say, by having a motorway built near them or an office block erected outside their garden — or whose area is decaying from planning blight, and at the same time to limit the benefits of the general increase in land values falling to individual owners, was to take land into public ownership. The implications of this would, of course, be enormous for planning. There would be the economic benefit to the local authority of the income from rents paid by profitable uses and from the increasing value of the land. This would provide resources to allow the local authority to begin to implement plans instead of relying on the private sector to do so. It could use the rental income as a resource to initiate development where it wanted it, and at the same time to ensure any piece of land was put to the best social use, including low-income housing, local shopping, theatres, social and sports centres or other 'non-profitable' activities, instead of being developed to recoup the maximum rental income. Of course, the problems of deciding which activities should go where would remain. Nevertheless, it would bring together the planning function and the economic and physical resources needed to implement it.

Rather than implementing the Uthwatt proposals, that all undeveloped land should be nationalized (this was, of course, highly controversial), the Labour Government in 1947 imposed a 100 per cent development tax, which, it was argued, recouped for the community the rise in land values which were created by the community.

This whole subject of land nationalization provides a very concrete illustration of one of the central issues in this book — how the encroachment of the public sector affects the private sector, and hence the dynamics of development. It has been suggested in Chapter 2 that in the national economy this encroachment can only proceed so far, with the public sector taking a very restricted role (running the basic services, etc. etc.), if the mixed economy is to remain viable in its present form. The balance between what the state contributes to the costs of production and what it takes away from the private sector for social consumption and other purposes (together with the balance between total private-sector production and the wages paid to labour) must be such as to preserve profitability in the private sector — otherwise development comes to a halt.

It has been suggested that there are limits to the growth of the public sector — and that there is a relatively large gap between all 'mixed' or market economies and the most market-orientated of the planned economies. Applying these arguments to the case of urban land, it would mean that the growth of public ownership or taxation could proceed only so far, before the private market was undermined. There is thus an incipient contradiction in a policy of public-sector intervention in the land market which will tend to restrict the policy, so that only small quantities of land are acquired, at the margin of existing development, ensuring that the land market as a whole continues to function. Otherwise a much more fundamental reform has to be instituted, so that as soon as the private market ceases to function it is replaced by public-sector-initiated development.

It can be argued that the reason why both the 1947 development tax and Land Commission failed was because this central and highly political issue was not faced. In the 1947 case, the development tax was effective and the land market dried up but it wasn't replaced by public-sector land development. This in itself is an argument against the taxation method; if the public sector does not own land it cannot so easily initate development to replace the land market. The case of the Land Commission shows the sheer logistical, legal and operational difficulties in embarking on land nationalization by piecemeal intervention in the land market. In the five years or so of its existence it acquired and assembled very little land indeed — a tiny fraction of the thousands of acres developed each year.

Public ownership of land is generally viewed differently from other types of nationalization. This stems from its peculiar nature as a commodity which cannot be produced, and also because of the difficulty of nationalizing a *commodity* (a physical object which is being used simultaneously by many different activities and industries) as opposed to the more common nationalization problem — the takeover of an *industry*, a group of *activities*. The latter can be done on a once-and-for-all basis; there is no question of piecemeal intervention in the market to acquire the industry factory by factory.

If piecemeal land purchase was embarked upon on a large scale it would not only undermine the land market in each urban area but would also cast the purchasing public authority in the role of arbitrary monopoly power holder buying up land at will — discriminating unfairly between those whose land was acquired and those who escaped. This in itself is an incentive to keep any such powers of public acquisition strictly limited to the easy situations, at the margins of development. Again it helps to explain why the Land Commission acquired so little land.

Piecemeal land-purchase schemes also put planners in a very difficult position, since a planning decision to designate land for development would also be a decision to take it into public ownership. In such a situation it would be very difficult to make a correct technical decision, separate from all the pressures of local politics and vested interests. The planning profession is unpopular enough already.

The key to a successful change in public powers over land lies in the 1947 act, which changed the rights of all landowners over their land *at a stroke*. The change was effected by a legal measure applied to all land instantaneously. Such an approach might possibly be used again to give the public authorities more control over land use and development; in a sense it would be an across-the-board change in the *system of land tenure* (see, for instance, Barras, Broadbent and Massey 1973). The aim would be to make the activity of land development a public-sector activity, but at the same time to ensure that all the various activities which use land and buildings for producing goods and services have access to the right buildings in the right places at the right time, and that all users of land and buildings have security of tenure.

In fact the distinction between *owners* and *users* of land is a crucial one. It is not always recognized that under public ownership, although the *owners* of land have to be compensated for loss of property rights, there is no reason to give them any further rights to the land in question unless they are also *using* and *occupying* the land. Once land has been taken into public ownership the only really significant agents from an economic viewpoint are the users and occupiers and the state as landlord. The previous owners serve no specific function by operating as middlemen. Of course, any such far-reaching change in the system of land tenure in the UK would have wide-ranging repercussions, and the legal reform would necessarily be complex. There is the immediate question of owner-occupiers, especially house owners; any scheme would need to make special arrangements to safeguard their rights. It might, for instance, be possible tc apply the measure first of all only to non-owner-occupiers — tenants, both commercial and private households.

There are many other issues inherent in a scheme of land-tenure reform. Two of the most important are the compensation aspects and the distinction between land and buildings. In the long run any viable scheme for public ownership would be bound to be profitable. This is because, as the economy grows and the total stock and value of capital equipment and buildings increase, the land, being fixed in quantity, tends to become more valuable and the public authority would benefit from the rising rentals. Any scheme, to be politically acceptable, and if it were not to step outside the traditional UK approach to nationalization, would need to compensate landowners. One way to make the scheme financially viable would be to compensate landowners over a period of time, out of the rental income being received by the state. In this way, the state rental revenue expenditure on compensation could be regulated so as to keep a favourable balance between the two. But this raises the thorny question of whether to compensate for the loss of the capital asset (in relation to some valuation), or whether simply to compensate for lost revenue — lost rental income. The latter would, of course, be much simpler to apply and there would be no need for a wholesale valuation exercise. One year the tenants would be paying a rent to their landlord, the next year, following the land reform, the rent would go to the state.

Once the initial capital investment has been paid off, a long-lasting asset such as a building really becomes incorporated into the land itself. Returns from the high-rent, commercial developments would be of major benefit in providing planning with more resources. In these cases public ownership of the *buildings* would provide resources for further urban development; but here again the issues are highly political. (An alternative way of bringing these large-scale developments into public ownership would be through acquisition of the largest property companies.)

To return to present realities, the Community Land Act of 1975 — although yet another attempt at piecemeal acquisition of land for development — does seem to recognize the existence of some of the inherent difficulties of a public-ownership-by-stealth approach. It recognizes that piecemeal acquisition will take a long time and therefore also proposes the parallel introduction of an 80 per cent tax on all development gains made from land sales. The idea is to begin by building up local authority land banks through land management and acquisition schemes to be submitted to the Department of the Environment. Private land sales (with many exceptions) will be eligible for the tax, but local authorities will buy land without the tax element — giving them an immediate incentive to begin 'dealing' in land. The idea is gradually to extend the land acquisition programme over all local authorities until on the 'second appointed day' all land which is to be developed or redeveloped is acquired by the local authority. But even in this long term the land will not necessarily be publicly owned, since it can be sold, leased or rented.

Note how this operation could quite easily become the traditional public-sector-run natural monopoly or basic service — providing a basic resource cheaply, and thereby reducing production costs for the private sector.

Not unexpectedly this scheme has been attacked as a bureaucratic field day. The official estimate is for at least 12,000 new staff to operate the scheme. It would seem that very little will happen in the short term under this legislation. Much land which will get developed in the next five years already has planning permission. It will be very complex to operate — it seems to be something of a retreat back to the detailed land-use maps of the 1947 act — since the detailed land management and acquisition schemes have to be submitted to central government. With all the political, administrative and legal problems of operating the scheme through local authorities with varying political control and planning expertiese, added to general difficulties of public-sector intervention in the land market noted above, it would seem that the scheme will not have much immediate effect on the major problems of urban planning in the UK — such as the polarization of land uses in situations like central areas where the competition for land is intense.

It will have little or no effect on the large-scale redevelopment of city centres and other sensitive sites where competition for land is intense (see, for instance, Ambrose and Colenutt 1975). There will be little money available to acquire the land, and even less to develop it. There will be considerable incentive for the local authority to sell quickly at a profit, or else to provide land on terms that the large financial institutions and property companies find acceptable. The Community Land Act of 1975, although a welcome recognition of the need for urban planning to have effective control over its basic resource, nevertheless offers little prospect of this being achieved in the foreseeable future.

For further reading on the land issue, see Hall (1965), Neutze (1973), and the institute of Economic Affairs (1974) for a free-market view.

Summary

The problems of the state at national level are even more severe at local level. Local government expenditure (though much of it is actually central government spending channelled through local authorities as agents) takes a growing proportion of public spending.

The largest proportion of the increase is taken up by direct social provision — 'social consumption' — with the local authority helping to 'produce' and maintain the labour force, through education, health and welfare, and environmental services. But with increasing growth comes increasing pressure for central control; local authorities are becoming more and more dependent on central government (and also on the private sector) for finance.

The public-sector debt problem becomes much more severe at local level where there is a larger gap between revenue raised locally and expenditure. So, while local expenditure and local state operations do not threaten the local private sector as directly as the national state might potentially threaten the private market, in many local authorities debt payments are a major item, particularly on housing accounts.

While the bulk of local government resources are used to provide social services, it is the urban planning function which illustrates very clearly the strict limits and rules imposed on state intervention in the mixed economy. It operates to regulate competition, and exhibits an increasingly sophisticated 'indicative' function which tries to provide guidance for private developers. Above all it highlights the gap between prepared plans and the powers and resources to implement them; the latter do not lie in the control of planners. Urban planning does not have powers to develop land, powers which would compete with or undermine the land market.

History shows the increase in the scale and sophistication of urban planning, which began with standards for the individual house and went on to concern itself with larger and larger areas. Public health ('social consumption') and housing were always the main concerns. Before the Second World War there was growing awareness of the nineteenth-century regional industrial problem, culminating in the Barlow Report, which, coupled with Keynesian national economic management, laid the basis for the planning system and philosophy existing today.

The early thinkers were not all utopians: Howard and Geddes in particular were concerned to examine real economic forces and trends and to harness them in constructive ways. Howard recognized that public ownership of land was essential to gain control of the returns which land-using productive activities generated.

The most important aspect of the current planning system, unchanged since 1947 (with its financial provision removed in the fifties), was that it changed property rights. It nationalized the right to develop land even though it did not obtain for the state the benefits of development — the rise in land values. It will be argued later that a further change in land-using property rights may now be required.

The 1968 act did nothing to change planning powers and resources. As increasingly powerful market forces invade the city, the 'negative control' powers to grant or refuse planning permission, although fairly effective in containing development in the countryside, are inadequate. Many significant social and economic changes of land use in cities are not subject to control. The negative development control powers largely operate to regulate competition at local level and to stop market forces getting out of hand. But as for strategic control of development, the new structure plans illustrate the classic divorce in the UK between planning on the one hand and lack of public resources and powers to implement the plans on the other.

The sad history of land policy — and attempts to collect the increase in land values for the community — represents a failure to give planning effective control over its basic resource.

5 Current theories of urban development

1 How ideas are used and produced: the function of theories in planning

Previous chapters have tried to show what urban planning sets out to achieve and why. Its function reflects in a very specific way some of the general limitations imposed on the public sector in the mixed economy. This chapter shows how and why urban planning thought draws on a wider universe of ideas and theories about society in order to help it to justify its role in principle and carry through its task in practice.

It is sometimes said that, when planners prepare their plans and perform their planning control function, they are exercising a specific and identifiable planning skill, and that this in part justifies the existence of a professional body to regulate training and job opportunities. (Some of the skills and techniques they use are discussed further in Chapter 6.)

This might imply that planning has developed its own brand of theory, specifically geared to understanding the key urban processes of urban development outlined in Chapter 3. In fact urban planning theory does not really exist separately from theories developed elsewhere, especially (and not surprisingly) from theories dealing with society. Planning has been very eclectic, drawing concepts and terms apparently arbitrarily from geography, economics, sociology, systems theory and other disciplines. This chapter tries to show that if we bear in mind the specific nature of planning outlined in Chapter 4, and the way this role is changing under the influence of economic trends discussed in Chapters 1 and 2, then the way theories influence planning is not so arbitrary as might first appear.

There are three ways in which a theory might he helpful to planning:

1 It might explain and justify the role of planning in general terms.
2 It might try to explain the processes of urban development and the competition for space and location.
3 It might be developed into an operational technique for preparing plans or otherwise aiding the planning task.

All these three influences overlap and any one theory or group of theories might do all three (what we have called 'mainstream' economics does this). But in what way does planning need this kind of help?

As described in Chapter 4, the purpose of planning is contradictory. On the one hand it has to undertake a very real task; although lacking any real power and resources to implement development, nevertheless the use of land has to be regulated. Both the community at large and the firms and agencies competing for space and location in the city (retailers, industrialists, property developers, construction companies) need planners and plans to provide some degree of certainty in the situation and to show how the pattern of supply and demand for goods and services is likely to change. Planning also helps to facilitate development by assembling land; the local authority itself has to know where and when to provide infrastructure and services — 'social investment' (roads and services) and 'social consumption' (public housing, social services, etc.) outlined in Chapters 2 and 4. This means there is a very real need to understand the processes of urban development.

On the other hand, since planning does not directly implement development, since there is a gap between plans and powers, and since the influence of the local authority over the main patterns of development is largely relative, there is a need to disguise some of the real processes of development and the ability of the public sector to influence it. This means that planning may be prone to draw on theories which give a true representation of some aspects of urban development and at the same time use others (or even aspects of the same theory) which give only a partial, onesided or even false idea of development processes and the role of the public sector. This is one reason for the well-known 'flights of fashion' in which planning tends to embrace and discard theories in successive waves.

The very notion of planning implies that the means of controlling development do exist and that this can be done somehow in the interest of the whole community. This means that more often than not planning will seem to be 'in crisis' and will tend to justify itself only in very abstract or unreal philosophical terms on the one hand, while on the other hand myopically getting on with the minutiae of day-to-day development control.

We try to show below how several different theories can be used either to aid the *understanding* of urban development and the role of planning, or to *mystify* and distort it. The theories all have their own assumptions (sometimes clearly stated, sometimes not), and we must try to distinguish between a theory wrongly applied (i.e. outside its own stated assumptions) and the question of whether the assumptions are reasonable in the first place.

Harvey (1973) identifies three types of theory: 'status quo' theories are 'rooted in reality' (i.e. provide some real understanding of urban development) but nevertheless put an interpretation on this reality which preserves vested interests; 'counter-revolutionary' theories also preserve vested interest but this time by distorting reality in some way; and 'revolutionary' theories are rooted in reality but interpret it in a way that points to a change in the status quo. Although there is some benefit in this three-way classification, it may imply that planning is free to choose its theories — for instance, to disguise its true nature at will. But it is one of the main points of this book that as urban processes develop, as production and consumption become organized on a larger and larger scale, and as planning becomes more explicit and more 'visible', then it has to draw on theories which really do explain what is going on in cities. In this situation it becomes more difficult to disguise the true nature of planning and the ability (or lack of ability) of the public sector to influence development.

2 Conventional economic theory

Introduction

Economics, the dominant social science, is 'the queen of the social sciences' (Samuelson 1973). Economists study 'how men and women obtain their livelihoods'; they believe themselves to be, as Keynes remarked, the 'trustees of the possibility of civilization'. Together with this universality of subject matter, economics often uses a relatively explicit logic, quantification, and mathematical and statistical analysis. The whole adds up to a pyramid of learning connecting high theory, philosophy and logic to real-life problems vital to everyday life and the 'survival of millions'. Although there are completely opposing schools of thought within economics, it nevertheless has a relatively well-defined body of intellectual concepts unparalleled in other social sciences. We try to show below and in Chapter 6 how the central concepts of conventional, 'mainstream' economics permeates thinking about cities and influences policies and plans. It is an influence sometimes acknowledged but often unrecognized.

The aim of this section is to outline in more detail the mechanics of neoclassical micro-economics and to establish more firmly some of its basic concepts, together with their critique. (As there is a strong critical element *within* mainstream economics, an attempt is made to distinguish this critique from the more 'radical' viewpoint which usually owes something to Marx.) This requires a brief outline of the detailed workings of the perfect market and the introduction of several jargon terms. (At a first reading the details in the section on 'Elaboration' can be omitted.) The aim is to show how the strengths and weaknesses of conventional economics are used by planning and how they influence planning thought. We shall then be in a position to show, in the following sections, how adding a geographical dimension to the conventional theory affects, and is affected by, these strengths and weaknesses, in order that its relevance to urban development can be assessed.

Fundamentals: the sovereign consumer, utility, the margin and the market

Introduction

As already stated, the micro-theory, which explains the way individual firms and consumers behave and compete, does not result in a satisfactory explanation of the way the economy as a whole develops: it cannot explain why there is a tendency to employment, or how fluctuations should be controlled and growth promoted. All this is left to the other main branch — macro-Keynesian theory, which is only imperfectly reconciled with the micro-theory.

The two main branches of modern mainstream economics were introduced in Chapter 2, in relation to state intervention in the economy.

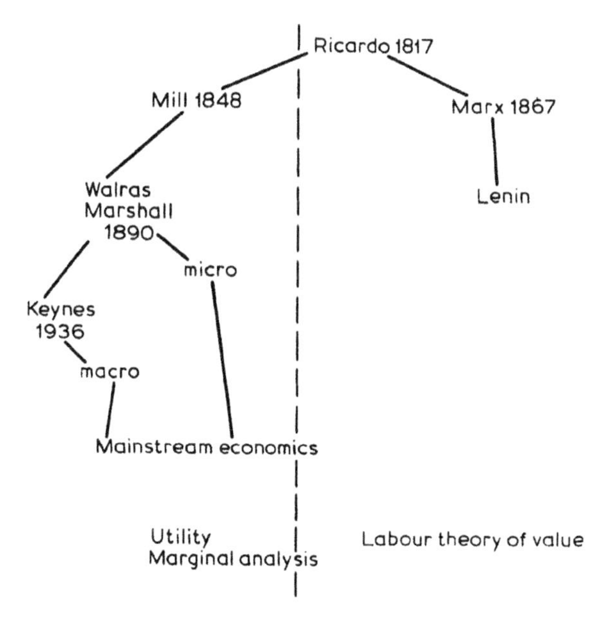

Figure 32 The development of economics
Source: After Samuelson (1973)

Samuelson's tree of economics (Figure 32) shows how the modern theory developed. The micro-theory of perfect competition involving infinitely small producers and consumers is concerned, as we have seen, with defining the conditions of a stable, static equilibrium in the economy. The essentials of this theory were laid down by the latter part of the nineteenth century, as represented, for instance, by the work of Marshall (1890). The central core of the change represented by this type of economics was the rejection of the central concept of 'classical' economics which asserted that the 'value' of all production was determined by the human effort or labour which went into it — 'the Labour Theory of Value' (Roll 1938; Dobb 1973). These earlier economists, culminating in Marx, felt that the immediately measurable economic indicators, such as the amount and mix of goods produced, their prices, and the wages and profits in the economy, were essentially 'surface' phenomena. The task of economics was therefore to tear away the 'monetary veil' and to investigate the real social forces underneath. Marx's view was that economic relations — that is, relations between goods and money (relations between *things*) — were really relations between *people* — social relations. He went on to define concepts such as 'exploitation', which were capable of exact definition and measurement. The 'value' of a good or service could be measured by the 'labour time' needed to produce it. The argument was that, effectively, the wages paid to the workers represented only part of the total value of the product they produced, the 'surplus value' being kept by capitalists who employed the worker, owned the capital and sold the goods produced by the worker's labour.

In Marx's view, then, profits ('surplus value') are not a reward to capitalists for services rendered in the production process, that is 'taking risks' or 'waiting' (Chapter 2), but are merely a part of the total output of society which they are able to appropriate because they own the means of production. Profits are then invested to reduce production costs in the continuing competition between capitalist firms. To an extent, Marx has been rehabilitated in recent years in some university courses (Deshai 1974; Samuelson 1973).

Utility

The neoclassical economists of the late nineteenth century looked at 'value' from the entirely opposite viewpoint (the view we are used to today), from the viewpoint of the consumer rather than the producer. How does the consumer 'value' the product? How much is it 'worth'? Conventional economics' measure of value is thus a subjective, psychological concept, measuring the *satisfaction* an individual obtains by *consuming* the product. This measure is called 'utility'. 'Utility' is a relative quantity, which varies from one individual to another, rather than an absolute quantity (as in the labour theory). This clearly causes problems if we are interested in the total increase in the utility of society as a whole (say, as a result of a new town), since it would entail adding the utility of many different individuals whose valuations of different products might be quite different one from the other, not least because of their different incomes. Nevertheless, utility is often used to evaluate the social benefit arising from urban development projects (see Chapter 6), although it is this very concept, so central to modern economics, which causes the most difficulty for planners when they try to use it in practice.

Consumer sovereignty therefore stands as the supreme guiding principle of micro-economics: 'production serves the needs of the people'. This idea may clearly be used (and often *is* used) to serve as a justification or rationalization for the market economy in general, and for whatever the pattern of production and consumption (or even level of unemployment) happens to be at a given time. The fact that utility or satisfaction is entirely relative and can only be defined for a single, isolated individual means that the total increase in utility or satisfaction of a group of people cannot be defined, because the separate relative utility of individuals cannot be added together. Even when we can define a quantity or quality for an individual which can be added to that of other individuals, there is still a great danger in doing so. The behaviour of a whole system (e.g. of consumers) is not merely the sum of its separate parts. This is known as the 'aggregation problem' (Blaug 1974).

Economists are often able only to observe how whole groups of consumers or firms behave in the aggregate (e.g. the way the consumption of food decreases with a price increase, or the supply of plastic gnomes increases with price). Consequently, because of the aggregation problem, it is extremely difficult to deduce the real behaviour of the individual consumer or producer, and therefore to prove that the market is really working as it should. We shall see in Chapter 6 how some very peculiar results can be produced if one follows the precept of satisfying the demand of this idealized, isolated, sovereign consumer in a situation where many thousands of consumers are involved (e.g. satisfying the 'demand' for urban transport).

The view of the urban system put forward in Chapter 3 clearly bears some affinity with the classical, labour view of the economy. We were not discussing questions of 'value' but were looking more or less directly at the *production process*. The city was viewed as a conglomeration of 'production' activities which competed for land, space and location. We were led into this 'production' view by some of the most obvious facts of the situation. Most production in the city involved exporting or importing goods from the national and international economy. If instead of production we were interested in the exchange of goods and services, the only truly local markets would be for land and labour, even though the city also encompasses the purchase of consumer goods by households. We took households as producers (of labour) because it was the needs of households as consumers *and* as producers which really defined the boundaries of the city, through the generalized home-work activity.

It is not possible to describe in detail how the labour theory of value was overthrown (Roll 1938; Dobb 1973), but the criticism was related to the fact that goods seemed to sell at prices unrelated to their labour content. It was felt that the 'value' of things in society must depend in some way on comparisons with other goods. There has been a long-running debate about the relationship between values and prices. At first sight, prices of goods and services seem to bear very little relation either to their labour content or to their utility. (In the Marxist theory this question is known as the 'transformation problem', where prices are derived from the original labour values, so that a good with a high value (high labour content) may in fact be sold at a low price.)

The price problem faced by the utility theory is rather different. Why does water, which surely must have a high utility, command a price lower than diamonds which are not so 'useful'? The way this classic and oft-stated problem is resolved is crucial to the usefulness or otherwise of mainstream economic concepts in urban planning. Quite simply, instead of trying to calculate prices from utility modern economics just avoids this problem altogether. Total utility of any good or service is not measurable; what is measured is the utility at the margin, the utility of the 'last unit consumed'.

The margin

Now this idea of the margin represents a crucial connection between ideas about the city and wider economic concepts and theories. We have already met the 'margin' — meaning the geographical edge of the city. We saw that activities can pay rent because they have lower production costs than activities on the edge of the city. Now classical economists such as Ricardo had used this idea to explain how rent was generated in an agricultural economy. The idea was that production costs become higher on land which is less fertile (or, in later versions of the theory, further from markets), and so less rent is paid. Von Thunen (1826), who is regarded as one of the founders of modern geography, also used this idea, in what begins to look like a more specifically 'urban' analysis. Instead of discussing fertility, he took a flat, fertile plain with a single isolated city in the middle, where the only differences in the cost of producing agricultural products for consumption in the city were the costs of transporting the produce to the city. He came up with a series of agricultural rings around the city, each ring producing a different product. This bears every similarity to the ring pattern for urban activities in later theories of urban economics (see below). Marshall's neoclassical economics then applied this 'marginal productivity' idea not only to land but also to other factors of production — labour and capital.

Wages and profits were determined by the 'marginal productivity' of the 'factor' — labour or capital (see below). Looking at consumption, 'marginal utility' was said to decrease in the same way as marginal productivity. The more of a good that is consumed by an individual, the lower the utility of the last unit consumed; the consumer becomes more and more 'satisfied'. The first pound of sugar consumed may well have a high utility (even though utility cannot be measured), but after consuming six pounds the utility of the next pound is very low: the consumer is already 'satisfied'. *Marginal,* not *total,* utility is the central economic factor in determining how much of a good is consumed.

We can see immediately why modern economics might be regarded by its critics as the economics of the status quo. It appears to take the existing pattern of consumption as in some way given, since at equilibrium everyone must, by definition, have maximized their own utility. The marginal utility of the last pound or dollar spent on each of all the goods and services consumed by an individual must be equal, otherwise the pattern of spending would be changed, and more would be spent on those goods that give more marginal utility.

The concept of *the margin* lies at the core of modern, mainstream micro-economics, and it also represents a key historical and theoretical link between the dominating economics of the time and the geography of cities. It is interesting that this central concept of economics was originally an application of a theory about space and land, a 'geographical' theory. Intuitively, we might feel that a view that concentrates on changes at the margin must have severe limitations. If we are interested in basic large-scale structural changes — in restructuring industry, building new towns, or reallocating resources — we might feel intuitively that an approach which looks purely at the margin of the existing pattern of development might not be too helpful. The concept of the margin is explored further below.

Supply and demand

But what is the overall orientation of the micro-economist? How does he or she tend to look at an economic problem? The modern economist assumes that ultimately, when all special, local and complicating factors are discounted, the problem is usually reducible to a network of *exchange transactions* in the market. Exchange is the dominant interest, not production as in the labour theory. The market is expressed through phenomena of supply, demand and price, first for a single good or service. People enter the system as 'consumers' and again as 'labour'. The economic system as a whole is then regarded as an enormous conglomeration of interdependent markets, involving many goods and services, rather than as a conglomeration of production activities as we have described. The central question becomes the explanation of the exchange process itself — and, in particular, the formation of prices.

Prices are determined by the interplay of supply and demand. The 'demand curve' is the cornerstone of the analysis. As the price of a good is increased, people are prepared to buy less of it, 'other things being equal' (*ceteris paribus*). The reason for this is the diminishing marginal utility described above: if the marginal utility of the last pound spent on each good is the same at equilibrium, then if the price of one good falls its marginal utility must also fall if the equilibrium is to be preserved. It is probably fair to say that very rarely has a demand curve been measured; other things rarely are equal, and prices do not usually vary over the full range. In the real world the demand curve itself can and does move. In Figure 33, for instance, the demand curve for cars can move upwards if the total incomes of the consumers increase.

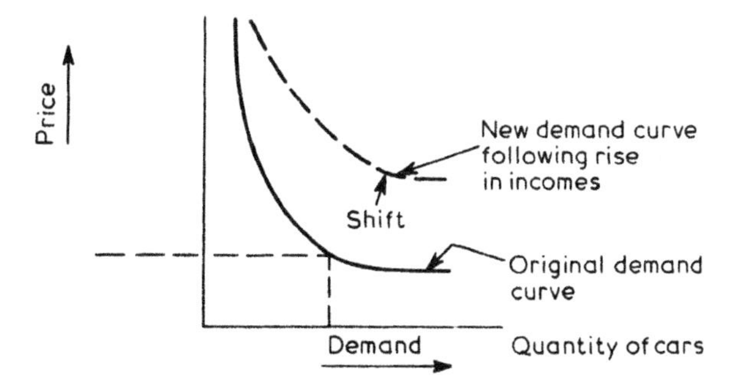

Figure 33 Demand curve

Keeping the market at the centre of the stage, modern economics then looks at the other side of the coin — supply. The demand curve is a theoretical proposition, stating how many cars *would* be bought at certain prices; but what is to say which prices are actually achieved? The *supply curve* (Figure 34) is used in the same way as the demand curve, but now looking at the *producers* of goods, to measure how much they would be prepared to produce at different prices. The higher the prices the more they will produce, therefore the supply curve slopes upwards usually exhibits *decreasing returns* to scale. This means that it slopes upwards ever more steeply as output

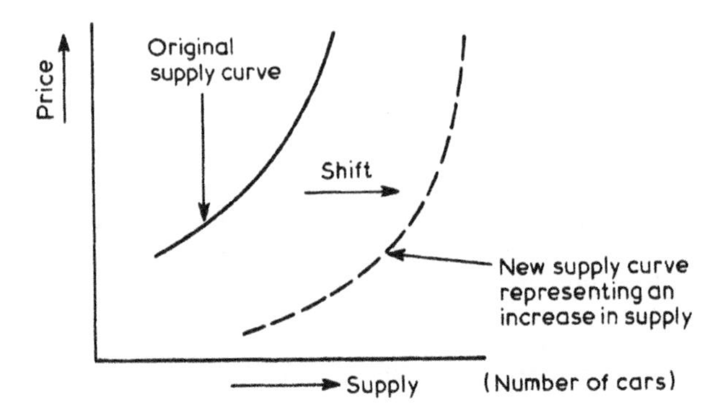

Figure 34 Supply curve

increases, because the *marginal cost* of production increases (in the same way as the *marginal utility* of consumption decreases). This concept of decreasing returns at the margin again echoes the idea of decreasing productivity at the geographical margin or edge of a city outlined in Chapter 3. Again the supply curve can shift: an overall increase in supply is represented by a shift to the right in Figure 30, so that more cars are supplied at any given price.

It is often said that urban planners restrict the supply of land and therefore increase its price (e.g. Simmie 1974), and so even this most elementary basic economic concept is used on a day-to-day basis in the debate about planning.

Notice that this detailed description of the exchange process in the market results in a very unclear representation of what is actually taking place in the *production process.* The supply curve describes production as an aggregate process whereby price calls forth 'supply' according to a mechanical formula. 'People' are not involved in production at this stage; 'firms' do the producing, people the consuming. This type of economics is dominated by smooth curves, with no irregularities. It implies that production can be increased or decreased smoothly from zero and that the various factors of production and commodities themselves are fully and smoothly *divisible* into smaller and smaller quantities. We have already mentioned the difficulty of deriving the behaviour of individuals from an observed aggregate curve. It might well be asked where are the big firms or the large factories which now dominate the production process at local and national level. We might feel that a smooth production curve cannot possibly represent the two or three large firms which dominate a city or even the UK economy. The smooth curves of neoclassical economics do indeed seem to draw a veil over the realities of the modern production process, and this is well recognized by urban economists (see below).

Equilibrium

Modern economics studies the interaction of supply and demand to show exactly what goods are produced at what price in what quantities. The

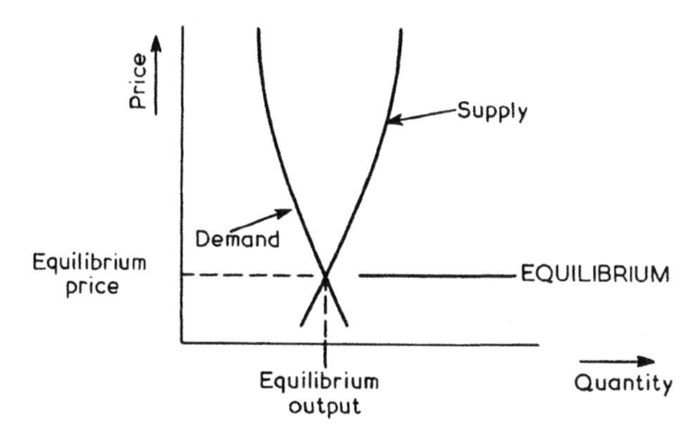

Figure 35 Equilibrium

equilibrium price is a price which 'lasts' because both consumers and suppliers are happy with it. There is only one price which occurs on *both* the supply curve and the demand curve. At anywhere else on the two curves except this intersection point (see Figure 35) there must be either a shortage or a surplus of goods on the market.

Producers will bring goods to the market if the price they obtain covers the cost of producing them. As the cost of production increases with output, they will stop bringing goods to market once the income received from sales does not cover the increased costs of production. Thus nearly all the firm's output is produced at a cost below the price received and this represents profit; if the firm continued producing even more goods once the decreasing returns to scale had pushed production costs above the prevailing price, total profits could be reduced. At maximum profit equilibrium *price is equal to the increase in the total cost of production due to the production of the last unit,* that is, the 'marginal cost' of production. (This is also equal to the rate at which total production costs of the firm are rising.) When all prices in the economy are equal to marginal costs, then society is at its most efficient; it is producing as much as it possibly can with its limited resources. If all prices were not equal to marginal costs, then the Pareto criterion we met in Chapter 2 is not satisfied. Production could be reorganized to make everyone better off — consumers with higher utility and producers with increased profits. The driving force of equilibrium which operates behind the supply and demand curves is the drive to maximize profits and utility by firms and consumers ('optimization behaviour').

Now, as Samuelson (1973) points out, a major difficulty in practice is that there are hundreds of goods and markets all functioning simultaneously — allocating given amounts of *land, labour* and *capital* (the three basic 'factors' of production) to a multitude of different productive activities. This is called the problem of *general equilibrium* and is clearly much more difficult than the partial analysis using demand and supply curves for single goods and services (see below). Most micro-economic exercises are partial in this sense.

Elaboration

What is the margin?
Economists are not always clarity itself when defining one of their central concepts. There appear to be at least three related meanings of the word margin and its derivative 'marginal'.

1 A *small absolute* increase (or sometimes decrease) in a quantity, e.g. a marginal change in demand or supply. This could be measured as a number of cars, houses etc.
2 A *rate of change,* e.g. the increase in utility per unit increase in ice cream consumed ('marginal utility') or the increase in production costs due to a unit increase in production (marginal cost). This could be measured in pounds per car.
3 An *absolute increase (or decrease) but the very last possible.* This definition implies some external reference and carries connotations of the 'edge' (e.g. of profitability). Marginal workers are those whom it is only just economic to employ; the marginal car might be the last car to be produced which still makes a profit.

Samuelson (1973, p. 214) — 'literally millions of readers' — seems to confuse the first two definitions; simply saying that marginal means 'extra' but then going on to illustrate the concept with reference to the second definition. Another use of the word margin means a *difference* between two quantities. A retailer's 'gross-margin' is similar to 'profit' — meaning the difference between the retailer's sale price and his/her purchase price. To add even further confusion, the second meaning of marginal (the rate of change) is also referred to as 'elasticity'. The 'price-elasticity' of demand is the rate of change of demand with respect to price.

It is relevant to recall at this stage, if we are seeking the link between prevailing economic theory and the control of urban development, that it was established in the preceding chapter that if urban planning in the UK has had any success, it is in controlling (or 'containing') urban development at the edge — the margin — of cities. Now we shall explain below (see also Chisholm 1970) that the economic margin (the third definition of the margin, i.e. the last unit in production) cannot simply be equated with the geographic margin. Nevertheless this crude parallel is not entirely accidental. It is true that market competition is at its most severe in city centres where profits are generated (or extracted) in the city. It is also true that planning, because of its lack of capital resources, has had to stand aside in these areas, while profitability and competition (i.e. supply and demand) decide which land gets developed or redeveloped. So the success of planning in containing urban development stands as a direct physical symbol of the failure to achieve its wider goal of allocating land and space *within* cities to uses according to social instead of market criteria. (Even under the conventional economic theory, 'market' criteria cannot be 'social' criteria in a situation where development capital is largely controlled by large institutions, or in an economy where large firms dominate.) The geographic role of urban planning both symbolizes and is partly determined by its economic role.

The two estates: wages and profits – 'what they deserve'

In Chapter 2 we briefly referred to the conventional explanation of why two of our three 'estates' gain a final income — profits for the private-market sector and wages for the workforce (population). These incomes are usually viewed as rewards for 'services rendered', e.g. 'risk' or 'waiting' for capital, and labour for wages. The detailed explanation of these 'factor incomes' is obtained by applying the concepts of supply and demand to the basic 'factors' of production — land, labour and capital. The demand for a factor depends on the demand for the products it is producing; it is a 'derived' demand. It also depends on technical factors, that is, how much can be produced from different combinations of the factors of labour and capital; the proportions can obviously vary, as they can *substitute* for each other. If labour becomes too expensive, then there is more incentive to use capital. Substitution can occur between any two goods — if they serve similar purposes. (This means that the demand curves for the two goods are not independent. If the price of one good rises, then the demand for the other will rise even though its price stays the same. We then move into a situation of more than one supply and demand curve, and more than one market. This makes the whole solution of the market problem immeasurably more complex. Incidentally the substitution between goods provides economists with a way of measuring marginal utility: it defines an *indifference* curve, where consumers are *indifferent*; they prefer increased quantities of both goods in the same proportion.)

If labour or land can be used for several alternative purposes, then a wage (or rent) may have to be paid that is equal to its price in its most productive use, in order to stop its transfer to this use. This *transfer payment* is the formal explanation given by conventional economics for that part of the rent paid by a user of land in the city in order to keep other activities away from using the same land.

The basic explanation for wages is still then the *marginal productivity* theory. In some sense the price of labour is equal to the marginal product — the increase in production resulting from a unit increase in the labour input. It pays a firm to increase its number of employees until the extra gain in income from sales is just equal to the extra wage. The market process — the interaction of supply and demand — again determines the price of labour wages.

The main radical criticism of this idea now becomes obvious. The conventional theory appears to be a justification of the status quo because people (labour) receive in wages just *what they deserve* — that is, what they produce in revenue from sales of the goods they produce.

The radical idea of exploitation of workers, as we have seen, is based on the notion that workers produce more than they earn — and that this is represented, at least in general terms, by the existence of profits received by owners of capital and/or investment in capital goods for future production. Conventional economics' definition of exploitation would be either the situation when workers receive less than their marginal product, or that in some sense the distribution of income between wages and profits is judged to be exploitative, based on outside values external to the straightforward economic analysis (Chapter 2, page 35). The radical critique would argue that even if workers do receive wages equal to their marginal product, capitalists still maximize profits by paying this wage (Sherman 1970), and that no productive services such as waiting or risks can be concretely defined to justify profit. The machinery and buildings needed for production are indeed productive, but the only service they represent is the effort (labour) previously expended by workers in the past to produce them. Here, then, is the kernel of the radical critique: that, even though the marginal productivity idea does not make any value judgements about the rate of wages and profits, it nevertheless provides an overall justification for the existence of private profit which is essentially spurious.

Big firms: unfair profits

As we have seen from Chapter 1, even looked at from the point of view of the national and world economy, firms are no longer small-scale microscopic entities. One of the main justifications for making the market (i.e. the exchange process) the centre of concern is that when firms are microscopic the most significant economic activity goes on outside the individual firm. The laws of supply and demand may then very well determine what is produced, for whom, at what price. But now that the world economy, the national economy and urban economy are being swallowed up inside large firms, the system really is beginning to operate the opposite way around — 'inside out' (see Figure 36). This means that what is happening inside firms assumes as much, or more importance, as what happens in the exchange relation between them. This again means that the *production process,* not only the *exchange process,* needs to be understood.

Mainstream economics does not ignore the problem of the big firms which so dominate the modern economy. But, not surprisingly, their operations are still largely described in terms of the market (supply and demand) and the way they deviate or cause deviations from the perfect-market equilibrium. The individual firm no longer has a vanishingly small effect on prices of products and the quantities produced; it can now affect both. One way of explaining the growth of large firms is when there are increasing returns to scale in production of goods. Unlike the situation in Figure 34, costs now decrease as output increases. Larger firms will therefore have a cumulative advantage over small firms, and the market will become dominated either by a few large firms (oliopoly) or, in the extreme case, by a single firm (monopoly).

Increasing returns can arise because there is a 'minimum economic size' of plant needed to produce certain products. (Some goods are now

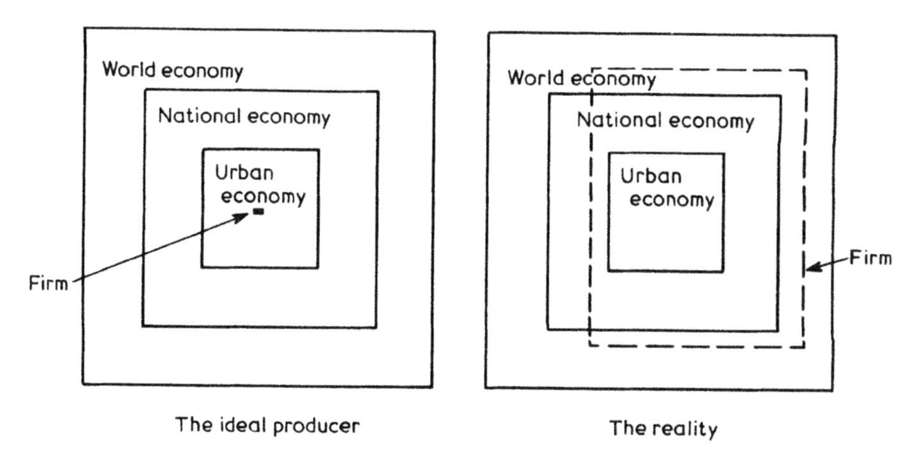

Fig. 36. The absorption of the economy inside large firms

manufactured in plants which produce a very large percentage of total UK output of the product (Pratten 1971).) The prices set by the larger firms are then higher than the marginal costs of producing the goods; but there is usually still a limit to such price increases, a limit set by the ability of consumers to pay. As the price increases, there may come a point where no one would buy, and the total income of the monopolist would fall to zero. So here again there will usually be an *equilibrium* situation, at which price is now higher than marginal costs, but the marginal increase in income from sales ('marginal revenue') is just equal to the marginal costs of production. With oligopoly, a few firms collude to maximize their joint profits and there may be a leading firm (e.g. the largest or most efficient) which decides prices. Oligopoly may result in a more 'rigid' price level than either perfect competition or monopoly. If one large firm increases its prices, all the others may hope to gain customers by maintaining a lower price. If a firm reduces its price, all the other firms may do the same in order to keep their share of the market. This is where 'game theory' comes into its own (von Neumann and Morgenstern 1953). This explains what alternative strategies are likely to be adopted by firms as they anticipate the likely response of their rivals.

Conventional economics thus defines monopoly or oligopoly (which as we saw in Chapter 1 is the norm in many industries) as an 'unfair' situation which has to be regulated. Monopoly is unfair because in a perfect-market equilibrium prices ought to be equal to marginal costs. By pricing their products above marginal cost, monopolists gain unfair super-profits. Neoclassic economics, when it is at its most consistent, is strongly critical of monopoly, since it undermines the inherent notion of equality or fairness which should operate between consumers and producers as they compete to maximize utility and profit. Monopoly contradicts the idea of 'balance' in supply and demand, and destroys the efficient allocation of resources to their most productive use. If a product really costs so much to produce, then its price should reflect this. This then justifies state intervention to control and reduce these unfair profits (see Chapter 2). Mainstream economics might now even justify strong international control over the multinational companies. Samuelson is appalled at the lack of control over oligopolies outside the USA.

Here, then, is another departure from the radical critique of profit which questions the basis of all profits and defines the existence of private profit itself as a misallocation of resources. Conventional mainstream economics rationalizes or justifies profits in general as a reward to owners of industry and capital for taking risks, but is strongly critical of 'unfair' monopoly profits. Indeed it defines this as 'exploitation' — not of labour but of the whole society (Samuelson 1973). Can mainstream economics then be used to question the whole structure of the UK economy where output is so dominated by big firms? Perhaps so, but usually the stock apologia for large firms is that they each produce many products, not just one, and therefore are often in competition with many other large firms for each product. Another excuse is that competition operates on a world scale, and in any given market there may be considerable international competition.

Many markets: general equilibrium

There have been many attempts to build large-scale 'general-equilibrium' models of economies involving a large number of commodities instead of the simple, partial supply-and-demand mechanism operating in a market for a single product. These models are usually said to derive from Walras (one of the founders of neoclassical theory) who attempted a mathematical formulation of this common but very complex situation. These models generally divide the economy into various sectors, each producing a given set of output commodities from a set of inputs. It is usually assumed that in each sector the relative proportions of inputs and outputs are always the same (that is, the models are linear).

It may seem an immediate drawback if the added realism of many markets can only be built in at the expense of making the production function for each product a straight line (capital cannot be substituted for labour as production increases). However, this does allow the 'aggregation problem' to be tackled head on; in fact reasonable *aggregate* production functions for whole industries (i.e. curves) can be derived, even if the underlying *individual* production functions of firms are straight lines (Blaug 1974). It has also been argued that the nearer we get to real individual firms and the real physical risks and processes which go on in production (real machines producing individual bits of products), the more 'linear' things become.

As an aid to thinking, and as a clarification of concepts, a general-equilibrium model is thus a great advance; it helps to clear up many of the confusions between aggregate concepts (i.e. for industries and whole economies) and truly micro-concepts (i.e. for individual firms), which abound in micro-economies (Blaug 1974). It has been said that it blurs many important mechanisms (McCormick *et al.* 1974), and that it seems to suggest a simultaneous static equilibrium across the whole economy with simultaneous determination of prices and quantities (supplies and demands) across the whole system. These authors prefer a partial approach which stresses the short-term mechanisms of non-equilibrium between different supplies and demands in different markets and also the different speed of response of different markets to changes in price. This argument seems to confuse two different purposes. One purpose of general equilibrium is to illustrate a *general tendency* and to explore general properties of a multi-sector economy rather than to tackle a specific short-term problem. If we are interested in what happens when the price of potatoes is raised and everything else remains equal, then clearly a single-market supply-and-demand analysis is the appropriate tool.

General-equilibrium models have been used to show how a static-equilibrium economy with perfect competition is identical to a centrally planned economy where all prices are calculated by the central planning board and fixed by law. Indeed Sherman (1970) suggests that the conventional neoclassical price theory is a useful practical tool for understanding how to set prices under a centrally planned economy, rather than a tool for understanding the workings of the modern market economy. The activity-analysis approach outlined in Chapter 3 is really no more than a general version of a multi-sector economy where the different 'sectors' are now 'activities' or activity groups (residential, service, industry, etc.).

General-equilibrium models have also been made operational through the input–output models of Leontief (1967). Their most widespread use has been to present the industrial structure of *whole countries* as an input–output table showing the flows of goods and services (purchases and sales) between all the different sectors of the economy. Most countries publish these tables, which have also been used to describe the industrial structure of regions and cities (e.g. Morrison and Smith 1976). They have not always been successful as practical forecasting tools, especially when used as exact mechanical models, but are most useful for examining general, underlying structures rather than examining the implications of small changes, since in the short term there are so many fluctuations which influence these input–output transactions. This was the point of using the activity-analysis presentation of the urban system in Chapter 3 to highlight underlying properties of the system of interlocking activities, especially in terms of real physical relationships involving people, land and buildings and not only the money flows which are subject to such fluctuations.

Summary of the market mechanism

We can now make an expanded list of the elementary concepts of modern micro-economics (modifying the list in Chapter 2) which underly many of the current theories of urban development.

1 The economic system is represented as a series of interacting supply and demand curves which determine how many of the different goods are

demanded and produced at what prices. All buyers and sellers have perfect knowledge of opportunities to buy and sell.

2 Relative preferences (or 'marginal utilities') of isolated individual ('incomparable') people lie behind the demand for goods. This is the 'value' theory of modern economics (the theory is all 'relative'). Individuals try to maximize their utility.
3 Marginal costs of production determine the supply curves of industries (i.e. how much is produced at what price).
4 Total output is determined by technical 'production functions' which show what output can be produced at what cost by individual firms from the basic factors of production (land, labour and capital).
5 Firms compete, trying to minimize costs (maximize profits) up to the point of equilibrium where their marginal costs are proportional to the price of their products. Thus the incomes of two of our three estates (Chapter 2) are determined by 'what they produce', and justified by the service they contribute to the production process. Factories are perfectly mobile; they can be transferred from one type of production to another.
6 The demand by firms for land, labour and capital again interacts with supply, to produce a set of 'factor prices' — respectively rent, wages, interest, etc. (Thus 'labour' is rather disembodied in this picture, being put on a par with 'inanimate' factors like land and capital.)
7 The primary resources (factors) are used to produce capital goods (machines and buildings) which ultimately determine the total amount produced in the economy. The economy itself measures the productivity of capital through the rate of interest. There is thus a capital market which determines where investment is made; capital seeks out the most profitable use — that which yields the best return.

As Dobb (1973) points out, there is a certain logic and simplicity in using neoclassical demand analysis for *short-term* situations when supplies are already fixed. Economists often describe their discipline as the science of choice — given a set of scarce resources. Using resources or supplies for one purpose means forgoing them for another; *'opportunity costs'* measure these lost production possibilities, the analysis of costs again being *relative.* But once supplies are no longer fixed, and the costs of production become important, then all the results of neoclassical economics depend on the given income distribution.

The staying power of neoclassical economics

While neoclassical economics is often attacked, it stubbornly refuses to go away. It has been applied to a whole range of problems in welfare, cost-benefit analysis, education, labour markets, the allocation and valuation of time, and so forth (Blaug 1974). 'For some reason of mathematical structure, the theory is highly manipulable; when faced with a specific issue it can yield meaningful implications relatively easily' (Arrow 1974). Clearly, then, when its assumptions hold, at least approximately, and especially when dealing with short-run situations of partial equilibrium, where a whole range of supplies and prices are already fixed and where the question of how to allocate scarce resources is paramount, neoclassical techniques have an important place (see the discussion on cost-benefit analysis in Chapter 6). This is not a value judgement; it simply acknowledges that conventional economic theory is being used on a day-to-day basis at many levels of industry and government to aid decision makers who have no interest at all in spurious 'justifications' or 'rationales'.

There is an interesting branch of literature which, instead of merely opposing neoclassical theory in a sterile way as some of the 'ultra-radical' thinkers appear to do, suggests that it may be possible to reconcile aspects of it with approaches based on the labour theory of value. The point made is that neoclassical economics does indeed provide the relevant tools for certain situations (e.g. economic management decisions in the short term) in the way which is perfectly compatible with a theory of long-term economic development based on the analysis of the production process, Marxist or otherwise (Sherman 1970; Johansen 1963). For example (paraphrasing): prices could be determined by the total input of labour even if one accepts the marginal-utility theory of consumers' behaviour. Even Marshall (1890) is said to have recognized that the conditions of production determine prices in the long term. If the two types of approach are compatible in this sense, then the 'mystification' or falsification of the economic development processes created by neoclassical theory arises when it is being applied to the wrong purpose at the wrong level, not because it is erroneous for the purpose for which it is particularly suited. But if the theory is used for a wrong purpose this in itself is no accident; it is because the contradictions of planning situation require it.

Before considering how conventional theory is used to explain the spatial distribution of economic activity (especially at the urban level) it is worth summarizing the way it is and can be used to justify or explain development and planning and to influence planning.

Justification

The conventional theory does not disguise the fact that the market is the main resource-allocating mechanism in the economy; it proclaims it, explains it and justifies it. It also justifies state intervention in very strictly limited cases (one might almost say 'marginal' cases). This seems to reflect the limited success of urban planning in controlling the development of cities only at the margin. Its idealization of the isolated, 'incomparable', individual consumer, together with the notion of 'fairness' in competition and 'balance' in supply and demand and its dislike of monopoly, are also often used to justify a market economy whose structure is far from the perfect ideal. It also provides a justification for wages — people get 'what they deserve' (as determined by what they produce) — and although monopoly profits are 'exploitation', normal profits are fair — a reward for risktaking.

Explanation and description
Conventional micro-economic theory has proved its usefulness in analysing partial (one-product), short-term problems where prices and supplies are fixed and where the choice as to how to allocate scarce resources is paramount. Many of its individual concepts (see the checklist in Figure 37) are useful aids to thought which help to clarify real-world problems, to analyse specific problems (Silbertson 1972), and to produce useful, general, empirical results which may have implications far beyond the strict conceptual formulation of the conventional theory. Because it focuses on the isolated individual consumer, it is conceptually incapable of providing unambiguous social criteria for measuring the benefit of production and consumption changes to the community as a whole. The macro-Keynesian theory has helped governments to control (or postpone) the incipient conflicts and crises of unemployment, inflation and instability in the national economy (see Chapter 2), but has done little to help understand the progressive decline of the UK economy and to suggest satisfactory ways of reversing it.

The conventional micro-theory gets into difficulty when it attempts to put its theories into practice. Whether the problem is partial (single-market) or general-equilibrium (several products), it is extremely difficult to measure and separate the various influences on price, supply and demand. It often cannot directly measure the behaviour of the individual consumer and producer on which its theory is based, and tends to retreat into complex statistical techniques (often unavailingly).

Once the assumptions of the micro-theory begin to break down, once large firms appear, long-term change is important, supplies variable, factors immobile, social consumption and total aggregate incomes and profits are of interest, the theory begins to seem more and more questionable as a reasonable description, and appears more like a justification. The following section examines how many of the assumptions become less justified at the urban level. Nevertheless, it has been suggested that its short-term, relative theory of consumer choice can be reconciled with alternative explanations of long-term economic development, e.g. those based on a study of production costs or even Marxist theories.

Planning techniques
In cost-benefit analysis, welfare economics provides a specific technique for deciding between alternative plans (see Chapter 6). It will be seen that this is appropriate only under very limited circumstances when the micro-assumptions hold and when the effects can be properly isolated and measured. However, the general philosophy of satisfying consumer demand is often implicitly or explicitly used as a justification for urban development which the private sector finds convenient. (See Chapter 6 for an example: satisfying 'demand' for roads.)

3 The urban economy: making conventional theory spatial

Introduction

We can now show briefly how conventional micro-economic theory, with its particular strengths and weaknesses is applied to a situation where production and consumption are no longer considered to occur at one point, but are spread out across geographical space. Building in a 'geographical dimension' tends to emphasize its weaknesses rather than its strengths. It can therefore produce broad pictures of urban development which correspond in general terms with reality (e.g. high profits in the city centre, declining density with distance from the centre, urban ghetto, etc.). However, its ability to explain problems, and in particular to verify its explanations in practice, is much more limited. Many of the perfect-market assumptions just do not hold in cities. Most problems are not 'partial'; there are very strong interlinkages, especially through the land market, as we have seen. This means that when the theory is used there is an even greater danger that it may be as a 'justification' rather than as a strict description or explanation of what is really going on in the city.

Not surprisingly, in view of the Chapter 3 discussion, the two specific problems in extending conventional economics into the geographical domain have been in the areas of (1) transport and transport costs; (2) competition for land (location and space). The basic split in economic theory between the aggregate macro-Keynesian theory of growth and equilibrium and the micro-static explanation of competition is also reflected in urban and regional studies. Micro-static concepts have been used to describe competition and the supply-and-demand process within or between cities. The theory has been used to explain an end-state, equilibrium pattern of activities located at various points in the country, of cities of different sizes at different places, and of residential location within the city. Concepts have been borrowed from international-trade theory, a theory which shows how and why different countries choose to produce certain goods, import others and export others. This leads to concepts of 'comparative advantage' and 'specialization' which are used to explain why cities and regions within an economy might choose to specialize in certain goods. On the other hand, macro-Keynesian multiplier ideas have been used to explain how a single city or region or a group of cities or regions grows. Again the two types of theory are very imperfectly reconciled; there is no overall explanation whereby the competitive process between firms and cities can also explain the way cities grow and decline.

Since the cost of transport to and from other places is a given characteristic of any given piece of land, then this must affect the costs and profits of the firm and hence its ability to make transfer payments to stop other activities from using the given location. Hence transport costs and competition in the land market are intimately linked and cannot be separated. However, we have already seen how competition for space is more important within cities than it is across a region, and so it is not unexpected that the

Utility	Subjective satisfaction (for an individual)
Marginal	Small increase in rate of change – 'least', 'last' or 'edge'
Elasticity	Rate of change
Demand curve	For a good or service
Supply curve	For a good or service
Relative preference	Marginal utility
Marginal costs	(Equals price at equilibrium)
Production function	(Shows the way in which capital and labour can be used)
Static equilibrium	(Of supply and demand)
Optimization	Maximization of utility, profit
Factor of production	(Land, labour, capital)
Mobility of factors	(Between different firms, producers)
Returns to a factor	'Income' of factors for 'services' provided, e.g. wages and profits
Substitution	e.g. of one good for another of one type of production for another
Opportunity cost	Cost of foregoing opportunities e.g. to consume another good
Transfer payment	Payment made for a reason other than the purchase of goods or services Examples: payment to landowners to stop others using the land; taxes
Scale	Returns to, economies of

Notes (a) Straightforward definitions are not in brackets
(b) Other remarks and examples are in brackets

Figure 37 Checklist of micro-economic terms (with some definitions and examples).

conventional theory has tended to compartmentalize. *Urban economics* concentrates on this competition and the allocation of land to uses (and especially residential uses), and is a separately identified subject from *regional economics* and *industrial location* which concentrate more on problems of transport costs in relation to other input costs for particular industries.

We discuss below how the concepts listed in Figure 37 become more limited and hedged about with qualifications, as we build in more spatial detail. But perhaps the first overall limitation is that of sheer complexity. As Isard (1956) points out, the usual type of general equilibrium, which involves a complex system of supply and demand for many goods and services, is only a special case of an even more 'general' case where all the *geographic* variations in inputs and outputs are recognized. Transport costs are no longer assumed to be zero and the factors of production are no longer mobile. The reader is referred to Chisholm's (1970) very useful account of how some of the key concepts of micro-economics can be applied geographically and how the basic assumptions of micro-theory have to be modified.

Urban micro-economics: assumptions breaking down

Inferior economies?

Urban economics contains and uses many of the elementary concepts of modern economics in modified terms. Perloff and Wingo (1968) describe the 'urban' versions of the two mainstreams of economic theory. Urban micro-economics has been developed from von Thunen (1826), by Weber (1968), Christaller (1966), Losch (1954) and Alonso (1964). All these authors emphasize the pre-eminence of micro-competition and optimization of costs and utility. Their work tends to result in a static picture of some end-state, equilibrium pattern of activity location within the city. Urban economics is regarded by some economists (e.g. Richardson 1971) as a very much underdeveloped branch of economics, even possibly as an inferior one. There is a frank acknowledgement in some standard works (e.g. Thompson 1968a) of the inadequacy of both the neoclassical resource allocation and the Keynesian multiplier approach. Thompson (1968b) feels that the economists 'run for cover' in situations where the supply side plays a major role as it does in urban development (cf. the production view of the city in Chapter 3). He reckons the conventional theory is too 'short-term', too dependent on a single-factor transport, and even growth theory explains only the fluctuations in income and output and not the underlying, longer-term determinants of growth. But there is also a great deal of controversy within economics, over what the limits of the theory really are when it is used in the urban situation.

Externalities

To Richardson (1971) urban areas are full of 'externalities'. We have already met this in Chapter 2. A given industry might be able to operate more or less effectively in a given city or region because some of its costs or benefits are borne by the wider system and not by the activity itself. If there really exist a large number of intangible 'externalities' which cannot be described or identified precisely, and which make the whole system so complex as to be impervious to any analysis, then this would indeed be a massive drawback, not only to the possible application of 'modern economics' but also to any kind of analysis. Williamson and Swanson (1966) found no definitive evidence of increasing external economies of urbanization in US cities, at least between 1820 and 1870, and Lampard (1961) feels that these economies remain as 'elusive' as ever.

Economists themselves are not always clear about what an externality actually is (Scitovsky 1954). Thompson defines externalities to be the effects of a transaction that affects other parties not directly involved in the transaction. He gives a list of urban 'externalities':

1 Increasing differentiation of firms whereby specialized services, specialized components previously manufactured within a single company, come to be performed by new auxiliary firms, which lowers the overall cost of production.
2 The availability of good transport facilities which lower the cost of production for firms producing commodities sold in national markets.
3 Large and flexible pools of labour.
4 Less need to retain large stocks and supplies within an individual firm because of the availability of supplies nearby in the urban area.

A common case often cited is that of an individual house owner who improves his own house by painting or remodernizing; the value of neighbouring houses is increased on account of this improvement of the 'environment'. We cannot pursue the problem of externalities in much detail here but it is at least worth trying to clarify when they exist and how they might be identified.

There are undoubtedly severe problems of measurement, but many modern economists would argue that, even though all the benefits and costs might not be retained by the participants in a given transaction, the overall effect is indeed transmitted through the economy to increased property values and other economic variables which are measurable. Most values, costs and benefits are tangible; they rarely arise from nowhere. Using a comprehensive and exhaustive classification of activities which fully cover the urban system and the use of resources within it (as in Chapter 3) should take us a long way towards being able to document most of the significant economic relationships in the system — including the effects external to any one firm or sector. The first of Thompson's external economies, the increasing differentiation and variety of specialist organizations, both public and private, are part and parcel of the increasing drive to reduce production costs throughout the whole economy, as noted in previous chapters. These activities can all be identified and described precisely as different types of production activity (see Broadbent 1973); Barras and Broadbent 1975; and Chapter 3).

Both Thomson and Haworth fail to distinguish adequately between external effects which are transmitted through the market from those which are not (Scitovsky 1954). *Non-market effects* include interdependent consumer utilities: spillover effects such as pollution, inventions which become available without cost to the firm, or a pool of labour created by one group of firms which may benefit another. (This is Thomson's case 3 above.) *Market effects* include most of those described in the 'development process' in Chapter 1, including the division of labour and specialization in production, where the different stages of a production process which was previously carried out in a single firm are now separated. This brings reductions in production costs which benefit all firms. Again this is a complex situation which requires a proper description of the production process and not only the exchange process dealt with by the conventional theory. Chisholm (1970) cites one simple way of looking at external economies and diseconomies, citing Neutze (1965), who measures the cost of providing services for cities of different sizes, obtaining an optimum city size of 500,000.

Alfred Marshall (1890, pp. 365-6) states that internal economies within an individual industrial firm are frequently very small when compared to those external economies which result from the general progress of the economic environment. Marshall notes that the increase in value of the particular site in an urban area is the value which results from the growth of a nearby rich and active population. This is the most striking example of the way in which the environment changes the cost of production in the individual firm. We have already seen how competition in the land market is the situation of externalities *par excellence.* If land is used for one activity it cannot be used for another. An increase of one activity in a given location implies a decrease in other activities at that place. The mere presence of other activities raises the rent payments of a given activity, as we have seen. Some of the confusion over externalities results from a failure to distinguish clearly between partial analyses (of an individual firm, household or sector of industry) and a comprehensive analysis which attempts to embrace all the interacting enterprises in the system — the city. In this latter analysis many of the so-called externalities are brought within the compass of study.

Lumpiness

Another generally acknowledged drawback is the inability of conventional economics to take account of the 'lumpiness' of many urban investments. Any one single enterprise may be a very major influence over the whole urban system. Big factories are clear 'indivisibilities': they cannot be subdivided into smaller and smaller units, and their production cannot be smoothly decreased as the smooth production curve might imply. Urban economists have long been aware of this problem; but, as we saw in Chapter 1, this is now becoming a problem for the national economy where single firms are also becoming important. (This might be another case where urban or spatial concepts are later applied to the national economy.) In many other fields of investigation (physics, biology, etc.) the most difficult situations to analyse are those where it is not possible to isolate and study the behaviour of one individual part of a functioning system separately from the functioning of the system as a whole. If at national level the growth of large firms is one reason why the ideal of the perfect market no longer applies, this is even more true at the urban level.

The sticky market

Conventional economics recognizes that in cities the market for goods and services tends to work far less well than it does within the wider national economy. Richardson notes the existence of monopolists in urban land ('planning authorities, property companies, and mortgage institutions'). Secondly, there are severe rigidities stemming from the very strong preferences of establishments to remain where they are; factors are just not mobile. This in turn arises because investment in plant and particularly in buildings is costly and is not easily recoverable in the market. Again there are 'supply-side' effects, which need a more explicit analysis of the production process than conventional micro-economics generally provides.

The location problem

Von Thunen, as we have seen, took a single, isolated city as a market centre and calculated the rent paid for agricultural land in relation to the cost of transport needed to reach the market. He showed that the different agricultural products would be produced at different distances from the centre of the city in a series of concentric rings. Anticipating the 'opportunity-cost' transfer-payment principle, he pointed out that the price of any product (e.g. milk) would have to be such that the land on which it was produced could not be used more profitably for any other product. In this theory, which is more or less similar to Ricardo and Marx's theory of differential rent, the point of production which is furthest from the market and where it is just economic to continue producing, is the place where the cost of production (including transport cost) is equal to the price at which it is sold.

The theory of industrial location was developed in 1909 (reprinted in Weber 1968), by Alfred Weber using many of von Thunen's ideas. It is a partial analysis and considers for *any given firm* all the different costs of production which will vary with different possible locations; the best location will be where all these costs are minimized.

A standard criticism of this cost-of-production approach has, of course, been that it does not deal adequately with the 'demand' side. It committed an unforgivable sin by not adequately considering the *market* for the goods being produced by the industry being located. Many industries tend to become 'market-oriented' (see Chapter 4) and hence to 'agglomerate' (cluster) together near this market rather than near their supplies. It might be argued that by looking at the whole system of indstries as in Chapter 3, including the cost of transport to the market as a cost of production, then any industry can become market-orientated, if the ratio between the transport cost of the outputs and the transport cost of the inputs so determines it.

Trade

Walter Isard (1956) has tried to develop location theory into a more general theory of the space economy by bringing it together with international-trade theory. Trade theory analyses the benefits accruing to two or more countries or regions which exchange those products that they each produce most cheaply. According to the principle of comparative advantage, trade may be mutually profitable to two regions (or cities) if each specializes in the products in which it has the greatest *relative* efficiency. This is the case, even if one region is in *absolute terms* more efficient in the production of all products than the other. This classic result is due to Ricardo and it shows why we might expect cities to specialize.

Once regions or cities are allowed to trade, an industry may be located in a place which in absolute terms is the worst possible for that industry (Chisholm 1970). Nevertheless, according to the theory, all consumers in both regions have higher incomes than if trade did not occur, because the total production and income is greater. The theory points to the benefits of a free market without barriers to trade — for example, tariffs and transport costs — trade is usually freer within a country than between countries.

Another reason for trade arises directly from a departure from the perfect-market assumptions — the physical *immobility* of a factor of production. Capital gets tied up in factories at specific places and labour does not always move spontaneously in response to the demand for jobs.

Monopoly again

The concept of monopoly also becomes important in spatial applications. Once the Weber analysis is extended to include the demand side, in the case of a single firm serving a uniform population distributed over a plain, there will be a finite *market area* for its product. Whether or not the transport costs are borne by the producer or the consumer, decreasing returns to scale and the increasing price of transporting the good will result in a boundary, even if the firm has no competitors, beyond which it will be unprofitable to transport the good. Once there are transport costs and market areas, the market is always at least *partially* a monopoly: a consumer only buys from the nearest single producer, according to the conventional theory. If consumers bear the transport costs, then price increases outwardly from the point of production. An increase in the size of the market area will yield a proportionately decreasing increase in sales, depending on the price-elasticity of demand. Once the firm is in competition with other firms to supply the market, the circular areas become squeezed into hexagons and profits are reduced to normal, non-monopoly profits. At the boundary lines between market areas the consumer is 'indifferent' between each producer.

Central places

Central-place theory, formulated by Christaller in 1933, and further elaborated and modified by Losch between 1940 and 1944, was mentioned in Chapter 3. This theory recognizes explicitly that urban centres provide a range of goods and services for their dependent populations according to a distinct, hierarchical-ordering system. Large cities are much more complex and encompass a very wide range of specialized functions and services which require a large, dependent population. In the original pure theory, a hexagonal system of cities and towns is proposed, so the idea is an extension of the concepts of spatial monopolies and market areas. The function of the highest-order city also includes the functions and services of the lower-order cities. It should be fairly clear that this elegant idea is also in principle applicable to central places within a city. Retail centres form a descending hierarchy from the city centre through district centres down to local centres. This theory produces a static-equilibrium picture once again, but it does allow for the continuing expansion and differentiation of industrial activities in urban centres. In terms of our Chapter 3 argument, it could be suggested that more and more goods and services are being provided from higher-order centres as the national economy continues its drive towards integration and specialization. Christaller recognized this in his brief discussion of the dynamics of the development of central places as the economy at large progresses. There have been many attempts to verify Christaller's theory, since Reilley (1931) argued for the existence of a hierarchical system of retail centres. Berry and Garrison (1958) discuss other possible explanations of the distribution of urban sizes. These authors have also demonstrated the existence of a central-place hierarchy in a United States county. In their study, three classes of towns seem to exhibit the discrete levels of population suggested by Christaller. Various authors have tried to correct observed patterns of central places to allow for the non-uniformity of the transport system and the physical terrain. They have purported to show that, when these corrections are added, the system of cities does indeed look hexagonal.

Angel and Hyman (1971), on the other hand, have argued that it is mathematically incorrect to transform geographic space in this way.

A review of the central-place idea can be found in Chorley and Haggett (1967), where it is noted that no overall theoretical interpretation and criticism of Christaller's work has been undertaken. Here again we see a branch of neoclassical theory reproducing in general terms observed spatial patterns, but again finding it difficult to test out the ideas in empirical detail. From the above remarks of spatial monopoly, it is clear that the hexagon idea refers to a situation of competition to divide up a *fixed market* between a number of firms. Clearly for many industries, firms and functions the idea just will not apply, since they serve the whole national market even though their plants are scattered about the country. Since the spatial economy is now largely inside firms rather than outside them, and since central-place theory concentrates largely on *exchange* and *competition* between plants and firms, it will be appropriate in fewer and more limited situations than it was before the growth of the large firm. It will be most useful in situations where there are a number of production centres competing to serve a uniform population — a local retail market is again the obvious example.

Describing the city: zones and rings

Evans (1973), in a distinctive work on the economics of residential location, traces the development of von Thunen's and Weber's early work and the various attempts to apply it in detail to the city. The original model of an isolated city has generally provided the framework for many of these further developments. Losch (1954) argued that the selection of a job in a different city was the most important factor in any individual's choice of residential location — and therefore, by definition, households within a given city would not be very mobile. Turvey (1967) felt that it would not be possible to use a static analysis to explain the layout of the town and its pattern of buildings, since the conditions which determined why these buildings were erected had long since changed; he did feel that the pattern of activities which use the buildings could be analysed much more readily.

An early attempt to explain the pattern of location within the city, put forward by E. W. Burgess in 1925, has since come to be known as 'concentric-zone' theory. In Burgess's view, von Thunen's successive rings specializing in different agricultural crops are now seen as rings specializing in different urban activities. In the centre there is the 'central business district'; surrounding this is an area 'in transition' between residential uses and newly invading business and light manufacture. The third ring is inhabited by workers who have escaped from the area of deterioration but who still desire to live within easy access of their work. Beyond this is a zone of high-class residential and apartment buildings, and still further out beyond the city limits is the commuter zone with surburban areas and satellite cities. As the city grows, each inner zone extends its area and invades the next outer zone. Thus any particular site serves a *succession* of uses. The reasons for all this are not clearly spelt out and certainly do not constitute anything like an economic theory of the city. It is implied that the oldest residential property will be near the city centre, the newest property will be further from the city and, since the highest income groups are able to afford the newest housing, they will be the ones located outside the city. Alonso in 1964 described this type of theory, therefore, as a 'historical' theory.

In another variant of the historical theory, Hoyt (1939) argued that successive groups of people, with progressively lower incomes and lower social standards, occupy a given residential structure which is deteriorating with the passage of time. However, he argued that the city does not tend to be circularly symmetrical and that the high-rent residential area is situated initially close to the retail and office centre on the side of the city which is furthest from the manufacturing district. Therefore, as the high-class residential population moves radially outward, the city as a whole adopts a sectoral pattern rather than a pattern of radially symmetric concentric zones.

Most economists now seem to agree that the overriding factor in determining the location of economic activities, and in particular the location of households within the city, is the cost of travel to the centre, which generally is the most important centre of employment. But most of the developments of the concentric-zone theory regard housing costs as externally given; each household is regarded as being involved in a trade-off situation, trying to balance the increasing cost of transport against the cost of space and land which, in these theories, tends to decline with distance from the centre. The exponents of this idea include Hoover and Vernon (1959), Alonso (1964) and Wingo (1963).

Thus in most of these theories the *buildings* of the city are ignored completely, although this omission is rectified in Muth's (1969) version of the theory; Evans's (1973) extremely careful analysis shows that with all the conventional neoclassical micro-market assumptions, including perfect competition, with both employers and employees following their respective goals of profit maximization and utility maximization, together with full employment both in the city and in the economy as a whole, nearly all the generally observed patterns of residential location in modern cities can be derived. He shows that the value of housing declines at a diminishing rate with distance from the centre and that the value of land and the density of housing units must also decline at a diminishing rate. Similarly with the population density, the analysis shows that an overall equilibrium is attainable in the city, but if people 'prefer' low densities then some government intervention may be needed to control density. He shows that under plausible assumptions it is quite possible for the rich to live *both* at the centre of the city and also at the very edge and that the poorest population may well live next to the rich in the city centre. This leads Evans to believe that the conventional, mainstream, neoclassical micro-economics can indeed to used to justify and explain the real pattern of residential location of different social groups within the city in a modern, mature economy.

The conventional theory distinguishes the location behaviour of residential households from that of industrial plants much more than we did in Chapter 3, where we regarded the household as a production unit. Most authors recognize that residential activity is far and away the most important user of urban land and that households tend to be less variable in their location behaviour than do industrial firms. But the Alonso version of the concentric-zone theory is also applied to industries. These activities can offset declining revenue and higher operating costs by lower site rent at locations increasingly distant from the city centre. He defined the 'bid rent' as the rent which an activity would be prepared to pay at each site, given the need to maintain the same profit level. The heterogeneity and complexity of industrial and commercial activities and the varying costs of different factors of production with the scale of output need to be analysed very specifically for each type of industry. For further reading on the location behaviour of industries, shops and offices, see Goldberg (1970), Vernon (1960), Cameron and Johnson (1969), Hamilton (1967), Berry (1967), Economist Intelligence Unit (1964), Location of Offices Bureau (1968), Dunning (1969), Wabe (1966) and Croft (1969). Mayer and Kohn (1959) give a useful selection of basic papers illustrating many of the most important concepts in urban geography.

Geographic margin and economic margin

Chisholm (1970) examines the concept of the economic margin in geography. He is at some pains to clear up some of the terminological confusion noted above but does not do so entirely. He points out that only in very special long-term cases (e.g. the von Thunen situation) will the geographic margin (the edge of cultivation, the furthest factory, the furthest city or the 'edge' of the city) coincide with the economic margin (i.e. type three, the least profitable firm or unit of production). Chisholm defines a further, rather confusing 'margin' which means the *highest* possible intensity (or density) of activity (rather than the *least).*

A geographic boundary between two land uses — that is, between two of von Thunen's rings — is in general terms a *substitution* boundary, where landowners are *indifferent* as to which of the two industries uses the land. We have already suggested that the 'containment' of urban development at the margin reflects the *economic* role of planning (i.e. restricted to the sidelines). The ideas of 'zoning' which underlies much of development control (see Chapter 4) reinforces and rationalizes the tendency of the market to produce concentrations of homogeneous land uses, whether in rings or ghettos. At the boundaries of the zones the market 'is weak' — it cannot decide between uses — and therefore here again planning steps in, in its usual 'negative-response' role, to draw a firm boundary between the different activities.

Chapter 3 showed quite clearly that there is polarization and specialization *within* a city, not only between activities but within the residential sector. The 'ghetto' areas tended to be concentrations of 'marginal' (type 3) workers who were most likely not only to have low incomes but also to be thrown out of work in economic downturns. In the competitive market economy, the conventional theory points out that there will *always* be a margin of type 3 — a least profitable unit of production or labour. Economic and planning policies designed to deal with a particular poverty group, whether or not it is located in one place (as the conventional theory itself tells us), are only really going to move the problem somewhere else — a 'boundary' change. The conventional theory finds it difficult to justify even a modest welfare programme safety net for the lowest income groups, since it will remove the incentive to participate in the market economy at all.

In the UK the tax threshold has descended to meet the level of the social security safety net — so there is often little economic 'incentive' to work.

If labour is in short supply (as it is with the full-employment equilibrium ideal) and is homogeneous (that is all workers have the same skills and can be transferred to *any* industry), then even the least productive industries will need to be paid a *transfer payment* to stop their workers moving to firms paying higher wages — all wages would be equal. If, on the other hand, labour is entirely non-homogeneous (one skill group or income group cannot be substituted for another) and workers in low-income industries do not have the abilities or skills to be transferred to high-productivity industries, then wages will depend, like land rent, on the demand for each industry's output. According to Samuelson, differences in skill are the most important cause of income differences (but why do skills continue to be different?). The final distribution of wage incomes is determined by a complex interaction between the supply-demand curves for labour of different types and the outputs of different industries. If labour is only partially substitutable, then in times of economic downturn there will be a tendency for some types of labour, especially unskilled labour, to appear 'marginal' and to be thrown out of employment, whereas others will retain their employment. When this process is mapped directly on to the residential map of the city and 'marginal' workers live in 'marginal' areas, then the geographical pattern clearly reveals how the economic process is generating income differences.

Urban macro-economics – the multiplier again: the economic base

Economic-base theory might be regarded as a reflection at the urban level of Keynes's ideas on the multiplier. The 'injection' into the local urban economy is now represented by the 'basic' sector of the economy, that is, that part of the local economy which exports goods and services to the national economy. The level of employment or income in these sectors ultimately determines the total level of all the other activities in the city. It is these activities which are therefore 'multiplied' to determine the total level of activity. The non-basic activities, by contrast, only serve the city itself. They provide goods and services for the local population and also for the basic sectors. Alexander (1954) traces the development of this idea from Au Rousseau (1921). The idea was developed further in the work of Hoyt (1939) and, as Perloff and Wingo (1968) note, economic-base theory is central to most projections of urban growth. There have been numerous efforts to identify and measure the economic base of different cities (Andrews 1953). At various times and in various cities the ratio between basic employment and non-basic employment varies widely, from a very small proportion of city activity up to 50 per cent and more. Richardson feels that base theory should be regarded rather as an extension of central-place theory (p. 83). Central-place theory identifies the source of a city's growth in its hinterland, base theory simply extends this notion of hinterland to cover the whole of the external world (Richardson 1971). But base theory has never been a comprehensive enough principle to provide a unifying theory of the development of an interlocking system of cities.

It has been subject to extensive criticism, not only because different methods of measurement give widely different sizes for the base, but also because it cannot be used to predict large-scale changes or long-term changes since the ratio between basic and service sectors changes as the city size increases. In addition, some urban growth and investment can occur autonomously, generated from profits within the area (Chapter 3). Blumenfeld (1955) has argued that attracting export industries to an area depends in the first place on having a dynamic service sector, and so a case can be made for turning base theory on its head and making the growth of the export sector dependent upon services. Weiss and Gooding (1968) and Tiebout (1956) have developed the idea further by dividing the export base sector into several different sectors and working out a multiplier for each. This can be developed further, and the service sectors themselves can also be divided so that the whole interlocking matrix of transactions within the service sectors and between the service sectors and the basic sectors can be identified (Morrison and Smith 1976); ultimately this turns the analysis into a disaggregated general-equilibrium model.

Assessment: description without explanation?

Conventional economics loses some of its conceptual elegance and explanatory power when it is applied to cities and regions. The reasons for this can be broadly divided into two. First, many of the perfect-market assumptions hold less well as we begin to look at the economy in finer detail. True, some things become easier — if we study the market for a single good we can use the traditional supply and demand curves of the partial theory. But other things become harder, since single industries and single cities are no longer made up of small firms, hence all the problems of indivisibility and immobility of factors. The theory says very little about some of the most important issues facing cities — the provision of development capital by large financial institutions, the effects of big firms, etc. etc. There is also the difficulty of what to assume about the rest of the economy *(ceteris paribus)* while we focus on an individual industry, city or region. It is very difficult to isolate the effect being studied, although with careful experimentation and statistical technique this can be attempted. All these then are the difficulties inherent in applying the conventional micro-technique to *any* part of a whole economy, in a situation where that part is linked to the rest of the economy through the demands and supplies of different goods.

Another general problem inherent in mainstream economics and which becomes more acute in urban and regional applications is the overall split between micro (competition) and macro (Keynesian) description. The whole mechanism of profit and utility maximization at the level of the firm and the individual consumer with its implied full-employment, static equilibrium, doesn't add up to an explanation of how the economy as a whole grows. Consequently it is no surprise that conventional theory does not link its explanations of the behaviour of firms and individuals inside cities with the behaviour of the whole system of cities and industries in the economy.

The second kind of difficulty does not simply arise from the innate limitations of the theory but from the specifically 'spatial' effects (i.e. of transport costs and the fact that activities are spread over, and compete for, space). This means, for instance, that there is 'imperfect' competition and other monopolistic effects which in any case the theory finds more difficult to handle. Because of these imperfections in the market, the theory might conceivably justify state intervention in land (public ownership?) and in the transport system, although its practitioners rarely pursue its logic this far. Its theory of comparative advantage, and the need for countries, cities and regions to trade and specialize in the commodities to which they are best suited, provides useful insights into why industries sometimes seem to locate in areas which are not the 'best' from the industry's point of view. But this again becomes less tenable when 'industries' in a city are really only a branch of a single firm. The theory implies that every region and city gains from specialization and free trade. However, even under all assumptions of the theory it may well be that these gains serve to increase income differences between rich and poor cities and regions. The theory is at best ambiguous and at worst inadequate in explaining unbalanced or uneven development within and between cities.

The geographic margin (or zone boundary) does not always coincide with the economic margin, but the general ideal of the margin continually reappears as marginal areas (ghettos) or as physical margins (boundaries of cities). Here again, reflecting the welfare-economic safety-net approach to public intervention, physical urban planning steps in to define boundaries more firmly where the market would be indifferent. Conventional theory can describe the pattern of residential location commonly observed in cities, high- and low-income ghettos, etc. It also shows how high-income, high-profit activities will force out less profitable activities from city centres. Households are microscopic entities where more of the micro-market assumptions might reasonably be expected to hold; but the fact that existing patterns can be predicted does not prove that the market is actually working, or that individuals are maximizing their utility. Indeed the zone boundary itself results from the activities of *many* firms or individuals (it is an (aggregate' smooth curve), and we have seen how difficult it is for the conventional theory to deduce from an observed smooth curve for a whole industry anything about the behaviour of individual firms or consumers that underlies it. This explains why the theory often appears 'non-operational' when it is used to study a particular city in detail. Its explanatory power disappears under a welter of statistics — searching for more and more detailed data on quantities and price to pin down more precisely what the idealized consumer actually 'wants'. 'How does the individual consumer value an extra bathroom, a garden, a clean street, etc. etc.?' This gargantuan appetite for data begins to obscure or mystify. The more detailed the data the less we seem to know about what is really going on. This 'data fetish' (together with the obsession with satisfying consumer demand) has had an influence on the techniques used by planners and has provoked a reaction against massive model-building exercises (see Chapter 6).

4 Pluralism: a market theory?

The correspondence between the theories of mainstream economics and the practice of UK social policy is really quite striking. It is not possible to say that the practice grew out of the theory or vice versa; rather, they developed together, and they suited each other. Practical policy makers and other more inchoate or less elegant theorists than economists do not always explicitly recognize that they are reflecting, interpreting or misinterpreting major economic theories. Yet the image of the perfect market, and the determination to restrict the field of inquiry only to the margin, can be discerned without too much imagination in the writings of many social theorists and planning practitioners.

'Pluralism' is really nothing more or less than the theories of mainstream economics, perhaps sometimes modified or distorted, applied to the practice of and legislation for social reform in the UK. Most of the planning legislation described earlier was a reflection of pluralist theory, as were the other great practical reforms like the Second World War education act and the National Health Service. It may seem crudely reductionist to lump together the great body of the UK social-academic establishment. The justification for doing so is that this school of thought, despite its heterogeneity, does in the ultimate hold to an underlying body of fundamental concepts which appear increasingly inadequate to address modern social and economic problems.

In the late 1960s, but perhaps rather less so recently, pluralist academics have been getting a rough ride from their students and from more radical theorists. The intellectual running in this attack is usually made by what might be called the 'ultra-radicals' (see below). The debate between these two schools is often so sterile, with the two camps forcing each other into their respective 'status quo' and 'destroy the system' entrenched positions, that the student or the erstwhile progressive planner may be at a loss to discern anything hopeful for the development of society.

It would be a hopeless travesty to attempt anything like a comprehensive classification and discussion of the different strands of thought within pluralist theory. We are mainly interested in the way it justifies changes and in its general perspective on society rather than in examining its enormous contribution to the empirical understanding of specific problems.

Simmie points out, that in contrast to economics, sociology has only a very limited number of useful theories. These theories are usually based on a combination of one or more of three basic notions — 'cooperation', 'anomie' or 'conflict'. Economics might be regarded as a superior social science, not least because its concepts are relatively well defined (and hence challengeable both theoretically and through measurement in the real world). But there is a more fundamental reason for the superiority of economics; any economic theory deals with the way individuals and groups interact and in this sense always embodies a social theory within it. Conventional micro-economic theory immediately seems to provide a fairly subtle social theory. It accommodates conflict at the micro-level of the individual, and yet comes out with a cooperative result for the whole society (whose utility — satisfaction — is maximized). (In contrast, the Marxist theory embodies conflict (profit maximization between individuals and firms), cooperation (of individuals within a class — the working class) and a new notion — contradiction between classes. Conflicts may well be reconcilable at a higher level of social organization, thus individual firms conflict but the whole class of firms (or those who own them) have a common interest (in preserving profits) *vis-a-vis* the working class. A contradiction, however, is not reconcilable in the ultimate, and therefore forces a radical restructuring of society.)

Pluralism is the dominant accepted social theory in the UK. It recognizes conflict (together with a concomitant diversity and complexity) in society and either states or assumes that the conflicts can be accommodated without fundamental social change. It parallels the neoclassical perfect-market ideas of 'fairness' (in competition), 'balance' (between different groups or classes) and 'diversity' (i.e. many entities having legitimate claims on the market). It tends to restrict its field of empirical investigation to the 'margin' — to investigating marginal groups (of the poor and deprived). It suggests new forms of support and state intervention to ameliorate poverty without questioning too deeply the resource-allocating mechanism of society which creates marginal groups.

Two contemporary authors in the urban social policy field who are relatively explicit about their pluralist ideology are Eversley (1973) and Donnison *et al.* (1975). Eversley points out how nineteenth-century utilitarian philosophy — 'the greatest happiness of the greatest number' — and such twentieth-century figures as R. H. Tawney provide reference points for many pluralists. Although reflecting the essentials of micro-market competition, the theory does appear to modify the market in two ways. *Groups* are brought in explicitly, in the place of mere individuals, but these are usually not analysed in terms of their role in the economy. They all have equal rights in their diversity. In his brief section on 'society', Eversley shows how diverse organizations can be (professional, business, union, academic, etc). The theory also diverges from the static equilibrium of the perfect market by recognizing that society has to change; but it imposes on social policy (the state) the task of ensuring that no single group ends up in an excessively privileged (monopoly?) position.

Simmie shows how other recent social theorists such as Dahrendorf (1969) have promulgated the pluralist idea as a direct refutation of the alternative, cooperative-functionalist theory of society, developed from the nineteenth century through Durkheim and Parsons (1951). Functionalist theory is not 'pluralist', but draws a parallel between society and a biological organism. Each group in society is like a distinct organ within a living being, having its own particular function which contributes to the harmonious development of the whole society. (See the discussion on systems and cybernetics below.) The Dahrendorf theory seems to go further than most pluralist formulations in recognizing the possibility of domination by some social groups over others. Rex (1970) goes even further in placing the conflict between *classes* at the centre of the stage, and thus begins to converge with Marxist ideas.

Pluralism begins to get into theoretical trouble with itself and with its micro-economic underpinning in considering one of its central preoccupations — equality. Equality is seen as a good thing, but in practice too much equality of the wrong kind (=greyness, =identity of interest, =class consciousness ?) can conflict with the need to maintain a desirable 'diversity' in society. This difficulty is often resolved by introducing the idea of equality of *opportunity*. Inequality preserves the 'incentive' notion of the market while equal opportunity preserves an element of balance and fairness in competition. Again the state is invoked in its usual welfare role to ensure that there is not too much inequality.

If the micro-economic theory and its pluralist derivation implies that the redistribution of resources can only take place at the margin (a small change in the position of overall equilibrium), some of its adherents are quite explicit about it. Eversley goes so far as to say that inequality in Britain has decreased. In discussing the resource implications of its suggested social reforms, the theory often avoids uncomfortable choices (taking away from one group to give to another) by relying on economic growth (redistribution at the margin). This is why the (non-growth situation seems to worry pluralists a great deal. (It was suggested in Chapter 1 that the 'new' non-growth situation is really the old problem in a more severe form, rather than a new phenomenon.) The pluralists idea that growth provides new resources which can be redistributed to the poor in order to decrease inequality conflicts in general terms with the underlying micro-economic, marginal-productivity theory. Growth is often promoted by high-productivity sectors which pay high wages, and so growth often increases inequality rather than decreases it.

A characteristic line of investigation by pluralists which reflects its 'marginalist' view of the world and also the poverty ('minimum standards') approach to social policy is discussed in Chapters 1 and 2. There is a never-ending search for the deprived — for new forms of poverty. Eversley recognizes the 'poverty industry': 'the old and the cold', 'the gypsies', 'the inner-city poor' are all new deprived groups discovered in recent years. Numerous studies demonstrate time and again pluralism's excessive concern with the margin, with connotations that the poor have to prove their rights to social welfare and that social provision is only for the poor, or for situations where the market breaks down.

It should be made clear that this kind of research can produce a wealth of factual material, much of it open to a wide range of theoretical interpretation.

Much of what we know about the UK has been produced by pluralists; often the theoretical orientation of the investigator may well not be relevant — facts are facts. Sometimes, of course, the excessive concern with the poor, reflecting as it does Victorian paternalism, often serves to divert the attention from the causes of social ills to the symptoms. But following the line of argument of Chapter 4 and Section 1 of this chapter, there are many occasions when, in so far as these studies serve a real practical purpose (e.g. developing legislation, new forms of social policy, new planning systems), they cannot avoid showing up things as they really are, even though the view may be onesided or the interpretation erroneous. Just as neoclassical economics cannot be dismissed out of hand, pluralist theory cannot either, and the ultra-radicals (see below) who do so do a disservice to the cause they purport to serve. They also help to reinforce the status quo by pushing the debate into a permanent, polarized, trench-warfare situation.

By initiating reforming legislation, the pluralists have helped to promote the increasing influence of the state, to advance planning generally and to provide the great bulk of basic empirical facts about city development. They are generally in favour of specific increases in state activity to ameliorate specific problems ('market imperfections'). But they do not usually ask the wider questions about the possible limits to the size of the state in the mixed economy or the possibilities for, or limits of, urban planning in the face of the increased forces of competition invading the city.

5 Ultra-radicals: planners as oppressors

The social upheavals in the USA in the 1960s and the increasing troubles of Western economics have helped to stimulate a new wave of radical social theorizing more or less overtly related to the other branch of Samuelson's economic tree — Marx. It would not be productive here to inquire which workers are truly 'Marxist'; studying the different tendencies in this field is a discipline in its own right. While all this type of work is concerned to point to the possibility, desirability or inevitability of more fundamental change in society, the version that has made most running in the urban planning field is that which exhibits more or less total antipathy to 'the state' (Broadbent 1975c). The thesis is that the state operates more or less directly in the interests of the 'ruling class'. As stated earlier, this assumes that the state's economic power is relatively independent of the rest of the economy, so that increased state activity, if required by the 'ruling class', can be achieved without threatening the profitability of the private sector. While pluralists tend to favour increased state activity in specific instances, the ultra-radical view of state intervention ranges from the deeply suspicious to the vehemently antagonistic. They often take the lead in conjuring up new invectives against planners.

Some of these authors use evidence from closely observed individual case histories, such as those of Dennis (1972), which find planners actively inhibiting public participation by withholding information or mystifying the real economic issues in urban development schemes. The more theoretical arguments (e.g. Simmie 1974) sometimes invoke conventional economics. Simmie uses the supply curve to show how planners reinforce and promote inequality. The argument is that planners restrict the supply of land, thereby increasing its price, and, since land is privately owned, benefit the landowner at the expense of the rest of the population. Over-eagerness to invoke a standard theory against the planners has led Simmie to overlook the fact that the planner's main activity is to *redistribute* development from one piece of land to another; planning does not impose an overall restriction on the total amount of land available nationally. Planners do restrict total development in particular places, the main effect of which is to benefit one set of landowners as opposed to another. Planning operates at the margin, and releases land at the margin. If land is not released at one site where a developer can make a profit, then it will have to be released at another profitable site, either in the same urban area or in another town, otherwise the development will not take place (see Chapter 4). Pahl (1975) has his own peculiar 'convergence' theory where he seeks to show that state planning operates in a similar way in both East and West, again essentially as an oppressor.

Combining a practical study of the city of Baltimore with some Marxist terminology about the theory of land rent, Harvey (1974) also casts planners in the role of exploiters. Here, imperfections in the market rather than market competition itself are seen as the major cause of urban inequality and oppression. People become 'trapped' in particular submarkets — in racial ghetto areas or poor-quality, high-priced private housing. It is the planners and other state bureaucrats who, by their restrictive zoning regulations and other legislation, stop people from escaping from these ghetto, submarket situations. There they remain, sitting targets for exploitation, while the landlords extract an extra payment of 'class monopoly rent'. Harvey presents a gloomy and terrifying picture of dominance and oppression by financial capital, backed by the power of the state. This one sided conclusion does not question or recognize the possible limits to the expansion of the state and its ability to 'back' the 'forces of oppression', or that there indeed may be some ultimate conflict between the growth of state power and the viability of the private sector. This picture fails to distinguish between what planners do in one *specific* situation, whether it be a development scheme or ghetto area, from the *total* effect of planning in the city or the economy as a whole. Planners do have to cooperate with the market, or development would not take place, but it is the market itself which produces whatever pattern of inequality and exploitation exists, rather than special monopolies produced by planners. The private ownership of land is in itself a monopoly because land cannot be produced in the economy. The growth of state planning is not an unambiguous good for the private sector, as we have argued in Chapter 2.

If practical action is a test of a theory, then the ultra-radicals tend to advocate grassroots, community action against both planners and developers. Although such resistance has had limited success against specific office developers in a few cases, it would seem a philosophy of despair to throw away all the possibilities and potential of existing institutions, including the state and planning. These institutions tend to be rejected without any concrete analysis of their role. They are defined to be oppressive rather than shown to be so. It is tempting to feel that the ultra-radicals' ideal would be something resembling the nineteenth-century situation where the people faced capitalists directly, and where the whole panoply of state institutions and welfare agencies did not exist.

This ultra-radical view seems to derive partly from Marcuse (1964), who sees the powers of the state and the corporation, with their instruments of cultural and economic domination, adding up to a comprehensive totalitarianism. There being no potential for radical change in existing institutions or the conflicts between them, the only response is the politically impotent one of 'total refusal'. Marcuse thus seems to propound a total 'dropout' strategy for change, which relies on the marginal groups, the poor, blacks etc., and not on the mass of the population and/or their pressure on the state.

How does this theory influence planning in the terms set out in Section 1?

Justification. It does not 'justify' planning at all as in any way serving social needs; there is no immediate or ultimate conflict between the state and the market — planning is simply on the side of big business. The theory may serve a totally negative end, the only appropriate action being to throw stones or dismantle the state.

Explanation. Its suspicion of the local planning bureaucracy may lead it to uncover the real issues at stake in a local planning scheme — and thus to play a real and positive role in participation or opposition to a particular development. Its mistake is to generalize from the specific case to a theory of a totally oppressive state.

It should be made clear that it is only the theory itself (the 'state as oppressor') which is identified as an ultra-radical view. It would be simplistic and unfair to attach such a label to particular individuals. Thus those who in some of their writings appear to subscribe to this view of the state at other times write or act in contrary directions, and have often contributed substantially to the development of understanding of the theory and practice of planning. Pahl, for instance, has been influential in the debate on the social effects of the Greater London Plan. Harvey (1973) has written an influential book *Social Justice and the City,* which describes how he, as a geographer, became dissatisfied with the new 'quantitative revolution' in geography in the 1960s with the stress on techniques, and scientific rigour (see Harvey 1969; Chorley and Haggett 1967), and arrived at a new questioning of the conceptual base of geography. Should it not be discussing the spatial implications of *social* relations, relations between people, rather than relations between 'things' or 'factors' like land, buildings and space? Much of his book bears on the issues discussed in previous chapters, but lack of space precludes a full critique of Harvey's work with its liberal and controversial use of Marxist terminology.

6 Systems and cybernetics: technique or social theory?

The author's own work has been in what is sometimes called urban systems theory or urban model building. The term covers the newer, 'more rigorous' scientific methods which have been used to study urban development processes. There is nothing remarkable about this; formal mathematical techniques and computers have invaded many non-quantitative fields of study. The 'systems approach' will be discussed in Chapter 6. By themselves mathematical relationships and cybernetic concepts are neutral. It is the social and economic interpretation of them and the use to which they are put which is crucial. Formal mathematics can be used to serve almost any type of social and economic theory — Marxist (e.g. Morishima 1973) or otherwise. It is when the mathematics or cybernetics itself is put forward as a social or economic theory that we have to worry. Very few 'model builders' are so bold as to make this claim directly. It is more usually a sin of omission where the social and economic interpretation of the mathematics is left unsaid.

Forrester's (1969) *Urban Dynamics* is an excellent example of a particular technique 'dynamic simulation', which has been raised to the level of an urban economic theory, developed to the point where specific policies (e.g. of providing fewer houses for the poor) are advocated. But it is the non-scientifically trained planners or sociologists who are inclined to embrace uncritically (or totally to reject) the neutral accounts of systems techniques such as those by McLoughlin (1969) or Wilson (1974). The critique of systems theory by sociologists such as Bailey (1975) and Simmie (1974) seems to embody a basic misunderstanding. They reject systems analysis and computers used by state planners in the same way as they react against planning itself. They seem overawed by the very terminology itself, so that it is assumed that the word 'system' necessarily implies some well-oiled complex engineering machinery in which every component has its assigned place and allotted function. By this means they equate systems analysis with the so-called 'functionalist' school of sociology propounded by Parsons (1951), where society functions because everything is in its allotted place, performing its own specific function in a harmonious, non-conflict, cooperative society

Buckley (1967) has shown how 'systems concepts' can be used to describe many different kinds of systems (i.e. societies), including those which *change* their structure. That technique is the servant of theory is an elementary but important point which cannot be emphasized too strongly. It is the *meaning* imparted to the results and to the variables and other elements of a technique which are the important thing. True, the existence of new tools means they are there available for misuse — and there was undoubtedly a tendency in the 1960s, especially in the USA, for model building to be given the status of a social theory. But the real question in this case should be to ask *why* has planning seized on yet another fashion. The answer lies in its particular role and its inability to tackle fundamental social issues (see Section 1). To attack systems analysis or mathematics as such is to tilt once again at a symptom rather than a cause.

But this is not to deny that the modelling and cybernetics field provides a new field of mystifying jargon of almost unlimited richness. Planners' use and misuse of systems analysis is discussed further in Chapter 6.

Summary

Out of the whole universe of ideas about society which influence urban planning, mainstream economics ('the queen of the social sciences') is the only really well-developed body of knowledge which covers both high theory and practical policies. Urban planners and social policy makers do not always acknowledge that they are reflecting economic constructs.

Mainstream economics views the city not as a system of production (as put forward in Chapter 3); it would regard this as a throwback to the labour theory of value. It views the city as a conglomeration of exchange transactions in a series of overlapping markets.

There are two key concepts that dominate ideas about urban planning:

1. The sovereignty of the *individual* as consumer and the idea that the value of a good or service (or a plan) is determined by the increase in subjective *utility* (or satisfaction) of the individual.
2. The idea of the *margin*: the main economic indicators of output, quantities consumed and produced and their prices are all determined by the *last unit* produced, consumed or spent. This central concept of modern economics was first developed by looking at cities — or at least at agricultural production *around* cities — and at its geographical margin.

Urban economics applies these ideas, developed into a comprehensive theory of competition, and also the more recent macro-economic theories of growth derived from Keynes, to the city. But it explicitly recognizes the difficulty of applying them to the urban scale, where the interaction between activities is intense (spillovers and externalities), and where the smooth curves of micro-economics are no longer appropriate — a single large factory (an 'indivisibility') influences the whole city. Location theory, trade theory and central-place theory all shed some light on the existing pattern of development in the UK.

Other schools of urban theory often reflect these economic theories — for instance the pluralists with their emphasis on conflict resolution, and reconciliation of competing claims on resources. The 'ultra-radicals' represent the state as an oppressor, and their arguments often coincide with *laissez-faire* economists who wish to dismantle the state. Sometimes a basic technique — such as 'systems analysis' — is mistakenly used as a social theory in its own right. But generally these new scientific approaches are the servants of theory, a point not always recognized by urban sociologists.

6 Planning practice

1 Introduction: what planners actually do

It might reasonably be asked why, in a wide-ranging discussion of the role of planning in society, a chapter on planning techniques is included. Why not leave this to another book geared to the needs of the specialist? The answer would be that this artificial separation between practical planning on the one hand and techniques on the other, is one reason why the technological revolution in planning in the 1960s went sour; there was a failure to see the requirements for new techniques in a wider socio-economic perspective. Also, it is possible to point out some of the key shortcomings and limitations of techniques by means of a non-technical discussion, which are often missed in books concerned only with the detailed workings of urban models (e.g. Lee 1973; Roberts 1974; Masser 1972). Furthermore, if we really want to understand what planning is about, it is worth trying to sketch out what UK urban planners actually spend their time doing. Most people are aware that many planners spend their time handling planning permissions (development control). This is largely, as suggested in Chapter 4, an *ad hoc* activity, separate from the business of preparing the strategic plan — the structure plan.

Structure planning, as we have seen, demonstrates in a piquant way the contradictions and dilemmas with which urban planning is faced. It has an implied wide-ranging social and economic objective: planners need to understand the processes of urban development outlined in earlier chapters. Yet planning has a chronic lack of positive power to implement development. Structure planning is sometimes regarded in the profession as a waste of time; and we have seen how few plans have yet been given statutory approval. Ironically structure planning is regarded abroad as another example (together with the new towns) of Britain's lead in planning and planning institutions.

This chapter tries to elaborate the generalizations made in Chapters 4 and 5 about the nature of the contradictions in planning. On the one hand, there is a very real day-to-day task to perform — to regulate competition, to help stabilize the urban system and to help the local authority in its wider role of providing social investment and social consumption. On the other hand, there is a decisive lack of public power to influence and implement development and produce a genuinely 'social' city. There is a need to disguise this, and to let 'present trends continue'. Chapter 5 described the way planning calls upon theory to support both sides of this contradiction; there is a genuine need to *understand* and a concomitant need to *mystify*, and a continuing tension between these two requirements.

Below we attempt to show how the need to *understand* — in a world of increasing scale of development and increasing competition described in Chapter 1 — has led to a demand for better technology and more up-to-date scientific understanding of urban processes. Much of this new technology did not fulfil its promise because, first, planning's lack of real power led it to invest the new technology (at least implicitly) with the power itself lacked. The new methods were expected to help to improve cities, to stop the over-concentration of activity in mono-use, high-profit central areas, to regenerate 'marginal' residential ghettos, and so on. That the new techniques did not meet this unrealistic objective meant that they were largely (though not entirely) rejected in favour of new, equally utopian concepts such as 'societal' planning, community action or merely better management — 'corporate planning'.

This swing of fashion pales beside that which occurred in the United States, where the attack on technique was taken so literally that some erstwhile leaders of the scientific and engineering community turned against the 'rational' approach and became involved in grassroots community action.

The way the new technology was developed and deployed to a great extent reflected a neoclassical economic view of the world. The emphasis was often on the individual consumer and the satisfaction of his or her 'demand'; there was an excessive preoccupation with the margin and changes at the margin. But there were also other special factors influencing the situation. First and foremost the influence of the United States: not only do the US multinationals have a disproportionate influence over the UK economy; in the mid- to late-1960s the USA also had a cultural and technical dominance, which has since abated. US planning technology was imported to explain and justify large-scale urban road building. This is discussed in Section 2, which goes on to show how central government imposed the methodology across the country. Section 3 outlines how present institutional structures, including professional restrictive practices, inhibits the development of expertise and often denies those with the most appropriate skills the opportunity of using them in planning. Section 4 tries to give a brief impression of what the typical team of structure planners actually does in preparing the plan, and shows how some of the better plans use fairly simple techniques to give a fair picture of the problems facing an area. Section 5 then outlines the legacy of the large-scale US land-use transport studies in the 1960s in more detail, and shows that much of their failure was due to an implicit micro-economic orientation. The section also shows how some of the methods, including those which have been applied in the UK, were often tautological — they could produce whatever answer was required. Another implicit embodiment of micro-economic concepts was the 'development-potential' method invented in the UK. This method was especially suited to justifying present trends or, at most, to containing development at the margin. However, the development and use of planning techniques is not all negative; and the last section shows how strategic urban planning itself has helped to show up the limits to and limitations of cost-benefit analysis — the technique developed by neoclassical economics for choosing between alternative plans.

2 The technical take-off: US transportation imports in the 1960s

Proving the demand for roads

We begin again, as in Chapter 1, with the growth of the national economy. If there is anything at all in Rostow's (1971) concept of leading sectors which cause take-offs in economic growth, it must be found above all in the car industry which helped to stimulate the relatively high growth of the UK economy in the 1950s and early 1960s. We have already considered how car production dominates the economy of the cities in which it is situated, and how increased mobility resulting from car ownership helps to decentralize all urban areas. We have also noted the relatively large state investment in new roads. Of course, this type of government spending creates immediate contracts and profits for the private sector — or at least for those who undertake the work (even though, as we saw in Chapter 2, this is financed by taxes, part of which were subtracted from what might have been potential profits of the private sector as a whole). Public opinion in the late 1950s and early 1960s tended to be neutral or even expansionist where new roads were concerned. A first-stage national motorway network of 1000 miles was planned and built. In the general climate of euphoria, urban road plans also mushroomed, many of them unsuited to the urban fabric in which they were sited, and manifestly unable to solve the urban transport problem. Despite the intense popular reaction against gargantuan road development, many of the plans still exist, even if in cold storage. The full story of the way these plans for massive 'social investments' were justified has yet to be told. The incipient pressure on government from a combination of car firms, the steel industry, component manufacturers, construction companies and road haulage companies must have been immense.

The immediate, 'naive' justification for urban road building by the pro-car academics (e.g. Day 1963) and also the 'environmentalists' (e.g. Buchanan 1973) was that the car was basically popular and had to be 'accommodated'. Day's was the overt 'keep the traffic moving' philosophy. Buchanan recognized that not enough urban roads could be built to satisfy the demand for the journey to work in central urban areas, but his starting point was that 'in view of the general demand for unrestricted use of the motor car, the greatest possible effort should be made to satisfy the demand.'

The proposed motorway network for London is a celebrated, long-running planning saga. The original form of the London structure plan — the Greater

London Development Plan — contained three concentric motorway rings. Given that urban-road-building costs (including land purchase) are on average ten times the equivalent costs out of town, the total cost of this system would be comparable with the entire national inter-urban network. The original scheme has been largely abandoned — and very little urban motorway has yet been built in London. Many of the routes originally designated still suffer from planning blight. In those days (the early 1960s) transport planning was almost entirely divorced from land-use planning — and still is, even though the functions were for a time combined in a single ministry. The road engineers and the planners maintained separate professional hierarchies inside the Department of the Environment but some attempt has been made to make the new-style, strategic 'transport plans and programmes' (TPPs) subordinate to the overall structure plan.

Road schemes had to be justified on some kind of planning grounds, and a series of special 'land-use/transport studies' was undertaken in all the big urban areas. Most of these studies did in the end come to justify the basic concept of massive road schemes, many of which were sketched out before these studies actually started. The most expensive example is, of course, the London Traffic Study which did indeed come to support the idea of three motorway rings for London.

Here we see demonstrated the lack of effective planning control, even of the negative 'guiding' kind, over these large-scale developments. This is not less true because the agent, pushing for these developments into the cities, was the national state. This shows how the relatively strong control of central government over local capital spending does not always act to *inhibit* spending but also to *impose* a specific sectoral pattern of capital development on the city and the special 'land-use/transport study' was part of the instrument for doing this.

It is not surprising, in view of the way other high-technology commodities have come to the UK, that the techniques for transport planning were imported into Britain from the USA. It often appears that in the USA there is a tendency to reduce what we recognize as complex social and economic problems to 'naïve' engineering solutions. See, for example, Krueckenberg and Silvers (1974) or Forrester (1969), although it is not fair to the former's useful treatise on planning techniques to equate it with the latter's exercise in cybernetic mythology.

These large-scale US transport studies also had an influence on the methods developed to forecast the pattern of land uses across the city. The land-use/transport models and the largely abortive studies in which they were developed also had an influence on UK planning techniques. This is discussed in Section 5 below.

The transport problem is, at first sight, amenable to the 'engineering' approach. The problem appears superficially to be relatively well defined. Trips to work, numbers of vehicles, and so on, are all immediately measurable, visible phenomena, especially when compared with the wider issues of social and economic development. It was not so much the actual methods used to study the transport problem as the meaning attached to the methods, the way the techniques were used and the interpretation of the results which can be criticized.

The 'engineering' technique of transport planning proceeds according to a well-defined series of rational steps and codified procedures which usually involves massive data gathering with household-survey exercises, roadside interviews and traffic surveys. Patterns of travel behaviour are examined down to a minute level of detail. The procedure is described in Meyer, Kain and Wohl (1965) and by Lane (1971). The whole exercise is dependent on external forecasts of the total growth in car ownership for all the different types of household in the city. Then the procedure focuses on the city in question, and four main stages in the study follow a logical sequence.

First, the city is divided into many (often hundreds of) very small zones. A household-interview survey of the whole area is undertaken to determine how many trips are made from each zone, per day or per week, for different purposes, such as work, recreation, shopping, and so on. This is called the trip *generation* stage.

The second stage is to determine where all the trips are going. From any one zone there are hundreds of choices; i.e. all the other zones. Some of these are further away than others; distant zones are less likely to be visited than the nearby zones. This is the *distribution* stage. This involves a computer calculation to find the shortest routes in the transport network between the various zones.

Having found out where all these trips are going, the process then looks at people's choices of travel mode. The basic decision for those who have cars is whether to travel by public or private transport, depending on the relative time, cost and convenience of the journey by each. This is the *modal split* stage. This stage is clearly crucial for the whole future of the city's transport system—and whether it is going to be orientated to private or public transport.

Finally, all these trips by public and private transport are loaded into public or private vehicles, and routed through the transport network. This is called the *assignment* stage.

The final result of this process is a road and/or rail network loaded with cars, buses and trains. To be more precise, in the UK transport studies, the networks were massively *overloaded* (once the growth of car ownership was projected into the future). We can immediately see the opportunities for justification of big road schemes once we understand some elementary facts about the journey to work in urban areas, especially the following two important characteristics.

The two urban activities (residential and workplace) at the origin and destination of the trip, have markedly different spatial distributions over the city, as we have seen. Residential is dispersed and extensive; employment is concentrated in city centres and at a few key points such as major manufacturing plants. Second, there is a morning and evening peaking phenomenon as workers travel from dispersed residential areas to concentrated workplaces. If all those with access to a car (over half the households now own cars) tried to use it for the work trip, this would require a road capacity which would far exceed any conceivable road-building programme. For London 90 per cent of trips to the centre are by public transport. No amount of road building could ever take more than a tiny fraction of these trips. So for large cities all solutions to the transport problem must be based first and foremost on public transport. These elementary facts explain why the results of these massive transport studies always showed *massively overloaded urban road networks.* Most of the studies used subsidiary procedures to 'restrain' traffic, once the demand was shown to exceed supply.

The methodology of these studies was based implicitly on giving priority to individual preference — the choice of destination and travel mode by the individual, rather than the design of the best and most efficient transport system for the city (and hence all its inhabitants) as a whole. This preoccupation with individual consumer choice does appear to reflect the micro-economic theories described in Chapters 2 and 5. In fact the very 'models' used at the traffic-distribution and modal-split stages can be derived directly from utility maximization and marginal demand curves (Neuberger 1971). The choice for the city is seen largely in the same way as a single individual choosing to make a travel decision. Thus, first decide when to make a trip, then decide where to go, and then how to travel. But we have already seen in Chapter 5 how the perfect-market model (where everything is *relative*) often comes out with global aggregate results (e.g. full employment) which are not valid. It is a fundamental principle of any system of interacting parts that the whole is not merely the sum of the individual parts. In the traffic system there are a large number of effects which take the system far away from the assumptions of the perfect market. Firstly the system is a general-equilibrium problem, involving many goods (trips) rather than a partial-equilibrium system — though this problem has been ingeniously formulated by several economists (e.g. Neuberger 1971). Second, the system is riddled with externalities; demand and supply interact through congestion on the road network and the demand curves are therefore not independent. This down-to-earth example of transport planning shows in a very immediate sense the limitations of the conventional economic theory — how an overriding concern for the individual consumer leads to an inability to plan effectively for the wellbeing of the city as a whole. To caricature Eversley's 'know them by results' philosophy, urban motorways are a product of a misapplication of micro-economics. (A *mis*application because it has sometimes been said that the proper application of neoclassical cost-benefit techniques would have shown many road schemes not to be socially viable. Nevertheless, since the market demand for road travel undoubtedly exists, any rejection of road building would need to be based on some kind of market valuation of those social diseconomics which the market did not value. This immediately leads to problems of indirect measurements and artificial manual adjustments ('fiddles') which are a feature of many applications of other types of new planning technology (see below). Cost-benefit analysis is discussed more fully in Section 6.)

Central control: imposing a method

This imported transport-planning technology was implanted into the existing structure of UK public-sector institutions described in earlier chapters. Central control was very strong, and no urban area or local authority was allowed to produce a transport plan unless it was justified by a study, of the type approved by the ministry. But if the state saw the need to employ this new planning technology, it did not see fit to invade a potentially profitable private-sector activity by actually applying the technology itself. The justification for this was that expertise in the use of complex computer programs was restricted to a few professional engineering consultancies, most of whom already had experience in the USA.

This again reflects Shonfield's 'separation of powers' tradition in the UK; the study consultants would provide independent advice — supposedly independent of government and the private sector. But we do not need to look very far for Mattick's 'superficial collaboration' between the state and private interests. The independence of the advice might have been open to question, since many of the transport-planning consultants were also consultants on the road-building and bridge-construction projects which their own studies helped to justify.

There is no suggestion that these arrangements were not made in good faith, nor that at the time, given the undeveloped state of transport planning in the UK and the lack of expertise in local authorities, there was any practical alternative. All we are trying to show is the concrete effect that these institutional arrangements actually had on the development of planning. There is a reasonable case to be made that the main effects of these studies were to justify road schemes which had often been decided beforehand on what were narrow or dubious grounds. The way the new planning methodology was applied, by a restricted circle of consultants, did not help to question or improve transport-planning methods as it should have, or increase the understanding of the implications of these transport plans on UK cities. In this sense, the studies did not justify the considerable sums spent on them. The largest was the London Transport Study, Stage 1 of which cost around £1 million at 1965 prices. Most of this money was spent on gathering detailed travel information on trip generation, distribution, modal split and assignment described above. London was divided into 1000 zones, and an army of survey teams was recruited to undertake household surveys and interviews, roadside traffic counts and roadside interviews. The resulting millions of data items posed a massive problem of storage, manipulation and retrieval, even for a sophisticated computer system. Much of the subsequent analysis involved trying to sort and manipulate these data rather than effectively testing a wide range of meaningful transport alternatives. At the end of the day, for this outlay, only the single 'motorway box' plan was actually subjected to a fully operational test using this transport-planning computer package, so complex was the procedure required to test a single transport plan.

Undoubtedly much of the experience gained has since been useful and has been of benefit to London transport planning in the longer run, especially after the policy emphasis swung away from roads. The consultants, Freeman, Fox and Partners, with American associates Wilber Smith, employed talented engineers on the project, many of them steeped in the US tradition of transport planning. They brought typically US skills of organization, attention to detail and rigorous technical competence. But far too much weight was given to assembling and manipulating data and the difficult problems of running the computer package, and too little attention was given to the overall policy implications and to seeing that a sufficiently wide range of plans would actually be tested at the end of the day.

All the other conurbation transport studies ran on very similar lines, and it did seem at the time that the basic methods and techniques advanced hardly at all, and did not justify the expenditure of several million pounds. Although it has been argued that the expenditure was small in relation to the transport investment involved, nevertheless it was very large compared with the total professional expertise brought to bear on new computer-based techniques for the non-transport side of urban planning.

Very few books on land-use planning or transport planning discuss the role of consultants in any detail. It is obviously a very sensitive area. Yet understanding their role is important to an understanding of the way urban planning and transport planning have developed during the last few years. Centres of professional planning expertise outside and independent of either central or local government are essential to the undertaking of specialist, one-off tasks involving new methodology or specialist studies with a limited time scale.

Consultants have to operate (at least in part) as private firms; they are in competition with each other for income and they have to sell their products on the market (in this case, the product is professional expertise), just as any other firm. This is, of course, limited by professional codes of ethics. In the transport planning of the 1960s, consultants had a monopoly of technical knowledge and expertise and were the only bodies allowed to carry out transport studies; local authorities were generally barred. There was therefore little or no incentive for consultants to share their knowledge or their computer programs or to spend time and money improving the methods. (Since then, however, new types of collaborations have been instituted between consultants and planning authorities — and authorities themselves have become more involved with the actual methods of transport planning.)

The overall lessons of the modern transport planning experience are as follows:

1 The use of high technology does not necessarily guarantee an improved understanding of the issues at stake in planning a city, and in this case the technology was used to produce plans that tended to benefit road users rather than the city residents.
2 The transport models essentially embodied an individual utility-maximizing approach, i.e. satisfying the 'demand for travel'. This is not the appropriate way to produce satisfactory results for a whole

interacting system such as a city. The expenses and excessive data needed to determine the 'real' behaviour of the individual, served to obscure many of the key strategic issues which a more aggregated approach would have revealed.

3 The institutional framework of central government control gave consultants almost a monopoly of the technology and of the power to carry out studies, and this helped to impose a uniform pattern of justification of urban road schemes across the country and inhibited a wider and broader-based critical attitude to these schemes.

(The Planning Research Applications Group (PRAG) was set up essentially as a pilot experiment to try to develop fruitful ways of applying new planning technology. This was to be done by direct collaboration with local authorities, so that policies, practice and research could be developed together (Broadbent 1975b). Some of these ideas were embodied in an earlier Ministry of Transport initiative in 1967. It was felt that there was considerable demand for a small, easily used computer program which would undertake the traditional transport study in one relatively easily used program which did not need massive data inputs, huge computer installations or large numbers of professional programmers and systems analysts to operate it. This was the so-called COMPACT program. It was subsequently released by the Ministry of Transport to all local authorities and transport-planning consultants, and used in several transport studies (Mackinder 1972; Broadbent and Mackinder 1973).)

3 Planning expertise: 'development control'

The traditional urban planning field (i.e. land-use planning rather than transport) being an older activity, not so dominated by technology, the role of consultants, while important, has not been as significant as that of the professional body, the Royal Town Planning Institute. Since 1947 the RTPI has had an increasing control over the job market for planners and over their education. We have already shown how the contradictory role of planning and its lack of power to implement development, leads to a need for better understanding of urban development processes as they become more complex, on the one hand, and to a need to disguise the lack of control over the processes on the other. Added to this, 'town planning' as a subject is a very ill-defined and elusive body of knowledge, theory and expertise. There is a point of view that much of what planners do (e.g. development control) could be done by anyone with a good general education and short specific training. Much of the rest of what planners do (e.g. economics or systems analysis) needs much more specialized skills than the standard town planning course provides.

The question set by the Institute in examinations show up some of the deficiencies in planners' training. The questions tend to be abstract and over-general, especially with regard to the true role of planning, and its powers, in relation to the market and to the national economic environment, as the following examples from a written examination set by the Royal Town Planning Institute show.

> Planning theory:
>
> 'Discuss problems of long-range forecasting in relation to physical planning'.
>
> 'The land-use plan reflects an analysis of urban activity systems' — discuss
>
> Planning techniques:
>
> '. . . what use is census data to a local planning authority and how would you decide upon the form in which you would wish to acquire the data?'
>
> 'The Development Plan Manual describes seven functions of Structure Plans. What are they and what interrelationship is there between them? What matters would you expect to see covered in the statement to the Structure Plan of a large rural County?'

Other professions, the civil engineers or the medical profession, have a much more established body of knowledge and a relatively well-defined standard of professional expertise which justify the existence of a professional body to advance and maintain standards.

Because of its relatively insecure foundations, the town planning profession has a tendency to be looking over its shoulder for possible encroachments on its own preserve and area of influence. It is extremely quick to latch on to new fashions and to adopt new procedures which may enhance its status *vis-à-vis* other professions, especially those in local government — treasurers, engineers, surveyors, social workers, etc. The 'systems approach' has been one such fashion. The more recent vogues for corporate planning, community involvement and 'societal planning' are others. The danger with fashions is that far too many expectations are built on to the fashion, and correspondingly there is some disillusion when 'urban models' are seen to resolve none of the fundamental contradictions of urban planning in the UK. But because the planning profession is strong and established in the local and central bureaucracy, and because the UK urban planning system, although having little direct power over development, is institutionally well developed, with uniform statutes and regulations and procedures from national to local level, each fashion as it comes along also tends to be absorbed and institutionalized. While the critical enthusiasm for the new fashion may be misplaced, the inevitable reaction does not therefore result in a complete abandonment of the new ideas, as is often the case with the even more violent swings of fashion in the USA. Thus the urban-model-building wave in the UK, while not so ambitious in the first place, has, at least in some respects, been consolidated and institutionalized. Lessons have been learnt and the achievements, though modest, are positive.

The Institute's grip on training, and the practice of planning skills is fairly complete, but the continuing tendency of planning to be in crisis has been reflected in continuing soul-searching by planners at large and by the Institute. In recent years, the Institute circulated its membership asking for reactions to alternative proposals for its future. Should it become more like a learned society, opening its membership (and by implication the job market) to those trained in other disciplines, and not lay down such stringent but narrow apprenticeship requirements — experience at the planning grass roots. Not surprisingly, the membership voted to retain its privileges. Another source of conflict has been with the various universities and colleges which run courses and require 'recognition' by the Institute to make their courses attractive to potential students. Interestingly, the main bone of contention has been over the inclusion of significant social and economic tuition in courses. This raises the essential point about the training of planners. Are they being trained to know or appreciate a wide range of topics in a general-knowledge sense; or are they being trained in the exercise of specific skills? At present it is very emphatically the former. It is argued in this chapter that much of what is wrong with recent work in structure planning and the development of new techniques is because the skills needed are in fact not being brought to bear. Experience in giving backyard planning permissions and a generalist planning education is no substitute for a proper training in a numerate, scientific subject providing identifiable skills which help the student to set about comprehending, understanding and measuring a complex system such as an urban economy (see also McLoughlin 1969). This problem grew from the serious to the severe when the demand for planners suddenly doubled on local government reorganization, and the top positions were restricted to members of the profession, many of whose skills and training were manifestly not up to the task.

Professional planners have their own hierarchical structure within central government. There is a Chief Planner who heads the profession inside the Department of the Environment; knowledge, practices and standards are at least partially monopolized by the profession operating at all levels of central and local government and in the colleges and universities. Part of the struggle to bring transport planning within the ambit of the structure plan (i.e. to make the case for roads subordinate to the wider socio-economic and land-use issues) has been a struggle between two professional groups (the planners and engineers) within government, each with its own specific control over knowledge and technique appropriate to its own subject.

Not surprisingly, given the origins of town planning sketched out in Chapter 4, many of the planning consultants were originally architects with a lack of engineering or scientific expertise, in contrast to the engineering-based transport-planning consultants. They were also often much smaller in size but, as in the case of the engineers, there have been periods of major growth and decline in the number of planning consultants in the UK, in the short-term booms or slumps in planning activity stimulated by central government. Since the war, there have been one or two major injections of cash into the consultancy sector. The design of the new towns in the 1940s and 1950s gave birth to several consultancy practices.

In the 1960s, when population growth was still expected, and the debate which finally resulted in the 1968 Planning Act had already started and the impact of the car on cities was beginning to be taken seriously, there was a new wave of subregional studies and also a series of special studies to examine the feasibility of building large new cities to cope with the expected increase in population. Some of these studies used consultants. The government of 1970–4, in line with the current vogue for channelling resources to 'where they are really needed', set up half a dozen special projects which were to concentrate on the key problem areas of the 'inner cities'. These again were largely undertaken by private consultants. (It is worth noting in passing that these studies reasserted the 'marginal' philosophy of micro-economics: 'deprived' areas are by definition on the economic margin — see the Coventry report noted in Chapter 3.)

Although in the past many of the prestige plans have been undertaken by consultants — such as the new city of Milton Keynes — the consultancy sector in the UK has been whittled away and it is now very small indeed compared to the public sector. It is estimated that the public planning system costs £100 million to operate at 1974 prices and employs about 6000 professional planners. The main private consultants probably employ far fewer than 1000 professionals all together. Salaries in the public sector escalated wildly as the demand for planners was almost doubled when local government was reorganized in 1974. The structure planning system itself has tended to reinforce the decline of the private sector by confirming the right of every local planning authority to produce its own 'consultants'-type report. British consultants, in both land use and transport, have recently turned abroad for much of their business, although there has been some resistance from the new customers to the apparently excessive enthusiasm of British transport planners for urban motorways.

4 Preparing a structure plan

Having noted that the main impetus behind the development of the new planning technology was generated by transport planning, and having described and compared the way knowledge, technique and expertise have been deployed and controlled, before moving on to discuss the details of the new planning technologies and their use and misuse, we now give a brief outline of the way a team of planners in a local authority undertakes the task of preparing the structure plan.

We have already used the South Hampshire Structure Plan to show that, however much of an advance in principle the written statement is over the old county maps, the new structure plans so far published are far from ideal. At the time of writing, some twenty-five such plans have been published. They are extremely variable documents. At one extreme — long and complex — is the South Hampshire plan already described. At the other, the East Sussex plan is brief and concise to the point of being almost telegraphic. This latter plan, or parts of it, is intended to be reviewed every year — a far shorter period, of course, than it takes the Department of the Environment to process the plans. There is already considerable worry that the new planning process is exhibiting as much time delay as the old system, and consequently circulars from the Department of the Environment (as well as the 'Dobry Report' — Department of the Environment 1975) encourage planners to speed up the whole structure planning process and to concentrate on relatively few key issues instead of covering a whole range of topics in depth. Unfortunately there is not much advice available as to how this should be done.

Our team of structure planners have an unenviable task. In a sense, it is a task they have to define for themselves. Because the structure plan itself exercises very little direct control over which land will actually be developed (other than in the sense of persuasion or negotiation), the approach adopted may depend largely on the whim of the particular Chief Planner.

We have seen in South Hampshire how tempting it is to retreat into obscurantism and techniques to justify a 'no alternative' — preferred strategy which lets present trends continue. It is also interesting to see some planners defining alternative development strategies in terms of 'old-fashioned' spatial blueprints such as linear cities, star shapes, radial cities, etc. There is relatively little concern with social and economic options — public versus private transport, rehabilitation versus renewal, service industry versus manufacture. These are now fashionable concerns, but again local politics probably limits planners' ability to state such options explicitly.

In our typical structure planning team of twenty professionals, most will have the typical planner's education as a generalist and will not have much statistical, economic or other specialism to bring to bear on the task. Many of the team will think it extremely important to use the structure planning exercise as an opportunity to spread the influence of the planning department throughout the local authority, and there may be a series of interdepartmental seminars or brainstorming sessions (as in Liverpool), where aims and objectives are discussed, policies reviewed and other departments urged to become interested in the plan. In fact, we can see here how the structure plan has become a concrete object under the control of the planning profession, making up to some extent for its lack of control over real resources compared with other local government professions. In this sense the structure plan has been a boon to the profession, extending, by implication, its influence over a wide range of subjects.

Most plans attempt to follow the stages in the structure planning (Figure 38) laid down by the Department of the Environment (1971).

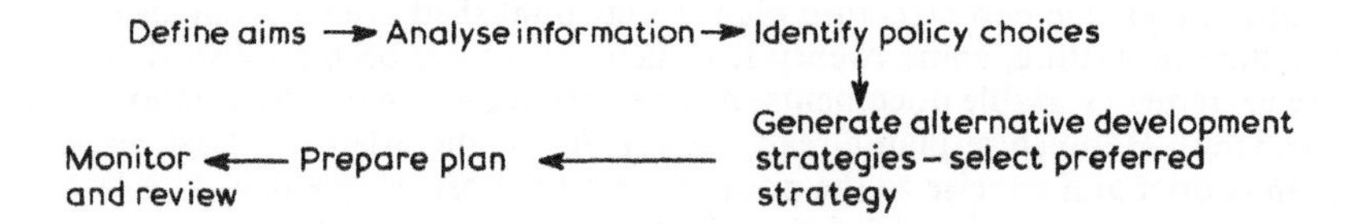

Figure 38 The stages in structure planning

These stages often proceed in parallel, and may be repeated several times — 'the cyclic planning process' (Boyce, Day and Macdonald 1970; Cordey-Hayes, Broadbent and Massey 1970). There is a requirement for public participation during the process, so that, in theory at least, the public can influence the choice of strategy. After the plan is published there is a formal legal examination in public, which again varies in content and quality. Sometimes key problems are thrown up (e.g. Staffordshire 1973) and more fundamental economic issues raised (e.g. by Coventry Community Development Project 1975). But often the examination in public is attended by no one else but planners.

The members of the planning team will be allocated to various tasks; they will often split into groups concentrating on separate stages of the plan or different subject areas and techniques. In view of what has been said, it is not surprising that the teams often become fragmented and explore different issues or subjects separately — often in far too much detail. The resulting plans do not show how all the subjects (i.e. the economic activities) affect one another (Jaffe 1976). The worst that can happen in this situation is that the whole exercise degenerates into a collection of rather academic *ad hoc* studies proceeding on their separate ways, and bearing very little on the central problems and issues of the area.

There is often a tendency to go in for large-scale information-handling exercises, for example, enumerating every single property in the area, or developing computer data files on every single employer in the area. This kind of information fetish also parallels the gargantuanism of the US transport and land-use studies discussed in Sections 2 and 5 of this chapter. The fact is that the very obsession with detail, and the effort to make sure that every last firm, property and individual is accounted for, absorbs all the attention, to the exclusion of any considered attention of the major problems. If 15,000 people work in one or two large factories there is very little use in chasing up the last 500 one-man businesses in the area.

But we have suggested that the structure plan is an advance in planning procedure. Despite its institutional and technical limitations, there are many positive aspects to these plans. It would be quite wrong to suggest that all plans have the same weaknesses or that they do not sometimes present some kind of cogent statement of the issues at stake in local areas. It was said in Chapter 4 that the best plans are those which deal with 'problems' rather than those which begin with abstract objectives. The best of these plans rely largely on the clear presentation of key facts, together with relatively simple techniques of analysis to explain and suggest what is right or wrong with the area and what might or ought to be done about it. The Birmingham (1974) plan and the Merseyside (1975) draft report of survey are good examples.

The Merseyside plan gives a coherent picture of the social and economic deprivation of the area, the decline of its traditional port-related industries, and the recent development of new industries which, with the help of national regional policy, has helped to bring the broad balance between manufacturing and service industry nearer to the national average. This has done little to alleviate the chronic social malaise in the area. The workforce is extremely 'unbalanced' with too many unskilled manual workers in comparison with non-manual and professional groups. There are few significant inter-industrial linkages which might offer prospects of further sustained growth in the future, or of stimulating a training programme to 'improve' the labour force. The condition of the housing stock is very poor, with up to one-third of the dwellings needing improvement or demolition. Unemployment in Liverpool is twice the national average (touching 20 per cent in early 1975 for non-manual workers who make up one-quarter of the labour force). The area continues to have a large out-migration flow, most of which involves the age group twenty-five to thirty-five, but because of its high birth rate it still has a young population. Within the county there has been a decentralization from the centre which (as outlined in Chapter 3) has left the central city and especially the inner areas with chronic unemployment and bad housing.

The county planners are especially concerned to attract further new industry, to convince central government of the need to invest more resources in housing and jobs in the area and to redirect investment from the neighbouring new towns which are seen as one reason for the continual drift of the 'best' people out of the area. There are gaps in this picture which could be filled with closer analysis, and some of the authority's analysis of the land-use problems and the effects of its own very large-scale demolition programme of the 1950s and 1960s might become more questionable on closer analysis. Nevertheless here at least there is a coherent starting point — a statement which is published and available and which gives a broad picture of some of the issues at stake. It amounts to what might be called a preliminary, aggregate, system-wide analysis.

Has the technological revolution in planning techniques helped this kind of presentation? On the surface perhaps very little, certainly not much, in the immediate sense of running complex, large-scale models of the sort discussed in Sections 2 and 5. Nevertheless, on closer examination a good number of formal (though simple) methods were used, and there is evidence of a good general capability of how to analyse a complex urban system — a capability which was probably not available before planning was subjected to influence from numerate disciplines. The planning team included several people with experience in the model-building field, gained not through the professional training of the planning schools but in research.

The methods used in this early stage of structure planning deal mainly with the relationships between the totals of the different types of activity and the broad patterns of land use across the area; it is an *aggregate* rather than a detailed, disaggregated approach. The 'elementary' techniques used for this purpose include a liberal use of simple totals (i.e. of activity levels, employment, population, land, houses, etc.). ratios (e.g. population density), averages of various kinds (e.g. average age of majority), complex ratios involving combinations of variables to measure the size of the economic base and the relative specialization and concentration in the economy, and finally more complex distributions of all these variables and combinations of variables, their trends over time and deviations from trends. Many of these elementary operations on data can be used to provide useful indicators, not only of the superficial structure of the system and recent changes in it, but also to provide indicators for different development theories. Obvious examples are the concentration and specialization measures (Palmer 1975) which can be related directly to theories of competition and the economic base. The point is that it is not so much the *level of detail* which is important, but rather the way even elementary and aggregate information can be used to *increase understanding*. 'Detail is not the same as 'understanding'.

When it comes to the future, by far the greatest effort goes into 'forecasting' what *might* happen, on the basis of either present trends or expected changes. 'Components of change' techniques, 'cohort survival' models, economic base, input-output, spatial allocation and transport models are all used to explore the future. True, they are sometimes used to test the implications of various actions by the local planning authority, but this is the exception rather than the rule. All this is a reflection of the lack of certainty in national economic and population planning, and the lack of direct powers available to the local authority to influence development in a direct manner in its own area.

Special mention should be made of the 'development-potential' technique, which has been popular in recent years for deciding the best locations for new development in areas where growth and expansion was still expected. This technique is very similar to the old 'sieve-map' techniques, whereby planners used to prepare a series of maps, each map displaying just one factor which inhibited development — areas of high slope, bad drainage, and so on. When all these maps were laid over each other, they showed the area which was suitable for development.

In the development-potential method, the area is divided into a number of small zones and each of the previous sieve maps is replaced by a so-called 'surface'. Each zone (and there may be several hundreds in a county) is given a score on each of several factors which might affect development. These will usually include accessibility to communications, services and employment, and the character of the land to be developed (e.g. good agricultural land). Each of these factors either enhances or limits the possibilities of development. Various weights can be attached to each. If the objective of the overall plan is to conserve agricultural land, then the agricural 'surface' will have a high 'negative weight' as a surface. The surfaces are added together using different combinations of weights to generate different strategies; the outcome is a combined development-potential surface or map which attaches a high or low number to each zone showing its overall potential for development. Not surprisingly, such a surface often shows that high potential exists where services are near, where employment is within reach and where there are good communications and where the land is not needed for agriculture. The best such areas are where development already exists (Coventry City Council 1971). This is not a very interesting result, and there is usually a final stage of analysis which subtracts all areas *already* developed. The indications are then that development is most suitable *at the margin* of existing development. Used in this way, the technique reflects almost precisely the present role of land-use planning — that is, restricted to guidance and control *at the margin* in the face of the market forces which drive development forward; it indirectly reflects the marginality theory of mainstream economics.

This method does not reveal the reasons why one activity should want to be near another. It involves no clear representation of the demand-and-supply relations for goods and services between activities, and it carries within it no explicit mechanism of competition for space and for location. The method also provides opportunity for 'manual adjustments', so that the strategy to emerge will be desirable — as with the Lowry-type model described in the next section. The way all the different factors are measured using combinations of different types of quantities — 'acres' of agricultural land, with 'average' distance to employment' and the *ad hoc* weighting procedures — provides opportunities for such adjustments.

On the positive side, this method does at least represent an attempt to consider *all* activities together, and relate them (through the weights) to the overall objectives of the plan, but the development potential method as it is now used, suffers from and tends to reinforce and justify the drawbacks of the negative, 'marginal' planning system.

5 The use and misuse of planning technology

US gigantism and its requiem

This section attempts to focus in more detail on the effects and after-effects of the technological revolution in planning techniques which occurred during the late 1960s and early 1970s. Although it was in and through the transport studies outlined in Section 2 that the new technology was applied with a uniform code of practice in true engineering fashion, 'land-use forecasting' techniques of enormous detail and complexity were also being developed, again largely in the USA. By the early 1970s disillusion had set in, and the modelling movement was widely regarded to have failed. In the US *Journal of the American Institute of Planners* a seminal article by Lee (1973) put the seal on this judgement. By this time a whole generation of talented US model builders, systems analysts and economists had taken to the streets in penance for their misdeeds. We have already tried to explain some of the factors involved in this — why planning was prone to seize on anything which might provide an answer to its agonizing dilemma, and even more prone to cast it brutally aside when this answer was (inevitably) not forthcoming. The belief that the diligent application of science and technology could solve all problems was widely prevalent in the 1960s, and was not solely confined to planning.

McLoughlin's (1969) book on a systems approach to planning crystallizes a view gaining widespread acceptance among the avant-garde at the time. It is a highly readable, enthusiastic (although somewhat subjective) account written almost in a spirit of optimism. It asserts the ability of 'man' to solve his admittedly enormous problems, by rational development and use of the resources to be found in the natural environment. This celebration of 'man in his ecological setting' reflects the widely held view at the time that systems analysis was only a logical development in an uninterrupted progress of man's increasing ability to control the environment. Few at the time believed that the long boom was over, and the convergence theories of Galbraith (1967) and the 'end of ideology' were widely proclaimed. McLoughlin's account of the American methods and models for forecasting different activities is one of the best and clearest introductions to the new methods developed in that decade. He talks of systems control, feedback, 'requisite variety' and many other basic cybernetic concepts, which are still being developed and applied in planning and elsewhere. Where the book reveals that it is a product of its time is in the absence of any concrete socio-economic context, or of any consideration of the social and economic barriers to the successful implementation of the systems revolution. With the benefit of hindsight we can now begin to assess these wider issues and also show some of the important technical and theoretical shortcomings of the various methods that have been revealed by experience. Other accounts which can be regarded as standard works include Wilson (1974), which is more conventionally academic and includes much of its author's own original research on a statistical theory to underpin models which had originally been developed as *ad hoc* empirical tools. Kreuckenberg and Silvers (1974) is a careful, step-by-step teaching book.

The landmark piece of literature in the reaction to the systems revolution was Lee's (1973) 'Requiem for large-scale models'. The more modest UK effort in model building had been marked by a sporadic, though consistent awareness that some degree of caution was required, especially with regard to model design problems, which the largely abstract accounts of the most ardent US advocates seemed to ignore. Some of this work (in which the present author was involved) pointed to these problems, in fairly neutral academic language (Broadbent 1970) which did not have the stridency or impact of Lee's comprehensive indictment. In summary, his case was that in practice none of the large-scale land-use prediction models had achieved any of the goals that they had set for themselves. On the rare occasions when they had been anything like fully tested, their results were unrealistic. They had assumed that, by trying to do several things at once, they would do each more efficiently than if each had been tackled separately. Thus they were over-comprehensive. They were also over-complex and over-detailed; the way they actually worked remained a mystery. They absorbed vast amounts of data and needed expensive surveys; there were often more numbers in the model than people in the system (cf. the London transport study described in Section 2). On the other hand the behavioural models they incorporated were only valid in general terms and the results produced were far too coarse to be of any use to local planning and development control in small areas. Finally most were not really derived from a cogent theory of location or behaviour. Above all, the models had to be 'massaged' into working order; this involved making many 'adjustments' so that the answers seemed reasonable. This meant that often the inputs determined the outputs, so there was a large degree of tautology inherent in their structure.

Most of these conclusions seem, superficially at least, to be valid, but some of them may lead to some misunderstanding. Far from not having a theory, nearly all the techniques incorporated a highly 'individualistic' neoclassical, micro-economic approach which presented the whole of the urban system rather as a large-scale projection of an individual consumer. More and more variables and data were built in to provide a more and more rigorous specification of all the relevant factors needed by the idealized consumer in the choice of location. This explains the relative neglect of the Keynesian-type aggregate approach which, as Lee points out, is the ultimate example of a relatively simple but useful and usable model. It also explains the excessive detail.

A land-use model in the UK

We now try to show, with reference to one specific and well-known urban model, the nature and implications of the new methods employed by planners in the UK. On the one hand there is the improved understanding — the explicit description of urban processes. On the other hand there is the considerable potential for mistakes and general abuse, and the possibility of using the technique to justify any policy — and most specifically a 'let present trends continue' policy. The basic model, invented by Lowry (1964), has been in existence for ten years and, although it has been much researched, there has still not yet been an adequate explanation of its relationship to urban economic theory, of how it behaves in a practical situation and of the possibilities for misuse. The best detailed scientific investigation of its characteristics and sensitivity is a recent application on Teesside (Cleveland) by Barras (1975).

Broadbent (1970) outlines some basic design problems of the model, and possibilities for misuse. It is possible to give here some idea of the implications and pitfalls of using this type of model, to which even the more overtly research-directed books sometimes give insufficient attention.

The function of the Lowry technique is to produce a geographical or spatial pattern of residential and service activity (or at least a 'demand' for that activity) across an urban area — given a fixed location of basic employment centres. In Chapter 3 we made a distinction between national-serving activities (usually big firms) and the other local activities which depend on these, principally the residential (labour-producing) activity and then the local consumer-serving activities (retail) and the local 'social consumption' services. This basic division appears also in the conventional theory of the economic base described in Chapter 5. This is exactly what the Lowry model does, but it takes the assumptions a stage further, assuming not only that the total size of these local activities depends on the nation-serving sector but that their spatial distribution also does. There is no transport cost-minimization process such as outlined in Chapter 3 for the generation of rent. The model makes an assumption similar to the traffic distribution technique in the transport model described above — namely that the further apart activities are, the less they interact. In spite of its relatively simple causal structure, the model can get quite complicated when written down in mathematics, with many lines of equations employing abstruse symbols. But the results of the model can be displayed extremely simply, to give an apparently meaningful and plausible picture of the pressure for residential and service development at different points in the city.

But this is where the main danger lies, for on the one hand the model is a set of equations which appear to represent nothing but a mystery, and on the other hand it is an extremely plausible set of results which a planner can put before his committee of politicians to justify putting in half a million square feet of shopping in the city centre. So a first requirement must be to expose a critical assessment the very innermost workings of such a model.

But even this is not sufficient. The inner workings of this model can be described in a very simple and plausible way (as, for instance, by Lee (1973) in what must be the most grossly inadequate and yet depressingly plausible presentation of urban models). Such an explanation can be presented as a series of boxes and arrows (see Figure 39), with a description which goes as follows.

First of all, the whole urban area is divided into small zones as with the traffic model. These zones can be either employment centres and/or zones of residence. All the distances or travel times between these zones (by road or public transport) can be measured and/or calculated from road distances using a computer. We first specify to the model which zones are the centres of basic employment. This gives us the total number of workers in the whole system who work in factories, producing goods for export outside the urban area (the economic-base concept of Chapter 5). To decide where these

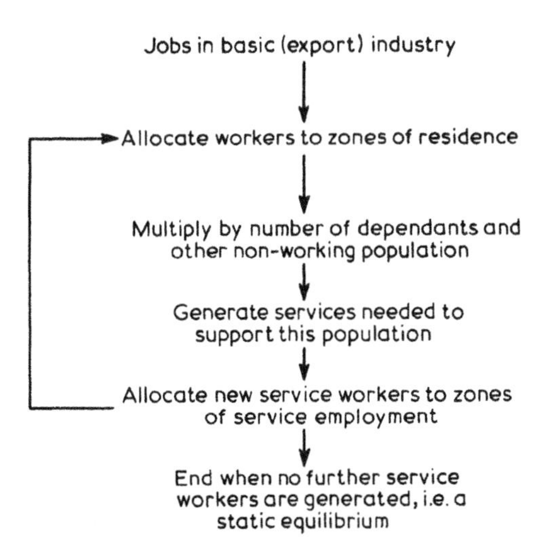

Figure 39 Elements of a residential and service allocation model

workers 'want' to live, given that they work in a fixed set of factories around the city, the model says that there are two factors involved:

1 The distance, time or cost of travel to work; the longer the distance from an employment zone to a residential zone, the less likely somebody would want to live there.
2 There may be some innate characteristic of the zone itself which makes a particular residential zone look 'attractive' to the worker.

On this basis, the model works out how likely each worker is to live in a particular zone. Each worker, now allocated to a residential zone, is multiplied by some average number of dependants (who thus also live there). This residential population then generates a need for local services — shopping and health facilities, local government, and so on. These in turn require a number of service workers to be employed in these service jobs. The service workers are then allocated to zones of service employment in the same way that 'basic' workers were allocated to zones of residence — by a combination of distance and the attractiveness of the zone. These new workers in service centres then also need allocating to zones of residence; they need to live somewhere, just as the workers in basic employment were allocated to residential zones. The process is repeated until equilibrium is reached. This is the essence of the so-called Lowry model which has been so much written about and been applied in several planning studies. (Examples include Bedfordshire — Cripps and Foot 1969; Notts-Derby 1969; Cheshire — Barras *et al.* 1971; Monmouthshire — PRAG 1973; Reading — Echenique 1973; Cleveland — Barras, Booth and England 1975.) This explanation, simple though it seems (and it has been given in several of the books cited above), is completely inadequate as a means of assessing whether this model is appropriate to a particular planning situation.

It is the *way the model is used and interpreted* which is crucial. The model itself can be as nothing if it is used or interpreted wrongly. It may even be better to have some evocative piece of journalism or perhaps to juggle a few simple statistics than to spend time and effort laboriously struggling with the computer programs and data inputs, in order to produce plausible-looking results which have little basic meaning.

There are a multitude of possible misuses of this model — all with the potential for arbitrary adjustments so that the ultimate answer coincides with whatever planning policies need justifying. There are several crucial choices which have to be made by the model builder/planner.

First, the whole system or urban area under study has to be defined. In our Chapter 3 terms, this means defining the *commuting watershed* — the labour pool within which all these different activities interact. If there is another urban area (labour market) nearby, there may be considerable work and service journeys between the two — which interferes with what is going on in the main city of interest. The author (Broadbent 1971b) outlines a hierarchical version of the model whereby two or more neighbouring areas can be identified separately, with an allowance for some commuting between them. This has since been applied in many studies (mainly by PRAG) including Monmouth, Gloucester, Cheshire and Cleveland.

Second, the way the area is divided into zones is absolutely crucial. Zones have to be small enough so that most journeys to work take place *between* them — the larger the zone, obviously the more work journeys there are which begin and end *within* the zone. In the early studies — like Notts-Derby — whole cities (e.g. Nottingham) were counted as a single zone and the great majority of work journeys occurred *within* the zone. This meant that most of the final results had little to do with the model itself, but rather depended on manual adjustments made to the results for the big cities. What should have been done was to give each of the separate labour pools — Derby, Nottingham and Mansfield — a separate Lowry model, and to divde each into much smaller zones (as was done on Teesside and Cheshire), so that the separate, distinct characteristics of each labour-market area would be revealed by having a series of linked sub-models for each separate labour-market area (e.g. Broadbent 1971b).

The most serious area of misuse is with the so-called 'attraction' factor (i.e. the factor, other than distance, which determines the number of workers likely to live in a particular residential zone). In order to see how well the model performs, most studies first of all compare the output of the model (i.e. the number of people living in each residential area) with the number who are actually living there, say as counted in the census. But the most commonly used attraction factor is usually *the number of people who live there already*. This is circular reasoning *par excellence;* Figure 40 shows how this tautology works for a single residential zone:

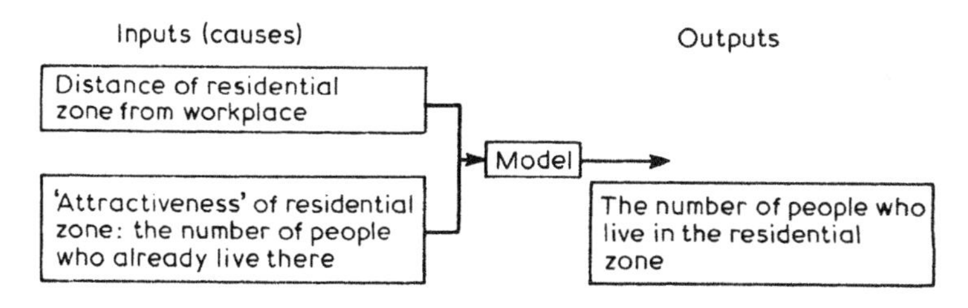

Figure 40 A model-building tautology

Elementary as this mistake may seem, many studies, involving a great deal of jargon and mathematics, nevertheless do not manage to understand that they are making it. The model fits the facts because the facts are fed into it in the first place. Lee (1973) doesn't recognize this mistake either, even when the model produced the 'correct' answer for a series of arbitrarily made-up population numbers.

This is the most obvious way in which the Lowry model can be used to justify a 'let present trends continue' strategy. Many modifications and adjustments can be made to the attraction factors (as they were in South

Hampshire and Notts-Derby), so that the results match present trends or indeed any other strategy which might seem desirable.

Lay writers on the new models like Hall (1974) are relatively unaware of the balance of advantages and dangers of these models. They tend to be content with admiring the new scientific techniques from a distance, defining the local authorities which use them as 'advanced'.

The history of the application of this residential and service activity allocation model in the UK can be interpreted in the light of the discussion in Chapter 5 and earlier sections of this chapter, and also in the light of Lee's discussion of the failures of large-scale models. The author's own work with others in the Planning Research Applications Group has involved trying to expose the basic workings of the model, its sensitivity to changes in inputs, how it works in practice, and to establish the most appropriate level of use — that is, to explore broad patterns of development. The Barras (1975) PRAG publication shows quite clearly that the model is sensitive to large-scale changes: it can show how the pattern of pressure for residential development is likely to change given a large-scale change in the distribution of employment such as a new steelworks. It is also clear that building in more and more detail in order to explain or predict small changes in small areas is not appropriate; it cannot solve this detailed problem. There is no purpose in attempting to 'massage' the model as in the South Hampshire study and many of the US models; this only increases the tautological element as more and more 'explanatory' factors are built in. However, once the overall behaviour and performance in practice have been established, there is further work to be done in understanding just what theory of household behaviour underlies the model. This again shows that *interpretation* is important. Another paper by the author (Broadbent 1973) shows how this technique can be portrayed as an urban production system similar to that described in Chapter 3.

An alternative and less fruitful line of work is to continue building in more and more detail, attaching more and more complex combinations of factors to explain the attractiveness of individual residential zones, adding in more time factors (so that the model seems to predict successive increments of development at five-year intervals), disaggregating the residential groups by income and social class, devising even more complex methods of 'fitting' the model to the real-world data. Now although such work often produces much useful spin-off in terms of new techniques (e.g. Batty and Mackie 1972) or new ways of organizing detailed information on a city (Echenique 1973); there is also a sense in which this work is proceeding down the disastrous road charted by Lee: a fruitless search for the 'real' behaviour of the microscopic individual, losing the understanding underneath the detail. Wilson (1968) has produced a whole school of unparalleled original work of high rigour on the derivation these techniques from basic statistical considerations. He has invented a general method for generating the most probable behaviour of large groups of people. He is the leading UK figure in theoretical research on urban allocation models, and his statistical work casts an interesting sidelight on the link between the conventional micro-market theory and the observed aggregate behaviour described in some of the land-use allocation models. Some of the models have been derived theoretically from utility-maximizing assumptions, but this does not 'prove' the market model (see Chapter 5). Wilson has shown that such models can be derived equally well using the 'neutral' properties of statistical physics. In summary, then, it is the meaning and design of a model or technique and the way it is used which is most important, and which determines whether or not it degenerates into a tautological instrument for justifying 'present trends'.

6 Cost-benefit analysis: the limits to micro-economics

The lack of a clearly identified choice of development strategy is a key shortcoming of many published structure plans. Clearly it is asking a lot of planners as a profession to be putting forward political policy alternatives at local level. At national level, these are settled by the political process itself: alternatives are presented to the electorate by political parties, not by the civil servants. But planners do use explicit methods to decide between alternatives, and it is here that their techniques diverge markedly from those derived from conventional micro-economic welfare theory.

The structure plan evaluation of alternative development strategies, however imperfect, is usually based on a preliminary statement of policy objectives, or problems to be solved. Objectives generally start from general ones about increasing overall wellbeing, and proceeds to define specific aims about providing adequate housing, jobs and services for the population. Ideally the planners try to show how one policy aim affects or conflicts with another. The development-potential method is an attempt to do this, although an extremely unsatisfactory one because the processes of urban development — in particular the way the different activities depend on each other, supply each other with goods and services and compete for space and locations — are not adequately described.

Micro-economics on the other hand has a very explicit, ideal model of the urban system — a model of many small, competing suppliers and consumers. It has a quite specific technique for deciding the economic worth of public investment projects — 'cost-benefit analysis' (CBA). The method develops quite naturally from the Pareto criterion outlined in Chapter 2. A planning project is worth supporting if some people gain and no one loses; the less restrictive Hicks-Kaldor principle suggests a project is worth supporting if the gainers from it can compensate the losers.

We saw how the modern form of welfare (micro) economics tries not to pass welfare judgements about the present distribution of income. It is impossible to add up the benefits of increases in the utility of different persons because utility is essentially a subjective concept. Cost-benefit practitioners do need to perform this addition, and therefore generally have to make some welfare assumption about the income distribution — e.g. that £1 is equally 'valuable' to everyone (Layard 1972). From then on a new planning project can be evaluated in two stages:

1 Value the total costs and benefits of the project in each year.
2 Discount all such future benefits and costs back to the present year — to produce a total 'present value' for the project.

Most work goes into Stage 1; the idea is to determine all the parties *directly* affected by the project. Thus, if the proposal is a new road bridge to replace an existing toll ferry, those directly affected include the taxpayers who pay for the bridge, ferry owners who lose their profits, existing travellers who no longer have to pay, and new travellers who did not previously cross the river at all. Cost-benefit analysts use the *demand curve* of Chapter 5 to evaluate the benefits arising from increased consumption (new travellers) which results from the lower price. They often assume a straight-line 'curve', so that the total benefit is equal to half the price fall multiplied by the quantity increase (the number of new trips). This is the area under the curve.

There is a massive literature on CBA. As seen in this example, it is more often than not used for a single sector, or a relatively small-scale project (such as a bridge) which supplies a single commodity — where a partial demand curve is most appropriate and where the immediate effects of the project can be isolated. (References include Layard 1972; Prest and Turvey 1965.)

CBA is the ultimate practical conclusion of the micro-economic theory (and some would say of the market economy itself). Everything can be shown to have a cash value, either because it is bought and sold or because people implicitly value it by consuming so much of it in relation to other things. This involves practitioners undertaking valuation exercises for non-traded commodities like the time spent travelling, the noise from aircraft, or historic buildings.

Such exercises are often highly controversial: it is possible to measure the value society as a whole puts on to any non-traded commodity — even a human life. But many alternative valuations can be discerned; society might spend millions to save someone stranded on a cliff, but relatively little to reduce the level of road accidents or deaths from smoking. Another valuation might be determined by calculating the contribution to total national output a person of a given age and socio-economic group would be expected to make over the rest of his or her life. The final choice of project may depend critically on the valuation of a particular benefit — itself stemming from some highly uncertain projection of demand for the new good or service provided. This is why, for instance, the value of travel time to the proposed third London Airport from central London, and the methods used to predict the decline in numbers using the airport as its distance from London increased, was subject to so much criticism.

There is no space here to devote to a detailed critique of CBA. There are two main types of criticism:

1 Accepting that all the underlying micro-economic concepts are truly representative of the real world, there are many limitations to the possible practical application of CBA.

2 The criticism that micro-economic welfare theory is not an adequate or meaningful description of the urban system, and therefore CBA carries with it all those limitations of the underlying theory from which it is derived.

No more needs to be said here on the second type of criticism, over and above what was said earlier in the discussion on micro-economic welfare theory in Chapters 2 and 5. The first type of criticism is widespread. Such critiques often concentrate on the way the costs and benefits are calculated, and the assumptions made about the incidence of costs and benefits, rather than on the conceptual underpinnings of the technique itself. For the third London airport, the difference in the total costs attributed to four alternative sites were overwhelmingly due to the so-called 'user costs' (i.e. essentially the costs of travelling to the airport from London). The difference in these costs between the four sites far outweighed the difference in construction costs. Only one of the many criticisms of this exercise was the fact that these 'user costs' were dependent on a long chain of reasoning involving many assumptions. The procedures required the use of a gravity model (similar to that described above) to estimate the reduction in the number of people travelling to an airport as the distance from London increased. All the above reservations of the use and interpretation of such techniques apply. There are clearly dangers when the cost-benefit decision is so highly dependent on a single model of travel behaviour, dangers that wrong definitions or 'manual adjustments' might be used almost unconsciously, to produce the required result.

Also of the first type of criticism is the fact that CBA cannot easily handle complex multidimensional large-scale projects, where there are many demand and supply curves and the separate effects of each cannot be identified. A large-scale urban expansion (or other 'strategic' structure plan situation) is one such example. Some of its logical inconsistencies stem from this 'partial' nature of cost-benefit analysis. Scitovsky (1941) shows that CBA might conceivably show a bridge to be worth building — and yet a repeated post-construction analysis could show that it would be worth pulling down again. This would be the result of changes in income distribution and prices resulting from the bridge. This is why simple or single (even though large-scale) decisions (e.g. about where to put London's third airport) have been subject to cost-benefit analysis, whereas very few large-scale regional, subregional or structure plans use it (Lichfield *et al.* 1975). The Planning Balance Sheet of Lichfield (1960) is an attempt to adapt CBA to a complex multisector situation; it tries to document all the benefits to the many different sectors of an urban planning proposal, such as 'occupiers of new developments, displaced industrialists, travellers on the road networks, etc.'. But this means making a large number of assumptions about the way all the partial demand curves interact and about the interrelationships (supply, demand and competition) between different activities and commodities across the whole system.

In practice, planners have found it inappropriate or impossible to apply

CBA to structure planning. They generally choose between alternative development strategies with reference to explicitly stared objectives and problems, often adapting versions of Hill's (1968) Goals-Achievement Matrix.

In this case, therefore, we see planning as no longer faithfully reflecting the prevailing ideas of conventional micro-economics. This is where planning practice (and the need to set up explicit procedures for deciding between large-scale alternatives involving many interactions between the different urban activities) transcends orthodox theory. It is evolving (albeit pragmatic and conceptually limited) procedures, which aspire to help the community to make a conscious choice of planning strategy with reference to explicit goals, rather than leaving the decision to the invisible hand of the market.

This is not to argue that planning strategies should not be based on a proper knowledge of costs and benefits; but the question is how are the benefits valued? The CBA technique appears to fail just at the point where it might hopefully be expected to be most relevant. It does so not only because its overall conceptual undepinning is questionable, but because conventional micro-economics is most at home in handling *partial* situations looking at the margin of development rather than at complex, strategic-planning situations which might significantly change the structure of a whole city over a ten- or twenty-year period.

It can be argued in this situation that what is needed is a more conceptually rigorous method for planners to describe urban problems and planning objectives, and to identify the interactions and conflicts between them. This in turn requires an explicit identification of the way urban activities interrelate — consume, produce and compete. A comprehensive production-process view of the city such as that outlined in Chapter 3 is one way of doing this; this is the technique we are exploring in a PRAG study for the Department of the Environment. There are no easy answers to this problem but an essential point is to develop procedures whereby the community may be able to make a more conscious choice for urban development, by showing clearly the way the different industries, services and population groups affect and compete with one another.

Summary

This Chapter shows how the new technology of urban planning arose in the 1960s — how transport planning provided the impetus and how planning technology, largely imported from the USA, came to be used in Britain. The use and abuse of technology and planning techniques (illustrated especially by the computer-based models of urban areas) is determined largely by the issues discussed in previous chapters — and not by specific features of the computer or the techniques themselves. The way many of these techniques were used reflected, for instance, the conceptual standpoint of micro-welfare economics — satisfying the demand of the consumer. The institutional arrangements whereby central government imposed a uniform transport-planning technology on all urban areas reflects the growing powers of the state in promoting investment. Professional restrictions also sometimes help to inhibit the effective use and deployment of appropriate expertise in planning.

The specific shortcomings of techniques now in use are discussed by reference to one or two specific methods. It can be shown that methods can be designed specifically to justify 'marginal' changes — to 'let present trends continue' — or, even more seriously, to be manipulated in arbitrary ways to produce the 'desired' answer.

The emphasis on 'forecasting' in local authority planning techniques reflects the impotence of urban planning and national planning to control the largest and most significant influences on urban development. Nevertheless the day-to-day practice of structure planning does in some ways transcend the dominant theory. This is particularly illustrated in the methods used to decide between alternative plans, where the most obviously relevant conventional economic technique, cost-benefit analysis, is very rarely used to decide between large-scale global alternatives for large urban areas.

7 Conclusion

1 The central dilemma

The main point of this book is to present some kind of plausible explanation for the way urban development occurs and why urban planning in particular takes the form it does. The pivot of the whole argument about planning lies in Chapter 2, which tried to portray the contradictory 'cleft-stick' position of the state in the modern mixed economy. Chapter 2 described the three interrelating, contradictory sectors: the state, the wage- and salary-earning labour force and the private (profit) sector — the three 'estates'. The private sector, the motor-generator of the economy, is in increasing difficulty in trying to maintain enough profit to reinvest and is apparently incapable of halting the overall relative decline of the national economy. The labour force is under increasing pressure to limit wages, to reduce consumption or to be unemployed, and variously attempts to resist such pressure. The state undertakes the crucial, short-term, steering role for the whole economy — trying to maintain some kind of balance in the economy, and to keep unemployment, inflation, the trade balance and other indicators within 'reasonable' levels. At the same time, it obtains its resources for doing this (and for providing for necessary social investment like roads) in part from deductions in tax from potential private-sector profits. The only way to preserve the dynamic of the mixed economy is to try to ensure that the private sector — the sector that sells marketable commodities — remains profitable and continues to invest.

For this reason the state does not invade possible profitable areas and tries not to take an entrepreneurial role; it remains on the sidelines, responding to the market — providing grants and incentives to firms (such as in regional policy) — and sometimes tries to put more of the burden of taxation on to the population or to reduce the costs of social services by reducing their quality. The constant theme, therefore, is 'how much (public sector) can we afford?' The image usually presented is one where the productive 'profit' sector produces useful products and generates the surplus needed to finance the whole inverted pyramid of 'non-productive' social services workers — local authority staff, social workers, teachers, doctors, civil servants, and so on. In the conventional view, we can afford these services only if we make enough profits in the private sector first.

While this picture serves to illustrate that the private-market sector provides the basic driving force for the whole system, even in a mixed economy the implied social priorities are at least questionable. Something that can be sold has social merit by the very fact that it makes profits and thereby provides taxes for social services. But surely health, housing and education could be regarded as immediate priorities for production in their own right. The social services involve 'production' in the same sense as car production: they involve the application of skilled labour and capital just as in industry, and they also increase demands for other industries (drugs, equipment, etc.). There is a universal demand for social services, even though they are not marketable goods. If a prime aim of modern production is to ensure a satisfactory life for the population, then education, health and housing would seem eminently suitable as a first priority for production. It would therefore seem a roundabout route to give overriding priority to the production of saleable products (plastic gnomes? hi-fi?), leaving health and housing and other social services to be financed from the resulting profits.

But we have seen that this lopsided approach to state planning is forced upon the government in the mixed economy. The economy is market-based — the market has to be sustained as a first priority. However different the initial intentions, all modern governments return willy-nilly to the old, oft-repeated, crisis-dominated, short-term planning formula of 'steering the economy'. Nowhere has this been more startlingly illustrated than in the speed of the retreat of the 1974 Labour government from its declared intention to seek new ways of breaking out of the economic straitjacket. The idea was to use the public sector in a new and positive manner to intervene positively, to invest directly in key sectors and to lead the economy by the nose out of its decline. The turnabout in strategy was completed *de facto* almost from the beginning, and was completed in terms of declared aims and intentions not long afterwards. The positive instruments of state planning initially proposed were divested of nearly all positive powers. The need to stimulate profit was re-emphasized, and the declared intention was to break out of the recession by holding back social services and public spending in favour of stimulating the private sector; this was to be coupled with a voluntary agreement to limit wage rises in an overall attempt to maintain the level of profit.

This recent history obviously does not prove any argument about state planning, but it is at least not inconsistent with the ideas expounded earlier about the role of the state. It is not inconsistent with the idea of a 'limit' or barrier to the growth of the state and to the way in which it operates. Understanding the reasons why the 'old' traditional type of state planning keeps on reasserting itself at least helps to guard against false expectations, and subsequent despair and disillusion about the intentions and good faith of governments. The same old films keep being re-run — speeches and plans for wage restraint and for limiting public spending are repeated verbatim over five, ten and twenty years — but at least we know why.

But some things do and have changed. The steering task is more difficult, the gap between goals and achievement has become wider, the decline of the economy has continued, and there is a wide-ranging search for new ideas and new solutions — even among the economic theoreticians who so influence planning.

If there really is a barrier to the growth of the state — and it was shown that even when the state does grow its growth is more apparent than real (often being merely a redistribution of the surplus produced in one part of the private sector to another (e.g. regional policy) or else a redistribution through taxes and subsidies between the wage- salary-earning labour force and the private sector) — and if in addition there is a large difference in the size of the state between the most capitalist of the planned economies and the most socialist of the market economies, there is likely to be no easy convergence of the two systems. If this barrier and gap exist it is an uncomfortable picture for those who believe in the gradual extension and growth of state planning. It means that we cannot necessarily have more and better planning without a more fundamental break with the existing economic system.

One central issue should be raised here. A reader has commented that the whole argument of this book points to the 'inevitable' conclusion that state planning and public ownership should be extended. The implication being that here is yet another 'weary' argument for increasing bureaucratic control. 'Bureaucracy improves nothing — why should planners or civil servants know better than individuals themselves what kind of cities they want? We are already overburdened with legislation and regulations and public spending is already too high.' Let us be quite clear what the central issue is in this argument about improving state planning. The issue is not simply that of extending the state, but of *changing* its function — emancipating it, breaking it out of the straitjacket. The fear of a non-responsive, monolithic state juggernaut is very real, and is often promoted by those who most benefit from state cooperation. Yet there is every reason to suggest that burgeoning bureaucracy is a result not of the growth in the state's direct powers but rather because of the failure to invest it with real powers. Two simple examples might illustrate this: if the state took the lead, for instance, in industrial planning at national level, there may well be less need for large numbers of planners to be employed in 'forecasting' what might happen to employment and industry in each local area. Similarly on the issue of public ownership of land discussed in Chapter 4 and below, a piecemeal approach — involving buying land laboriously in the market — means employing thousands of planners, surveyors and valuers. A straightforward reform of the land-tenure system to change the relationship between owners and users of land and planners' powers to affect the use of land through a once-off legal measure would make wide-ranging land purchase and the associated bureaucracy unnecessary.

We are not therefore proposing a bureaucratic field day. Those who argue for limiting the role of the state to the sidelines on the grounds of limiting bureaucracy are confusing ends and means. Bureaucracies tend to be hierarchical, impersonal and specialized and tend to involve 'systems of control based upon rational rules — rules which try to regulate the whole organizational structure on the basis of technical knowledge and with the aim of maximum efficiency' (Mouzelis 1967). This is the means; the discussion here is concentrated on the ends, on what the state bureaucracy should be trying to achieve.

It would be neither feasible nor desirable to have every facet of life controlled from the centre by the state. There is a good case to be made for the State 'changing places' with the private sector — that is, that it should take the leading role in the economy. The market could operate just as freely within this framework as it does now. It would of course be operating at a lower level. The decision to have a steel industry of a given size or a house-building programme of a certain size would be one for the state. The decision to have a mixture of activities in central cities — including housing — would be one for the local state. But within this broad framework, various activities and small businesses, householders, etc., would all have scope to compete for markets and for locations.

2 The main themes summarized

All the various themes discussed in this book, about the national economy, the role of urban centres within it, about urban planning and the theories and techniques that underpin it, are most clearly seen in relation to this central argument about the state and the three estates in the economy. It is worth briefly reiterating some of these main themes.

According to all superficial indicators Britain is in decline *vis-à-vis* other Western economies. The fundamental process which makes the economy tick, and which determines what factories are built, what is produced and for whom, is the market process in the private sector. The essential aspect of this process is the production of commodities for sale on national and world markets; to keep the process going these sales must generate enough profits — profits which are later reinvested within firms or else circulated through other financial institutions in the national economy and later still ploughed back into further investment in productive capital. Britain is the most 'mature' market economy: it has more wage and salary earners than similar economies, and it has no room left for new extensive growth, for exploiting new resources (notwithstanding oil and other energy reserves); in fact, there are a series of barriers to the expansion of the economy. The only way forward is through internal restructuring and reorganization of industry. This is taking place through increasing competition and the drive to reduce costs of production. The economy has become dominated by big firms; some sectors grow at the expense of others without improving the overall position of the economy. This points to the need for a more conscious attempt to control the economy — to the state.

Following the central three-estates argument, Chapter 2 described the conventional theoretical justification for state intervention in the economy (to correct imperfections in the market, to steer the economy and to maintain stability). The central question remains: how to overcome the classic separation between planning and powers, which limits the battery of Treasury controls to the short-term control of the economy.

Chapter 3 showed how the pattern of urban centres in the UK has been changing, how geographically concentrated the population is and how the centralization and decentralization of cities and the relative decline of the inner areas of cities have been proceeding. It tried to show briefly how this national urban structure is affected by the national economy — the big firms and the financial institutions which have invested in urban development, especially in the commercial, profit-making, central areas. The individual city is first and foremost a pool of labour — and the city itself is not an economic system in its own right; there are few links within the city economy other than those that flow from the employment of labour, and from the consumer spending of the labour force. These flows pass through households, which should themselves be regarded as 'producers', as small factories. The second key process in the development of urban areas is a reflection of the competition between firms and other agencies in the economy within limited resources and where there is continual displacement of one activity by another. This process is the competition for space — for land and for key locations like city centres where profits can be generated.

Chapter 4 showed how the problems of the local state (local authorities) are even more severe than those of the national state: the scope for action is more circumscribed, and the dependency on the central government is growing, despite the apparent increase in size of the local state. While most of the real resources channelled through the local state are devoted to direct services like education and housing, it is the urban planning function which illustrates very clearly some of the limits imposed on state intervention, both nationally and locally. Over time, planning has become more pervasive, more visible, and concerned with larger geographic areas. Nevertheless, in the face of increasing competition between activities competing for space and location, planning remains on the sidelines; it cannot implement its own plans. Its greatest success (and even this is disputed) is at the 'margin' of cities, not in the crucial inner-urban areas where competition between activities is severe. The continued failure of successive attempts to take land into public ownership — an aim which can be traced back to Howard and other pioneers of urban planning — also symbolizes the 'barrier' to state intervention at the urban level.

Chapter 5 argued that the function of urban planning theory was to help explain the role of planning and understand the real forces and processes shaping urban development, and to underpin planning techniques. Planning itself is contradictory: it has a very real need to understand what is going on in cities, but at the same time it needs to disguise its lack of real influence to allocate urban resources in socially optimal ways.

Economics is the dominant social theory. Of its two branches (micro and macro), neoclassical micro-economic theory, with its central concepts of the individual, isolated consumer and the 'margin', is appropriate for analysis of specific economic problems under limited circumstances. It is often misapplied as an overall philosophy, to justify satisfying consumer demand and the restricted role of planning. Other schools of theory often reflect economic ideas. Pluralism — which emphasizes the conflicting and competing claims on resources, with planners ensuring fair play and supporting oppressed groups at the margin — can also be seen as an echo of market ideas. The ultra-radicals represent the state largely or solely (and erroneously) as an oppressor, and seek to dismantle it; they do not see the potential in existing institutions. Urban sociology tends to make a recurring and elementary mistake, in taking a 'technique' such as systems analysis as a social theory in its own right.

Chapter 6 showed how the new technology of urban planning arose in the 1960s, how transport planning provided the impetus and how planning technology, largely imported from the USA, came to be used in Britain. The use and abuse of technology and planning techniques (illustrated especially by the computer-based models of urban areas) is determined by the issues discussed in previous chapters — and not by specific features of the computer or the techniques themselves. The way many of these techniques were used reflected, for instance, the conceptual standpoint of micro-welfare economics — satisfying the demand of the consumer. The institutional arrangements whereby central government imposed a uniform transport-planning technology on all urban areas reflects the growing powers of the state in promoting certain types of investment. Professional restrictions also sometimes help to inhibit the effective use and deployment of appropriate expertise in planning.

The specific shortcomings of techniques now in use are discussed by reference to one or two specific methods. It is shown that methods can be designed specifically to justify 'marginal' changes — to 'let present trends continue' — or, even more seriously, can be manipulated in arbitrary ways to produce the 'desired' answer.

The emphasis on 'forecasting' in local authority planning techniques reflects the impotence of urban planning and national planning to control the largest and most significant influences on urban development. Nevertheless the day-to-day practice of structure planning does in some ways transcend the dominant theory. This is particularly illustrated in the methods used to decide between alternative plans — where the most obviously relevant conventional economic technique, cost-benefit analysis, is very rarely used to decide between global alternatives for large urban areas.

3 Positive suggestions?

What are the positive suggestions coming out of this argument? In essence these grow almost concomitantly out of the main themes outlined above.

Taking first of all the central, pivotal discussion about the role of the state at national and local level, the key aim should be to *emancipate* public planning, to change the rules under which it now operates, and to use the state to achieve *directly* some of the national economic and social goals, and the local urban goals. This would entail the state's building new factories, carrying out redevelopment or rehabilitation; the state could build houses, employ people who are otherwise unemployed, and use all its potential purchasing power in an active and positive manner to achieve social goals directly, rather than relying first and foremost on the private sector. This means bridging the gap between 'planning' (so often an exhortatory exercise on the sidelines) and 'powers', especially Treasury powers, so often geared to entirely short-term, sterile goals which do nothing to halt the decline of the economy.

But there have been many suggestions to this effect, and they have not been implemented. We have already discussed above the reasons for this — the 'barrier', the threat to the essentially market-based structure of the economy. There is no point in ducking this uncomfortable fact; the question of national state planning has to be left there. There can be no smooth increase in the effectiveness of state planning; the choice will eventually become a stark political alternative. Bridging the gap between plans and powers will eventually mean not simply improving arrangements in the public sector, but must mean a more or less fundamental leap from a market-based and market-led economy to a public-sector-led economy.

Whether or not there remain some useful limited ways in which state planning can be made more effective and given more teeth and more powers, even within the present market set up, remains an open question, needing detailed study. If such effective ways of tackling the decline of the economy were available, they would surely have been discovered and tried by now. The uncomfortable choice is slowly but inexorably coming closer. Even within their present straitjacket, both national and urban planning continue to become more pervasive, more visible, more long-term, more strategic, drawing more and more on sophisticated techniques and theory. In this sense (i.e. in order to perform even its limited existing task) there will be a continual demand for improved planning, for better theories and techniques which show how planning can influence the processes driving the national economy and urban development. (Similarly there will be a continuing need to disguise why planning cannot solve fundamental problems directly.)

More concrete prescriptions for limited reforms and improvements in technique and planning can be suggested: one or two are outlined below, but they should be done so in the full awareness of the underlying dilemma outlined above.

Urban planning is a part of the overall state planning activity, with a restricted scope and limited effect on the whole economy. Is urban planning therefore a case where public-sector planning can be improved, where reforms to close the gap between plans and powers, to put teeth into the urban planning system, can be effected without raising the central dilemma of national planning outlined above?

In Denmark there appears to have been some success in a system of urban planning briefly touched on in Chapter 4, based on 'channelling the economic forces'. This approach accepts the increasing scale of many developments such as retail centres — that is, the forces promoting decentralization or urban areas — and tries to put these forces to positive use, for instance to take pressure away from central areas and to make integrated provision for large-scale developments, with 'bundles' of motorways, campuses and hypermarkets all sited well away from city centres. With this approach, as much activity (e.g. shopping) as possible is concentrated at the local level. It is a philosophy which begins from the bottom up, by focusing first on local centres, putting as much activity there as possible, and then building a sufficient number of large centres to absorb the *surplus* shopping demand. This contrasts with the UK approach whereby as much activity as possible seems to be forced into the single town-centre development, through comprehensive clearance schemes which also help to remove small retailers and other activities.

Clearly, the 'channelling the forces' approach cannot halt the overall trend to larger developments. There is surely an argument to be made that this trend itself provides more scope, more elbow room to choose and to decide quite consciously *where to develop.* A certain minimum level of local services can be preserved, large comprehensive developments can be sited away from the historic town centres, and such centres can then retain a mixture of activities, including housing.

But why hasn't this been attempted already? Clearly in some sense it is not profitable for the investing institutions and the large finance and property companies to do so. Presumably the production costs (which it has been argued underlie the whole process of the decisions to develop) are somehow too high. It may, for instance, be cheaper to knock down an existing centre for a new large-scale comprehensive development and utilize the existing town-centre 'infrastructure' (roads, sewers, bus stations, etc.etc.) than to develop a whole new infrastructure of access roads and services away from the centre. If at the same time the comprehensive clearance is administratively quicker and if it helps to speed the removal of competition, so much the better.

But from the point of view of the whole community this approach may actually be the most expensive, since costs of disruption, longer journeys, and so forth, are borne by the population or by the local authority, not by the investing institutions. Strengthening the planning system would then mean making the cost-of-production calculation from the point of view of the whole community, *and having the powers to implement* a plan which minimizes these total social costs, not merely the private-sector, financial and institutional costs.

But supposing we try to tackle the central failing of urban planning head on — and independently of the rest of the economy; that is, we try to put teeth into urban planning — to bridge the gap between plans and powers, between plans and resources, by giving the public sector at the urban level direct power over land and resources to carry out development directly. How could this be done?

It is arguable that to do this would require a fundamental change in property rights over land — just as the 1947 act changed once and for all the legal position and rights of landowners over their land. Let us suppose for a moment that all land became publicly owned, so that local authorities would have the benefit of income from land and rental income from major development schemes. How would this affect the rest of the economy?

Depending on how far this control also extended to the properties themselves, the major financial institutions would be derived of some of their income from rent; and for these institutions, property represents a significant proportion of their total assets, as we have seen. Such a removal might undermine a large part of the financial sector's economic base and so there would be great difficulty in undertaking such an exercise.

The effect on the ground would be twofold. First, it would remove the rental element from the decision to develop, so that the competition between activities (which now pushes up rents for less profitable activities and drives them away from favourable locations) would be removed. Second, local authorities would be able to use the rental income to implement at least part of their own plans; in other words, planning and powers would be no longer separated. The overall rationale of the 1947 act — which assumed that most urban development would be made by the public sector — would be restored.

The aim would be to allow all the *users* of urban land — industry, retailers and even possibly developers — to make their normal profits from the direct revenues and costs of their operations; they would not have to pay excessive rents to property companies or financial institutions. Nor would they need to rely on the appreciation in the value of their own properties.

This picture is far removed from the existing situation. There have been several attempts to remove the rental element from the land market; as we have seen, none of these attempted an overall legal change in the owner's right to land as did the 1947 Planning Act. The attempts to buy up urban land gradually over a period under the present proposals (the Community Land Act) is a classic illustration of the 'barrier' to the increase in the public sector once the market and profits start to be undermined. In 1947-56, the market in land dried up when 100 per cent tax on the increased land values was imposed.

A similar thing will happen now if local authorities are seen to be acquiring large amounts of land in an urban area. The new proposal for a tax on increases in development value will also help to reduce the supply of land as owners wait for a change of government. If authorities try to acquire sufficient land by compulsory purchase, they will run into tremendous controversy — as large monopoly land buyers discriminating between landowners. The creeping-encroachment approach can only be pursued if it is consistent with maintaining profits, and in this case this means generally only acquiring small quantities of land at the margin; it cannot be carried out in major problem areas *inside* the cities, which are already developed, over-developed or decayed.

If, however, the land-tenure system was changed in a comprehensive way, so that *all land* in specific categories (or uses) came into public ownership at once, *all land users* — retailers, householders, etc. — could have full rights to use the land for a given purpose, with security of tenure. They would hold these rights direct from the state or local authority. Such a measure would aim to cut out pure land *owners* from the picture; there should be no pyramids of owners and subowners or lessees and sublessees intervening between the state and the user or occupier.

This is a much more fundamental measure than anything that has recently been contemplated; but here again the failures of the partial, 'marginal'-reform approach is at least not inconsistent with the idea of a barrier to state planning and intervention. It seems to indicate that even in the limited sphere of urban planning, easy partial-reform answers are not so easy to come by.

But if this is a major conclusion, what about a list of constructive suggestions for improvement, which have some chance of being implemented? All the discussion throughout this book has suggested that even within the present planning straitjacket there will be a continuing demand to improve planning to make it broader in scope, based on it better and generally more effective techniques, in order that it might fulfil its central function of providing a framework of certainty for the private-market sector in cities. There is, in other words, every scope for considering limited reforms — to help give urban planning more teeth, more resources and more powers — provided it is done in the knowledge that the more fundamental changes and options are more and more likely to be raised.

Ambrose and Colenutt (1975) give a mixed list of proposals, some of which are fairly limited, such as extending the development-control powers so that wider social and economic criteria could legitimately be considered in planning decisions. Their other suggestions are much more sweeping, including, for instance, the public ownership of all land and property. They do not make an adequate distinction between the fundamental reforms which could threaten the basis of the market economy, and the minor reforms which do not.

If we are looking for limited reforms for narrowing the gap between plans and powers at the urban level, then a national development agency, coupled with a reform of the financial institutions, should be early candidates. These are reforms at the national level, where urban investment funds are marshalled and from where they are distributed. At the local level the structure plan itself could become the active instrument for implementing development. The centrally directed public funds and the national building activity would be channelled through the social-cost-based structure plan.

If the structure plan itself were made into a part of a complete global local authority development plan for a whole city or group of cities, then it would be easier to bring all public capital spending on the different sectors — education, health, roads, sewage, etc. — into bundles specifically directed at the problems of a particular area. This would help to make plans more political (in the proper sense); there would be more direct and open debate over the structure plan. There would no longer be so much excuse for a 'let present trends continue' strategy, since planners would have real powers to affect these trends.

Another gap which might be narrowed or closed in the short term is that between the strategic plan itself and the day-to-day exercise of (negative) planning control. There is every reason to distinguish between the strategic large-scale developments which affect whole areas from small-scale improvements to houses. The round of negotiation and discussion with property companies, public-sector institutions and large-scale developers should become part of the strategic structure plan activity. There is no reason why these major developments should not be discussed and negotiated in public together with possible alternatives and presented and evaluated over a period of a year or eighteen months. This would entail attaching a 'current developments' element to the structure plan, one that would be subject to regular short-term publication — for instance annually in association with the various annual reports and budgets. Even here, though, there are grounds for pessimism. The Dobry Report which suggested a more or less purely administrative change to distinguish large-scale developments from small developments was not implemented. Whether this was a result of professional restrictive practice or not, it is difficult to see how this proposal would threaten the present planning system.

By the same logic, the wider range of social and economic issues now discussed in the written structure plans could be brought into development control itself, so that the control of activities which are allowed to use specific land and buildings becomes more effective. As of now, the major land-use classes are so broad that the character of whole areas can change — a car factory can be converted into a toothbrush plant, food shops into boutiques, rented accommodation into owner occupation — and yet still remain within the same official land use. These user classes could be made much finer and more specific, so that planning permission would need to be sought to change between these activities or use classes. This type of reform, which would in fact need no new legislation, would strengthen planning powers considerably.

Many other minor reforms could stem directly from the discussion in preceding chapters. A reform of the planning profession and its control over education could help to bring in a wider range of expertise into planning, and national and regional planning agencies and centres of expertise could cut out duplication of effort and minimize bad practice and misuse of techniques in justification of present trends.

These limited reforms and extensions of direct planning at the urban level, aimed at replacing at least some of the market element in the allocation of land uses in cities by direct planning, may well be consistent with the continued operation of a profitable private sector at national level, even though the implications for the financial institutions may be serious. But a major extension of public ownership and planning at the national level could pose much starker questions for the survival of the market economy than has hitherto been recognized.

Bibliography

Abercrombie, P. (1945) *The Greater London Plan.* London, HMSO.
Adelman, A. (1961) *Theories of Economic Growth and Development.* Stanford, Stanford University Press.
Adelman, I.G. (1958) A stochastic analysis of the size distribution of firms. *American Statistical Association Journal* 53 893—904.
Alden, J., and Morgan, R. (1974) *Regional Planning: A Comprehensive View.* Leighton Buzzard, Leonard Hill Books.
Alexander, J.W. (1954) The basic-non basic concept of urban economic functions. *Economic Geography* 30, 246—61.
Allen, G. C. (1970) *British Industries and Their Organization.* London, Longmans.
Alonso, W. (1960) A theory of the urban land market. *Papers and Proceedings of the Regional Science Association* 6, 149—58.
Alonso, W. (1964) *Location and Land Use.* Cambridge, Mass., Harvard University Press.
Althusser, L., and Balibar, E. (1970) *Reading Capital.* London, New Left Books.
Ambrose, P., and Colenutt, R. (1975) *The Property Machine.* Harmondsworth, Penguin.
Amery, C., and Cruickshank, D. (1975) *The Rape of Britain.* London, Elek.
Andrews, R.B. (1973) Mechanics of the urban economic base. *Land Economics,* 22.
Angel, S., and Hyman, G. (1971) *Transformations and Geographic Theory.* Working Paper 72. London, Centre for Environmental Studies.
Arrow, K. (1974) Limited knowledge and economic analysis. *American Economic Review* 64, March.
Ashby, W.R. (1956) *An Introduction to Cybernetics.* London, Chapman and Hall.
Ashworth, W. (1954) *The Genesis of Modern British Town Planning.* London, Routledge and Kegan Paul.
Atkinson, A. B. (ed.) (1973) *Wealth, Income and Inequality.* Harmondsworth, Penguin.
Au Rousseau, M. (1921) The distribution of population: a constructive problem. *Geographical Review* 11, 574.
Bacon, R. (1970) An approach to the theory of consumer shopping behaviour. *Urban Studies* 8, 55—64.
Bailey, J. (1975) *Social Theory for Planning.* London, Routledge and Kegan Paul.
Bain, J. S. (1966) *International Differences in Industrial Structures of Eight Nations in the 1950s.* New Haven, Yale University Press.
Ball, R. J., and Doyle, P. (1969) *Inflation.* Harmondsworth, Penguin.
Banham, R., Barker, P., Hall, P., and Price, C. (1969) Non-plan: an experiment in freedom. *New Society,* 20 March, 435—43

Bank of England (1974) Overseas sterling balances 1963—73. *Bank of England Quarterly* 14 (2), June, 162—75.

Baran, P., and Sweezy, P. (1968) *Monopoly Capital.* Harmondsworth, Penguin.

Barker, T. S., and Lecomber, J. R. C. (1969) *Economic Planning for 1972: An Appraisal of the Task Ahead.* London, Political and Economic Planning.

Barlow Report (1940) *Report of the Royal Commission on the Distribution of the Industrial Population.* Cmnd. 61530. London, HMSO.

Barras, R. (1975) *A Spatial Interaction Model of Cleveland.* Technical Paper 11. London, Planning Research Applications Group.

Barras, R., Booth, D. J. W., and England, J. R. (1975) *The Use of Models in Structure Planning: Applications in Cleveland.* Technical Paper 10. London, Planning Research Applications Group.

Barras, R., and Broadbent, T. A. (1974) The development of an activity-commodity representation of urban systems as a potential framework for evaluation. In Cripps (1974).

Barras, R., and Broadbent, T. A. (1975a) *A Framework for Structure Plan Analysis.* Technical Paper 8. London, Planning Research Applications Group.

Barras, R., and Broadbent, T. A. (1975b) *Elements of an Activity-Commodity Formalism for Representing Socio-Economic Systems.* Paper presented at European Regional Science Association Conference, Budapest. Research Paper 18. London, Centre for Environmental Studies.

Barras, R., Broadbent, T. A., Cordey-Hayes, M., Massey, D., Robinson, K., and Willis, J. (1971) An operational urban development model of Cheshire. *Environment and Planning* 3, 109—242.

Barras, R., and Catalano, A. (1975) *The Financial Structure of Property Companies.* Paper presented to Conference on Urban Economics, Keele, June.

Batty, M. (1971) Design and construction of a sub-regional land use model. *Socio-Economic Planning Sciences* 5, 97—124.

Batty, M. (1974) Spatial entropy. *Geographical Analysis* 4, 1—32.

Batty, M., and Mackie, S. (1972) The calibration of gravity, entropy and related models of spatial interaction. *Environment and Planning* 4, 131—250.

Batty, M., and Masser, I. (1975) Spatial decomposition and partitions in urban modelling. In E. C. Cripps (ed.), *Regional Science - New Concepts and Old Problems.* London Papers in Regional Science No. 5. London, Pion.

Baxter, R., and Williams, I. (1972) *The Second Stage in Disaggregating the Residential Sub-Model.* University of Cambridge, Centre for Land Use and Built Form Studies.

Becker, G. S. (1965) A theory of the allocation of time. *Economic Journal* 75 (299), 493—517.

Beckerman, W. (ed.) (1972) *The Labour Government's Economic Record 1964–1970.* London, Duckworth.

Beckerman, W., and associates (1965) *The British Economy in 1975.* Cambridge, Cambridge University Press.

Beer, S. (1966) *Decision and Control.* New York, Wiley.

Bell, C., and Bell, R. (1972) *City Fathers.* Harmondsworth, Penguin.

Berry, B. J. L. (1967) *Geography of Market Centres and Retail Distribution.* Englewood Cliffs, N. J., Prentice-Hall.

Berry, B. J. L., and Garrison, W. L. (1958) Alternate explanations of urban rank size relationships. In Mayer and Kohn (1959).

Bertalanffy, L. Von (1971) *General System Theory.* London, Allen Lane.

Best, R. (1959) *The Major Land Uses of Great Britain.* Wye College, University of London.

Best, R., and Champion, A. G. (1970) Regional conversions of agricultural land to urban use in England and Wales 1945—67. *Transactions of the Institute of British Geographers* 49, 15—32.

Best, R., and Coppock, J. T. (eds) (1962) *The Changing Use of Land in Britain.* London, Faber.

Beveridge Report (1940) *Social Insurance and Allied Services.* Cmnd. 6153. London, HMSO.

Blair, J. M. (1972) *Economic Concentration.* New York, Harcourt Brace Jovanovich.

Blank, S. (1973) *Industry and Government in Britain. The FBI in Politics 1945–1965.* Farnborough, Hants, Saxon House.

Blaug, M. (1968) *Economic Theory in Retrospect.* London, Heinemann.

Blaug, M. (1974) *The Cambridge Revolution: Success or Failure?* London, Institute of Economic Affairs.

Blumenfeld, H. (1955) The economic base of the metropolis: critical remarks on the basic-non basic concept. *Journal of the American Institute of Planners* 25 (2), 114—32.

Blunden, W. R. (1972) *The Land-Use Transport System.* Urban and Regional Planning Series, 1. Oxford, Pergamon.

Board of Trade (1968) *The Movement of Manufacturing Industries in the United Kingdom.* London, HMSO.

Bolton Committee (1971) *Committee of Inquiry on Small Firms.* Cmnd. 4811. London, HMSO.

Boswell, R. J., and Lewis, E. W. (1970) The geographical distribution of industrial research activity in the UK. *Regional Studies* 4, 297—306.

Boyce, D. E., Day, N. D., and McDonald, C. (1970) *Metropolitan Plan Making.* Monograph No. 4. Cambridge, Mass., Regional Science Institute.

Brand, J. A. (1965) Ministry control and local autonomy in education. *Political Science Quarterly* 36, 154—63.

Bratten, C.. Dean, R. M., and Silberton, A. (1965) *Economics of Large Scale Production in British Industry.* Cambridge, Cambridge University Press.

Bray, J. (1970) *Decision in Government.* London, Gollancz.

Brittan, S. (1971) *Steering the Economy.* Harmondsworth, Penguin.

Broadbent, T. A. (1970) Notes on the design of operational models. *Environment and Planning* 2, 469—76.

Broadbent, T. A. (1971a) *An Activity Analysis Framework for Urban Planning.* Working Paper 61. London, Centre for Environmental Studies.

Broadbent, T. A. (1971b) A hierarchical interaction model for a two-level spatial system. *Regional Studies* 5, 23—7.

Broadbent, T. A. (1973) Activity analysis of spatial allocation models. *Environment and Planning* 5, 673—91.

Broadbent, T. A. (1973) *A Resource Allocation Model for Households and Housing* Paper presented to the Operational Research Society.

Broadbent, T. A. (1975a) *An Attempt to Apply Marx's Theory of Ground Rent to a Modern Urban Economy.* Research Paper 17. London, Centre for Environmental Studies.

Broadbent, T. A. (1975b) An approach to the application of urban models in the planning system of the UK. In Snickers and Karlqvist (1975).

Broadbent, T. A. (1975c) How to put teeth into planning. *Municipal Review* 46, 137—8.

Broadbent, T. A. (1975d) Planners' plight. *New Society* 20 November, 426—7.

Broadbent, T. A. (1977) *Urban Production Activities.* Technical Paper Forthcoming. London, Planning Research Applications Group.

Broadbent, T. A., and Catalano, A. (1975) *Notes on Office Development by Major Property Companies 1960–1975*. Paper presented to conference of the North-West European Regional Science Association, September. Working Note 426. London, Centre for Environmental Studies.

Broadbent, T. A., and Mackinder, I. H. (1973) Transport planning and structure planning. *Traffic Engineering and Control* 14, 536—7.

Buchanan, C. D. (1963) *Traffic in Towns*. (Shortened edition of the Buchanan Report.) Harmondsworth, Penguin.

Buckley, W. (1967) *Sociology and Modern Systems Theory*. Englewood Cliffs, N.J., Prentice-Hall.

Bullock, N., Dickens, P., Shapcott, M., and Steadman, P. (1974) Time budgets and models or urban activity patterns. *Social Trends* 5. London, HMSO.

Cairncross, A. (ed.) (1970) *The Managed Economy*. Oxford, Blackwell.

Cambridge Political Economy Group (1974) *Britain's Economic Crisis*. Spokesman Pamphlet 44. London, Bertrand Russell Peace Foundation.

Cameron, G. C., and Clark, B. D. (1966) *Industrial Movement and the Regional Problems*. University of Glasgow Social and Economic Studies, Occasional Papers No. 5. Edinburgh, Oliver and Boyd.

Cameron, G. C., and Evans, A. W. (1972) The British conurbation centres. *Regional Studies* 7, 47—55.

Cameron, G. C., and Johnson, K. M. (1969) Comprehensive urban renewal: the Glasgow case. In Orr and Cullingworth (1969).

Cameron, G. C., and Wingo, L. (eds) (1973) *Cities, Regions and Public Policy*. Edinburgh, Oliver and Boyd, for the University of Glasgow and Resources for the Future.

Carney, J., *et al.* (1974) *Regional Underdevelopment in Late Capitalism*. North-East Area Study. University of Durham.

Central Statistical Office (1968) *National Accounts Statistics: Sources and Methods*. London, HMSO.

Chadwick, G. (1971) *A Systems View of Planning*. Oxford, Pergamon.

Champion, R. G. (1974) *An Estimate of the Changing Extent and Distribution of Urban Land in England and Wales*. Research Paper 10. London, Centre for Environmental Studies.

Chancellor of the Exchequer (1975) *An Approach to Industrial Strategy*. Cmnd. 6315. London, HMSO.

Chenery, H. B. (1965) Comparative advantage and development policy. In *Surveys of Economic Theory*. American Economic Association, Royal Economic Society. London, Macmillan.

Cherry, G. (1974a) *The Evolution of British Town Planning*. Leighton Buzzard, Leonard Hill.

Cherry, G. (ed.) (1974b) *Urban Planning Problems*. Leighton Buzzard, Leonard Hill.

Chisholm, M. (1970) *Geography and Economics*. 2nd edition. London, Bell.

Chisholm, M., and Manners, G. (1971) *Spatial Policy Problems of the British Economy*. Cambridge, Cambridge University Press.

Chorley, R. J., and Haggett, P. (1967) *Models in Geography*. London, Methuen.

Christaller, W. (1966) *Central Places in Southern Germany*. Englewood Cliffs, N.J., Prentice-Hall.

Clawson, M. (1971) *Suburban Land Conversion in the United States: An Economic and Governmental Process*. Baltimore, Md., Johns Hopkins University Press.

Coates, B. C., and Rawstron, E. M. (1971) *Regional Variations in Britain*. London, Batsford.

Coleman, A. (1961) The Second Land Use Survey; progress and prospect. *Geographical Journal* 127 (2), 168—86.

Coleman, A., and Maggs, K. R. A. (1965) *Land Use Survey Handbook*. Second Land Use Survey. Department of Geography, King's College, London.
Colston Papers (1970) *Regional Forecasting*. London, Butterworth.
Commission of the European Communities (1971) *Regional Development in the Community: An Analytical Survey*. Brussels, EEC.
Commission of the European Communities (1973) *Report on the Regional Problems in the Enlarged Community*. Brussels, EEC.
Cordey-Hayes, M., Broadbent, T. A., and Massey, D. B. (1970) Towards operational urban development models. In Colston Papers (1970)
Cordey-Hayes, M., and Gleave, D. (1973) *Migration Movements and the Differential Growth of City Regions in England and Wales*. Paper presented at the European Regional Science Association Congress, Vienna. Research Paper 1. London, Centre for Environmental Studies.
Cordey-Hayes, M., and Gleave, D. (1974) *Dynamic Models of the Interaction Between Migration and the Differential Growth of Cities*. RR—74—9. Vienna, International Insititute for Applied Systems Analysis.
Cordey-Hayes, M., and Wilson, A. G. (1970) Spatial interaction. *Socio-Economic Planning Sciences* 5, 73—96.
Cornwall, J. (1972) *Growth and Stability in a Mature Economy*. London, Martin Robertson.
Coventry City Council (1971) *Coventry-Solihull-Warwickshire: A Strategy for a Sub-Region*.
Coventry Community Development Project (1975) *Final Report: Coventry and Hillfields: Prosperity and the Persistence of Inequality*. London, Home Office.
Cowan, P. (ed. (1973) *The Future of Planning*. London, Heinemann.
Cowan, P., *et al.* (1969) *The Office: A Facet of Urban Growth*. London, Heinemann.
Cripps, E. C. (ed.) (1974) *Space-Time Concepts in Planning*. London, Pion.
Cripps, E. C., and Foot, D. H. S. (1969) The empirical development of an elementary residential location model for use in sub-regional planning. *Environment and Planning* 1, 81—90.
Croft, M. J. (1969) *Offices in a Regional Centre*. Research Report 3. London, Location of Offices Bureau.
Cullen, I., and Godson, V. (1975) Urban networks; the structure of activity patterns. *Progress in Planning* 4, Pt 1. Oxford, Pergamon.
Cullingford. D., Flynn, P., and Webber, R. (1975) *Liverpool Social Area Analysis (Interim Report)*. Technical Paper 9. London, Planning Research Applications Group.
Cullingworth, J. B. (1972) *Town and Country Planning in Britain*. London, Allen and Unwin.
Cullingworth, J. B. (1973) *Problems of an Urban Society*. Vol. 1: *The Social Framework of Planning*. Vol. 2: *The Social Context of Planning*. Vol. 3: *Planning for Change*. London, Allen and Unwin.
Dahrendorf, R. (1969) *Class and Conflict in Utilitarian Society*. London, Routledge and Kegan Paul.
Day, A. (1963) *Roads*. London, Mayflower Books.
Deane, P. D., and Cole, W. A. (1969) *British Economic Growth 1688-1959*. Cambridge, Cambridge University Press.
Denison, E. F. (1967) *Why Growth Rates Differ – Postwar Experience in Nine Western Countries*. London, Allen and Unwin.
Dennis, N. (1972) *Public Participation and Planners' Blight*. London, Faber.
Department of Economic Affairs (1965) *The National Plan*. Cmnd. 2764. London, HMSO.
Department of Economic Affairs (1969a) *The Task Ahead*. London, HMSO.

Department of Economic Affairs (1969b) *The Intermediate Areas*. Cmnd. 3998. London, HMSO.
Department of the Environment (1965a) *Town and Country Planning (Use Classes) Order*. No. 229. London, HMSO.
Department of the Environment (1965b) *The Future of Development Plans*. Report of the Planning Advisory Group. London, HMSO.
Department of the Environment (1967) *Management of Local Government*, 1 and 2 (Maud Report). London, HMSO.
Department of the Environment (1969a) *People and Planning. Report of the Committee on Public Participation in Planning*. London, HMSO.
Department of the Environment (1969b) (1969 onwards) *Development Control Policy Notes*. London, HMSO.
Department of the Environment (1970) *Development Plans. A Manual on Form and Content*. London, HMSO.
Department of the Environment (1971) *Management and Networks: A Study for Structure Plans*. London, HMSO.
Department of the Environment (1973a) *Using Predictive Models for Structure Planning*. London, HMSO.
Department of the Environment (1973b) *Town and Country Planning*. General Development Order No. 31. London, HMSO.
Department of the Environment (1974) *Structure Plans*. Circular 98/74. London, HMSO.
Department of the Environment (1975) *Review of Development Control System*. By G. Dobry. London, HMSO.
Derry, T. K., and Jarman, T. L. (1970) *The Making of Modern Britain*. London, Murray.
Desai, M. (1974) *Marxian Economic Theory*. London, Gray-Mills.
Diamond Commission (1975) *Royal Commission on the Distribution of Income and Wealth*. Report No. 1. Cmnd. 6171. London, HMSO.
Dobb, M. (1973) *Theories of Value and Distribution since Adam Smith*. Cambridge, Cambridge University Press.
Donaldson, P. (1974) *Guide to the British Economy*. Harmondsworth, Penguin.
Donnison, D. V. (1973) What is the 'good city'? *New Society* 13 December, 647—9.
Donnison, D. V., *et al.* (1975) *Social Policy and Administration Revisited*. London, Allen and Unwin.
Dorfman, R., Samuelson, P. A., and Solow, R. J. (1958) *Linear Programming and Economic Analysis*. New York, McGraw-Hill.
Dow, J. C. R. (1970) *The Management of the British Economy 1945–1960*. Cambridge, Cambridge University Press.
Drake, M., McLoughlin, J. B., Thomson, R., and Thornley, J. (1975) *Aspects of Structure Planning in Britain*. Research Paper 20. London, Centre for Environmental Studies.
Drewett, R., Goddard, J., and Spence, N. (1976) *British Cities, Urban Population and Employment Change 1951–71*. Research Report 10. London, Department of the Environment.
Dunning, J. H. (1969) The city of London: a case study in urban economics. *Town Planning Review* 40, 207—32.
Dunning J. H. (1975) *Economic Analysis and the Multi-National Enterprise*. London, Allen and Unwin.
Echenique, M., *et al.* (1973) *A Disaggregated Model of Urban Structure*. Working Paper 8. University of Cambridge, Centre for Land Use and Built Form Studies.
Echenique, M., Crowther, D., and Lindsay, W. (1969) A spatial model of urban stock and activity. *Regional Studies* 3, 281—312.

Economist Intelligence Unit (1964) *A Survey of Factors Governing the Location of Offices in the London Area.* London, Location of Offices Bureau.
Eddison, A. (1975) *Local Government: Management and Corporate Planning.* Leighton Buzzard, Leonard Hill.
Evans, A. W. (1973) *The Economics of Residential Location.* London, Macmillan.
Eversley, D. E. C. (1973) *The Planner in Society.* London, Faber.
Eversley, D. E. C. (1975) *Reform of Local Government Finance: The Limitations of a Local Income Tax.* Evidence submitted to the Committee of Enquiry into local government finance (Chairman: Frank Layfield, QC). Research Paper 16. London, Centre for Environmental Studies.
Expenditure Committee (1971—2a) *Public Money in the Private Sector.* Sixth Report. HC 347. London, HMSO.
Expenditure Committee (1971—2b) *British Regional Policy: A Critique.* Trade and Industry Sub-Committee. London, HMSO.
Expenditure Committee (1972—3) *Urban Transport Planning.* Vol. 1: *Report and Appendix.* London, HMSO.
Faludi, A. (1973) *Planning Theory.* Oxford, Pergamon.
Fitch, B. (1968) A Galbraith reappraisal: the idealogue as gadfly. *Ramparts* May, 73—84.
Florence, P. S. (1972) *The Logic of British and American Industry.* London, Routledge and Kegan Paul.
Forbes, J. (ed.) (1974) *Studies in Social Science and Planning.* Edinburgh, Scottish Academic Press.
Forrester, J. W. (1969) *Urban Dynamics.* Cambridge, Mass., MIT Press.
Foster, C. D., and Beesley, M. E. (1963) Estimating the social benefit of constructing an underground railway in London. *Journal of the Royal Statistical Society* Series A, 126, Pt 1.
Friedman, M. (1959) Monetary theory and policy. In *Employment, Growth and Price Levels.* Hearing before the joint Economic Committee, 86th Congress, 605—12. US Government Printing Office.
Frost, M. E. (1975) The impact of regional policy. *Progress in Planning* 4, Pt 3. Oxford, Pergamon.
Galbraith, J. K. (1962) *The Affluent Society.* Harmondsworth, Penguin.
Galbraith, J. K. (1967) *The New Industrial State.* London, Hamish Hamilton.
Geddes, P. (1968) *Cities in Evolution.* London, Benn.
George, K. D. (1975) A note on changes in industrial concentration in the United Kingdom. *Economic Journal* 83, 124—8.
Gibson, N. E. (1974) Monetary, credit and fiscal policies. In Prest and Coppock (1974).
Glyn, A., and Sutcliffe, R. B. (1972) *British Capitalism, Workers and the Profits Squeeze.* Harmondsworth, Penguin.
Goddard, J. B. (1973) Office linkages and location. *Progress in Planning* 1, Pt 2. Oxford, Pergamon.
Goddard, J. B. (1975a) *Office Location in Urban and Regional Development.* London, Oxford University Press.
Goddard, J. B. (1975b) Organisation theory, industrial location and the urban system. In Massey and Morrison (1975).
Goddard, J. B. (1975c) *Organizational Information Flows and the Urban System.*
Goldberg (1970) An economic model of intra-metropolitan industrial location. *Journal of Regional Science* 10, 75—9.
Goodman, R. (1972) *After the Planners.* Harmondsworth, Penguin.
Gottmann, J. (1961) *Megalopolis.* Cambridge, Mass., MIT Press.
Gower Economic Publications (1974) *The 100 Centre Guide.* Epping, Gower

Economic Publications.
Gray, H. (1968) *The Cost of Council Housing.* London, Institute of Economic Affairs.
Green, R. J. (1960) *Provincial Metropolis.* Manchester, Manchester University Press.
Griffith, J. A. G. (1966) *Central Departments and Local Authorities.* London, Allen and Unwin.
Hall, A. D. (1962) *A Methodology for Systems Engineering.* New York, Van Nostrand.
Hall, P. (ed.) (1965) *Land Values.* London, Sweet and Maxwell.
Hall, P. (1974) *Urban and Regional Planning.* Harmondsworth, Pelican.
Hall, P., Gracey, H., Drewett, R., and Thomas, R. (1973) *The Containment of Urban England,* 1 and 2. Harmondsworth, Penguin.
Hamilton, F. E. I. (1967) Models of industrial location. In Chorley and Haggett (1967).
Hammond, E. (1967) Dispersal of Government offices: a survey. *Urban Studies* 4 (3), 250—75.
Hansen, N. (1972) *Growth Centres in Regional Economic Development.* New York, Free Press.
Harris, N. (1973) *Competition and the Corporate Society.* London, Methuen.
Harrod, R. F. (1948) *Towards a Dynamic Economics.* London, Macmillan.
Hart, P. E., and Prais, S. J. (1956) Analysis of business concentration. *Journal of the Royal Statistical Society* Series A, 119, 150—81.
Hartwick, J. M., and Hartwick, P. G. (1975) The activity analysis approach to urban model building. *Papers of the Regional Science Association* 35, 75—85.
Harvey, D. (1967) Models of the evolution of spatial patterns in human geography. In Chorley and Haggett (1967).
Harvey, D. (1969) *Explanation in Geography.* London, Edward Arnold.
Harvey, D. (1973) *Social Justice and the City.* London, Edward Arnold.
Harvey, D. (1974) Class monopoly rent, finance capital and the urban revolution. *Regional Studies* 8 (3—4), 239—55.
Hayward, J., and Watson, M. (1975) *Planning Politics and Public Policy.* Cambridge, Cambridge University Press.
Heap, D. (1973) *An Outline of Planning Law.* London, Sweet and Maxwell.
Heap, D. (ed.) (1959—75) *Encyclopaedia of Planning Law and Practice.* London, Sweet and Maxwell.
Heclo, H., and Wildavsky, A. (1974) *The Private Government of Public Money.* London, Macmillan.
Hepworth, N. P. (1971) *The Finance of Local Government.* London, Allen and Unwin.
Herbert, J., and Stevens, B. H. (1960) A model for the distribution of residential activity in urban areas. *Journal of Regional Science* 2, 21—36.
Hicks, J. R. (1965) *Value and Capital.* Oxford, Clarendon Press.
Hill, M. (1968) A goals-achievement matrix for evaluating alternative plans. *Journal of the American Institute of Planners* 34, 19—29.
Hird, C. (1975) *Your Employers' Profits.* Workers' Handbook No. 2. London Pluto Press.
Hobsbawm, E. J. (1969) *Industry and Empire.* Harmondsworth, Penguin.
Hoch, I. (1972) Income and city size. *Urban Studies* 9(3), 299-328.
Holland, S. (1972) *The State as Entrepreneur.* London, Weidenfeld and Nicolson.
Holland, S. (1975a) *Strategy for Socialism.* Nottingham, Spokesman Books.
Holland, S. (1975b) *The Socialist Challenge.* London, Quartet Books.
Holland, S. A. (1976) *The Regional Problem.* London, Macmillan.

Holliday, J. (ed.) (1973) *City Centre Redevelopment.* Tonbridge, Kent, Charles Knight.

Hoover, E. M., and Vernon, R. (1959) *Anatomy of a Metropolis.* Cambridge, Mass., Harvard University Press.

Hoskins, W. G. (1970) *The Making of the English Landscape.* Harmondsworth, Penguin.

Howard, E. (1965) *Garden Cities of Tomorrow* (1902). London, Faber.

Hoyt, H. (1939) The future growth and structure of cities. In Weimer (1972).

Hudson Institute, Europe (1974) *The UK in 1980.* London, Associated Business Programmes.

Hunt, E. K., and Schwartz, J. G. (1972) *A Critique of Economic Theory.* Harmondsworth, Penguin.

Hyman, G. (1969) The calibration of trip distribution models. *Environment and Planning* 1, 105—12.

Ilersic, A. R. (1974) Financing Local Government. *The Accountant*, 30 May, 170, 698—700.

Institute of Economic Affairs (1974) *Government and the Land.* London, Institute of Economic Affairs.

Isard, W. (1956) *Location and Space Economy.* Cambridge, Mass., MIT Press.

Isard, W. (1960) *Methods of Regional Analysis.* Cambridge, Mass., MIT Press.

Isard, W., and Langford, T. W. (1971) *Regional Input-Output Study*, Cambridge, Mass., MIT Press.

Jacobs, J. (1972) *The Economy of Cities.* Harmondsworth, Penguin.

Jaffe, M. (1976) *A Review of Some Published Structure Plans.* Technical Paper 15. London, Centre for Environmental Studies.

Johansen, L. (1963) Labour theory of value and marginal utilities. *Economics of Planning* 3, 89—103.

Johnson, M. B. (1971) *Household Behaviour.* Harmondsworth, Penguin.

Jones, G. P., and Pool, A. G. (1966) *A Hundred Years of Economic Development in Great Britain 1840–1940.* London, Duckworth.

Kaldor, N. (1966) Causes of the Slow Rate of Economic Growth of the United Kingdom. Cambridge, Cambridge University Press.

Keeble, D. E. (1968) Industrial decentralisation and the metropolis: the North-West London case. *Transactions of the Institute of British Geographers* **44**, 1—54.

Keeble, L. (1969) *Principles and Practice of Town and Country Planning.* London, Estates Gazette.

Kennedy, M. C. (1974) The economy as a whole. In Prest and Coppock (1974).

Keynes, J. M. (1936) *The General Theory of Employment, Interest and Money* London, Royal Economic Society.

King, M. A. (1975) The United Kingdom profits crises: myth or reality? *Economic Journal* , March, —

Klir, J., and Valach, M. (1967) *Cybernetic Modelling.* London, Iliffe Books.

Kozlowski, J. (1972) *Threshold Analysis.* London, Architectural Press.

Krueckeberg, D. A., and Silvers, A. L. (1974) *Urban Planning Analysis: Methods and Models.* New York, Wiley.

Labour Part (1973) *Labour's Programme.*

Labour Party (1974) *Paying for Local Services.* Evidence to the Layfield Committee.

Labour Research Department (1973) *The Two Nations: Inequality in Britain Today.*

Lambert, C., and Wier, D. (1975) *Cities in Modern Britain*, London, Fontana.

Lampard, E. E. (1961) The evolving system of cities in the US. In Perloff and Wingo (1961).

Landes, D. S. (1969) *The Unbound Prometheus.* Cambridge, Cambridge University Press.
Lane, R. (1971) *Analytical Transport Planning.* London, Duckworth.
Lange, O. (1965) *Wholes and Parts.* Oxford, Pergamon.
Langley, P. (1975) *Structure Plan Techniques.* Internal Note. Department of the Environment.
Layard, R. (ed.) *Cost Benefit Analysis.* Harmondsworth, Penguin.
Layfield Report (1976) *Local Government Finance.* Report of Commission of Inquiry. Cmnd. 6453. London, HMSO.
Lean, W. (1969) *Economics of Land Use Planning: Urban and Regional.* London, Estates Gazette.
Le Corbusier (1971) *The City of Tomorrow.* London, Architectural Press.
Lee, C. (1973) *Models in Planning.* Urban and Regional Planning Series 4. Oxford, Pergamon.
Lee, C. H. (1971) *Regional Economic Growth in the UK.* New York, McGraw-Hill.
Lee, D. B. (1973) Requiem for large scale models. *Journal of the American Institute of Planners* 39 (3), 163—78.
Leicester City Council and Leicester County Council (1969) *Leicester and Leicestershire Sub-Regional Planning Study.*
Leontief, W. (1967) *Input-Output Analysis.* London, Oxford University Press.
Levinson, C. (1971) *Capital, Inflation and the Multinationals.* London, Allen and Unwin.
Lichfield, N. (1960) Cost benefit analysis in city planning. *Journal of the American Institute of Planners* 26, 273—9.
Lichfield, N., Kettle, P., and Whitbread, M. (1975) *Evaluation in the Planning Process.* Oxford, Pergamon.
Liggins, D. (1975) *National Economics Planning in France.* Farnborough, Hants, Saxon House.
Little, I. M. D. (1967) *A Critique of Welfare Economics.* Oxford, Clarendon Press.
Livingstone. J. M. (1971) *Britain and the World Economy.* Harmondsworth, Penguin.
Livingstone. J. M. (1974) *The British Economy - in Theory and Practice.* London, Macmillan.
Local Government Financial Statistics (1971—2) Department of the Environment. London, HMSO.
Location of Offices Bureau (1968) *Offices in a Regional Centre.* Research Paper 2. London, Location of Offices Bureau.
Losch, A. (1954) *The Economics of Location.* New Haven, Yale University Press.
Lowry, I. S. (1964) *Model of Metropolis.* Memorandum RM—4035—RC. New York, Rand Corporation.
McCormick, B. J., *et al.* (1974) *Introducing Economics.* Harmondsworth, Penguin.
McCrone, G. (1969) *Regional Policy in Britain.* London, Allen and Unwin.
Mackinder, I. H. (1972) Compact: a simple transportation planning package. *Traffic Engineering and Control* 13, 512—16.
McLoughlin, J. B. (1969) *Urban and Regional Planning - A Systems Approach.* London, Faber.
McLoughlin, J. B. (1973) *Control and Urban Planning.* London, Faber.
Manners, G., Keeble, D., Rodgers, B., and Warren, K. (1972) *Regional Development in Britain.* New York, Wiley.
Marcuse, H. (1964) *One-Dimensional Man.* London, Sphere.
Marshall, A. (1974) *Principles of Economics* (1890). London, Macmillan.

Masser, I. (1972) *Analytical Models for Urban and Regional Transport Planning.* Newton Abbott, David and Charles.
Massey, D. B., and Morrison, W. I. (1975) *Industrial Location: Alternative Frameworks.* Conference Paper. London, Centre for Environmental Studies.
Mathias, P. (1969) *The First Industrial Nation.* London, Methuen.
Mattick, P. (1969) *Marx and Keynes.* London, Merlin Press.
Maud Committee (1967) *Report of the Committee on the Management of Local Authorities.* London, HMSO.
Mayer, H. M., and Kohn, C. F. (1959) *Readings in Urban Geography.* University of Chicago Press.
Meade, J. E. (1975) *The Intelligent Radical's Guide to Economic Policy.* London, Allen and Unwin.
Meadows, D. H., *et al.* (1972) *The Limits to Growth.* London, Earth Island.
Medio, A. (1972) Profits and surplus value. In Hunt and Schwartz (1972).
Merseyside Metropolitan County Council (1975a) *Draft Structure Plan: Report of Survey.*
Merseyside Metropolitan County Council (1975b) *Consultants' Report on Sub-Contracting in Merseyside.* Undertaken for the structure plan.
Meszaros, I. (1970) *Marx's Theory of Alienation.* London, Merlin Press.
Meyer, J. R., Main, J. F., and Wohl, M. (1965) *The Urban Transportation Problem.* Cambridge, Mass., Harvard University Press.
Miliband, R. (1969) *The State in Capitalist Society.* London, Weidenfeld and Nicolson.
Mills, E. (1974) *Recent Developments in Retailing and Urban Planning.* Technical Paper 3. London, Planning Research Applications Group.
Mills, E.S. (1972) *Urban Economics.* Glenview, Ill., Scott, Foresman.
Minns, R., and Thornley, J. (1976) Municipal enterprise boards. *New Society* 37 (726), 499.
Mishan, J. (1969) *The Costs of Economic Growth.* Harmondsworth, Pelican.
Moore, B., and Rhodes, J. (1973) Evaluating the effects of British regional economic policy. *Economic Journal* 83, 87—110.
Moore, B., and Rhodes, J. (1974) Regional policy in the Scottish economy. *Scottish Journal of Political Economy* 21 (3), 215—35.
Morishima, M. (1973) *Marx's Economics: A Dual Theory of Value and Growth.* Cambridge, Cambridge University Press.
Morishima, M., and Murata, Y. (1972) *An Input-Output Analysis of Disguised Unemployment in Japan 1951-1965.* Cambridge, Cambridge University Press.
Morrison, W. I. (1976) *Economic Model of Merseyside.* Technical Paper . London, Planning Research Applications Group.
Morrison, W. I., and Smith, P. (1976) *Input-Output Methods in Urban and Regional Planning: A Practical Guide.* Technical Paper 6. London, Planning Research Applications Group.
Moser, C. A., and Scott, W. (1961) *British Towns.* Edinburgh, Oliver and Boyd.
Moses, L., and Williamson, H. F. (1967) The location of economic activity in cities. *American Economic Review* 57, 211—22.
Mouzelis, N. P. (1967) *Organisation and Bureaucracy.* London, Routledge and Kegan Paul.
Mowat, C. L. (1955) *Britain Between the Wars 1918-1940.* London, Methuen.
Mumford, L. (1961) *The City in History.* Harmondsworth, Penguin.
Musgrave, R. A. (1959) *The Theory of Public Finance.* New York, McGraw-Hill.
Muth, R. (1969) *Cities and Housing.* Chicago, University of Chicago Press.

Myrdal, G. M. (1957) *Economic Theory and Underdeveloped Regions.* London, Duckworth.
Nanton, V., and Sharpe, L. J. (1975) *Conference on Local Government Performance.* London, Social Science Research Council.
National Economic Development Office (1975) *Finance for Investment.*
National Industrial Conference Board (1969) *Concentration and Productivity.*
National Institute of Economic and Social Research (1974) *National Institute Economic Review.*
National Institute of Economic and Social Research (1975) *The UK Economy.* Commission of European Communities.
Nelson, H. J. (1955) A service classification of American cities. *Economic Geography 31, 189–210.*
Nettleford, J. S. (1914) *Practical Town Planning.* London, St Katherine Press.
Neuberger, H. N. (1971) User benefit in the evaluation of transport and land use plans. *Journal of Transport Economics and Policy* 1, 1—24.
Neumann, J. Von, and Morgenstern, O. (1953) *The Theory of Games and Economic Behaviour.* Princeton, Princeton University Press.
Neutze, G. M. (1965) *Economic Policy and the Size of Cities.* London, Cass.
Neutze, G. M. (1973) *The Price of Land and Land Use Planning.* Organization of Economic Cooperation and Development.
Nevitt, A. A. (1966) *Housing, Taxation and Subsidies.* London, Nelson.
North, G. A. (1975) *Teesside's Economic Heritage.* Cleveland County Council.
Notts-Derby Sub-Regional Planning Unit (1969) *Notts-Derby Sub-Regional Study.*
O'Connor, J. (1969) Scientific and ideological elements in the economic theory of government policy. *Science and Society* 33, 385—414.
O'Connor, J. (1973) *The Fiscal Crisis of the State.* New York, St Martin's Press.
Organization of Economic Cooperation and Development (1974) *National Accounts for OECD Countries 1962–1973.*
Orr, S. C., and Cullingworth, J. B. (1969) *Urban and Regional Studies.* London, Allen and Unwin.
Pahl, R. E. (1975) *From Urban Sociology to Political Economy.* Paper presented to British Sociological Association.
Palmer, D. (1975) *Planning and Forecasting Employment and Economic Development in Structure Planning.* Technical Paper 13. London, Planning Research Applications Group.
Palmer, J. A. D. (1972) Introduction to the British edition. In Goodman (1972).
Parsons, G. (1972) The giant manufacturing corporations and balanced regional growth. *Area,* 4, 99—103.
Parsons, T. (1951) *The Social System.* London, Routledge and Kegan Paul.
Peacock, A., and Wiseman, J. (1967) *The Growth of Public Expenditure in the United Kingdom.* London, Allen & Unwin.
Peaker, A. (1974) *Economic Growth in Modern Britain.* London, Macmillan.
Pen, J. (1971) *Income Distribution.* Harmondsworth, Penguin.
Perloff, H. S., and Wingo, L. (1968) *Issues in Urban Economics.* Baltimore, Md., Johns Hopkins Press.
Phillips, A. W. (1958) The relationship between unemployment and the rate of change of money wage rates in the United Kingdom, 1861—1957. *Economica* 25, 283—99.
Pigou, A. C. (1920) *The Economics of Welfare.* London, Macmillan.
Pigou, A. C. (1933) *Theory of Unemployment.* London, Cass.
Planning Research Applications Group (1973) *Monmouthshire: A Strategic Planning Model.* Technical Paper 2. London, Planning Research Applications Group.
Political and Economic Planning (1968) *Economic Planning and Policies in Britain, France and Germany.*

Poulantzas, N. (1975) *Classes in Contemporary Capitalism*. London, New Left Books.
Pratten, C. F. (1971) *Economics of Scale in Manufacturing Industry*. Cambridge University Press.
Pred, A. (1973) The growth and development of cities in advanced economies. In A. Pred and G. Tornquist, *Systems of Cities and Information Flows*. Lund Studies in Geography (B) 38.
Prest, A. R.. and Coppock, D. J. (eds) (1974) *The UK Economy: A Manual of Applied Economics*. London, Weidenfeld and Nicolson.
Prest, A. R.. and Turvey, R. (1965) Cost benefit analysis: a survey. *Economic Journal* 75, 685—705.
Pryor, F. (1970) The extent and pattern of public ownership. *Weltwirtschaftlichen Archiv* 104 (2), 159—88.
Quandt, R. E. (1970) *The Demand for Travel: Theory and Measurement*. Lexington, Mass., Lexington Books.
Quarmby, D. E. (1967) Choice of travel mode for the journey to work. *Journal of Transport Economics and Policy* 1, 273—314.
Rasmussen, D. W. (1973) *Urban Economics*. New York, Harper and Row.
Rasmussen, D. W., and Haworth, C. T. (1973) *The Modern City*. New York, Harper and Row.
Rees, J. (1972) The industrial corporation and location decision analysis. *Area* 4, 199—205.
Rees, M. (1973) *The Public Sector in the Mixed Economy*. London, Batsford.
Reilley, W. J. (1931) *The Law of Retail Gravitation*. New York, Putnam.
Report of an Investigation into Conditions in Certain Depressed Areas (1934) Cmnd. 4728. London, HMSO.
Report of the Commission for the Special Areas (1935—8) London, HMSO.
Rex, J. (1970) *Key Problems of Sociological Theory*. London, Routledge and Kegan Paul.
Richardson. H. W. (1970) *Regional Economics: A Reader*. London, Macmillan.
Richardson. H. W. (1971) *Urban Economics*. Harmondsworth, Penguin.
Richardson. H. W. (1972a) *Regional Economics*. London, Weidenfeld and Nicolson.
Richardson. H. W. (1972b) *Input-Output and Regional Economics*. London, Weidenfeld and Nicolson.
Richardson. H. W. (1973a) *Regional Growth Theory*. London, Macmillan.
Richardson. H. W. (1973b) *Economics of Urban Size*. London, Saxon House.
Roberts, M. (1974) *An Introduction to Town Planning Techniques*. London, Hutchinson.
Robinson, J. (1966) *An Essay on Marxian Economics*. London, Macmillan.
Robinson, J. (1969) *The Economics of Imperfect Competition*. 2nd edition. Reprinted 1972. London, Macmillan.
Rodgers, A. (1966) Matrix methods of population analysis. *Journal of the American Institute of Planners* 32, 40—4.
Roll, E. (1938) *A History of Economic Thought*. London, Faber.
Rostow, W. W. (1971) *The Stages of Economic Growth*. Cambridge, Cambridge University Press.
Round, J. I. (1972) Regional input-output models in the UK: a reappraisal of some techniques. *Regional Studies* 6. 1—9.
Rowthorn, R., and Hymer, S. (1971) *International big business 1957-1967*. Department of Applied Economics Occasional Papers 24. Cambridge University Press.
Roy, D. (1974) *State Holding Companies*. Young Fabian Pamphlet.
Samuelson, P. A. (1973) *Economics*. New York, McGraw-Hill.
Schumacher, E. F. (1974) *Small is Beautiful*. London, Sphere.

Scitovsky, T. A. (1941) A note on welfare propositions in economics. *Review of Economic Studies.* Reprinted in K. Arrow and T. Scitovsky (eds), *Readings in Welfare Economics.* London, Allen and Unwin.

Scitovsky, T. (1954) Two concepts of external economies. *Journal of Political Economy* 17, 143—51.

Self, P. (1971) *Metropolitan Planning: The Planning System of Greater London.* London, Weidenfeld and Nicolson.

Self, P. (1972) *Administrative Theories and Politics.* London, Allen and Unwin.

Senior, D. (1969) *Royal Commission on Local Government in England.* 2: *Memorandum of Dissent.* Cmnd. 4040. London, HMSO.

Shanks, M. (1972) *The Stagnant Society.* Harmondsworth, Penguin.

Shepherd, W. G. (1961) A comparison of industrial concentration in the USA and Britain. *Review of Economics and Statistics* 43, February, 7—75.

Shepherd, W. G. (1966) *Changes in British Industrial Concentration 1951–1958.* Oxford Economic Papers. London, Oxford University Press.

Sherman, H. J. (1970) The Marxist theory of value revisited. *Science and Society* 34, 257—92.

Shonfield, A. (1965) *Modern Capitalism.* London, Oxford University Press.

Silbertson, A. (1972) Economies of scale in theory and practice. *Economic Journal* 82, 369—91.

Simmie, J. M. (1974) *Citizens in Conflict: The Sociology of Town Planning.* London, Hutchinson.

Smith, P., and Morrison, W. I. (1974) *Simulating the Urban Economy.* London, Pion.

Smith, T. (1975) Britain. In Hayward and Watson (1975).

Snickers, F., and Karlqvist, A. (1975) *Dynamic Allocation in Space.* Farnborough, Hants, Saxon House.

Solesbury, W. (1974) *Policy in Urban Planning.* Oxford, Pergamon.

South Hampshire Structure Plan (1973)

Staffordshire County Structure Plan (1973)

Stalley, M. (1972) *Patrick Geddes.* New Brunswick, N.J., Rutgers University Press.

Stamp, L. D. (1962) *The Land of Britain: Its Use and Misuse.* London, Longmans.

Stewart, M. (1967) *Keynes and After.* Harmondsworth, Penguin.

Stone, P. A. (1970) Urban development in Britain: standards, costs and resources 1964—2004. *Population Trends and Housing* 1. National Institute of Economic and Social Research. Cambridge University Press.

Stone, P. A. (1973) *The Structure, Size and Cost of Urban Settlements.* Cambridge, Cambridge University Press.

Stone, R. (1967) *Mathematics in the Social Sciences.* London, Chapman and Hall.

Structure Plan (1973) Hampshire County Council.

Structure Plan (1975) Cleveland County Council.

Structure Plan (1975—6) Greater Manchester County Council.

Structure Plan (1975—6) Merseyside County Council.

Structure Plan (1973) Birmingham City Council.

Thompson, W. R. (1968a) *A Preface to Urban Economics.* Baltimore, Md., Johns Hopkins Press.

Thompson, W. R. (1968b) Internal and external factors in the development of urban economics. In Perloff and Wingo (1968).

Thunen, J. H. Von (1826—50) *Der isolierte Staat.* Edited by P. Hall, *Von Thunen's Isolated State.* London, Pergamon, 1966.

Tiebout, C. M. (1956) The urban economic base reconsidered. *Land Economics* 32, 95—9.

Titmuss, R. M. (1974) *Social Policy: An Introduction.* London, Allen and Unwin.

Town and Country Planning Act (1971) London, HMSO.

Townroe, P. M. (1971) *Industrial Location Decisions: A Study in Management.* Occasional Paper 15. University of Birmingham, Centre for Urban and Regional Studies.

Townroe, P. M. (1973) Some behavioural considerations in the industrial location decision. *Regional Studies* 6 (3), 261—72.

Turner, G. (1971) *Business in Britain.* Harmondsworth, Penguin.

Turvey, R. (1957) *The Economics of Real Property.* London, Allen and Unwin.

Unwin, R. (1909) *Town Planning in Practice: An Introduction to the Art of Design in Cities and Suburbs.* Fisher Unwin.

Uthwatt Report (1942) *Final Report of the Expert Committee on Compensation and Betterment.* London, HMSO.

Utton, M. A. (1970) *Industrial Concentration.* Harmondsworth, Penguin.

Vernon, R. (ed.) (1974) *Big Business and the State: Changing Relations in Western Europe.* London, Macmillan.

Wabe, J. S. (1966) Office decentralization: an empirical study. *Urban Studies* 3, 35—53.

Warneryd, O. (1972) *A Systematic Approach to the Study of Interdependence in Urban Systems.* Third Annual Conference, Regional Science Association, British Section 1969. London Papers in Regional Science No. 2, ed. A. G. Wilson. London, Pion.

Warren, B., and Prior, M. (1974) *Advanced Capitalism and Backward Socialism.* Spokesman Pamphlet 46. London, Bertrand Russell Peace Foundation.

Webber, R. (1974) *Social Area Analysis of Greater London.* Technical Paper 7. London, Planning Research Applications Group.

Webber, R. (1976) *Liverpool Social Area Study. 1971 Data: Final Report.* Technical Paper 14. London, Planning Research Applications Group.

Weber, A. (1968) *The Theory of the Location of Industries.* University of Chicago Press.

Weimar, A. M. (1972) *Real Estate.* 6th ed. New York, Ronald Press.

Weiss, S. J., and Gooding, E. C. (1968) Estimation of differential multipliers in a small regional economy. *Land Economics* 44, 234—44.

Westway, E. J. (1974) The spatial hierarchy of business organizations and its implications for the British urban system. *Regional Studies* 8, 145—55.

Williamson, J. G., and Swanson, J. A. (1966) The growth of cities in the American North East 1820—1870. *Explorations in Entrepreneurial History* 4, 3—101.

Wilson, J. (1969) *Population Growth and Movement.* Working Paper 12. London, Centre for Environmental Studies.

Wilson, A. G. (1968) *Entropy in Urban and Regional Modelling.* London, Pion.

Wilson, A. G. (1972) Behavioural inputs to aggregative urban system models. In *Papers in Urban and Regional Analysis,* 71—90. London, Pion.

Wilson, A. G. (1974) *Urban and Regional Models in Geography and Planning.* New York, Wiley.

Wilson, J. Q. (ed.) (1970) *The Metropolitan Enigma.* Cambridge, Mass., Harvard University Press.

Wingo, L. J. (ed.) (1963) *Cities and Space.* Baltimore, Md., Johns Hopkins Press.

Worswick, G. D. N., and Blackaby, F. T. (1974) *The Medium Term, Models of the British Economy.* London, Social Science Research Council and National Institute for Economic and Social Research.

Yamey, B. S. (1973) *Economics of Industrial Structure.* Harmondsworth, Penguin.

Young, S., and Lowe, A. (1974) *Intervention in the Mixed Economy.* New York, Croom Helm.

Index

Index of Names

Abercrombie, P., 149, 215
Adelman, A., 13, 251
Adelman, I. G., 251
Alden, J., 74, 251
Alexander, J. W., 201, 251
Allen, G. C., 251
Alonso, W., 193, 199, 251
Althusser, L., 251
Ambrose, P., 163, 169, 251
Amery, C., 124, 155, 251
Andrews, R. B., 201, 251
Angel, S., 198, 251
Arrow, K., 251
Ashby, R., 88, 251
Ashworth, W., 44, 143, 251
Atkinson, A. B., 23, 251
Au Rousseau, M., 201, 251

Bacon, R., 6, 251
Bailey, J., 210, 251
Bain, J. S., 19, 22, 251
Balibar, E., 251
Ball, R. J., 53, 251
Banham, R., 251
Bank of England, 251
Baran, P., 12, 18, 45, 82, 252
Barker, P., 252
Barker, T. S., 71, 252
Barlow Report, 150, 151, 252
Barras, R., 67, 106, 115, 230, 232, 234, 252
Batty, M., 234, 252
Baxter, R., 252
Becker, G., S., 114, 118, 252
Beckerman, W., 15, 252
Beesley, M., 257
Beer, S., 88, 252
Bell, C., 143, 252
Bell, R., 143, 252
Berry, B. J. L., 197, 200, 252
Bertalanffy, L. von, 86, 252
Best, R., 252, 253
Beveridge Report, 151, 253
Blackaby, F. T., 265
Blair, J., 253
Blank, S., 253
Blaug, M., 146, 176, 186, 253
Blumenfeld, H., 202, 253
Blunden, W. R., 253
Board of Trade, 77, 253
Bolton Committee, 18, 253
Booth, D. J. W., 106, 232, 252
Boswell, R. J., 105, 253
Boyce, D. E., 225, 253
Brand, J. A., 253
Bratten, C., 253
Bray, J., 55, 58, 61, 62, 253
Brittan, S., 48, 49, 51, 52, 253, 254
Broadbent, T. A., 26, 31, 108, 115, 119, 220, 229, 232, 234, 253
Buchanan, C. D., 214, 254
Buckley, W., 210, 254
Bullock, N., 254

Cairncross, A., 58, 117, 254
Cambridge Political Economy Group, 2, 254
Cameron, G. C., 200, 254
Carney, J., 78, 254
Catalano, A., 67, 108, 252, 254
Central Statistical Office, 254
Chadwick, G., 254
Champion, R. G., 96, 253, 254
Chancellor of the Exchequer, 254
Chenery, H. B., 254
Cherry, G., 2, 28, 143, 254
Chisholm, M., 192, 194, 196, 200, 254
Chorley, R. J., 109, 198, 254
Christaller, W., 193, 254
Clark, B. D., 254
Clawson, M., 25, 254
Coates, B. E., 254
Cole, W. A., 7, 255
Coleman, A., 163, 255

Colenutt, R., 163, 169, 251
Colston Papers, 255
Commission of the European Communities, 255
Coppock, D. J., 21, 27, 263
Coppock J. T., 253
Cordey-Hayes, M., 77, 100, 225, 255
Cornwall, J., 255
Coventry City Council, 227, 255
Coventry Community Development Project, 106, 225, 255
Cowan, P., 255
Cripps, E. C., 232, 255
Croft, M. J., 200, 255
Crowther, D., 256
Cruickshank, D., 124, 155, 251
Cullen, I., 114, 255
Cullingford, D., 120, 255
Cullingworth, J. B., 2, 28, 151, 255

Dahrendorf, R., 205, 255
Day, A., 214, 255
Day, N. D., 225, 253
Dean, R. M., 253
Deane, P. D., 7, 255
Denison, E. F., 13, 255
Dennis, N., 141, 207, 255
Department of Economic Affairs, 70, 71, 78, 255
Department of Environment, 158, 159, 160, 224 256
Derry, T. K., 256
Desai, M., 176, 256
Dickens, P., 254, 256
Dobb, M., 34, 35, 36, 76, 175, 177, 188, 256
Donaldson, P., 19, 58, 256
Donnison, D. V., 23, 205, 256
Dorfman, R., 256
Doyle, P., 53, 251
Dow, J. C. R., 49, 58, 64, 65, 256
Drake, M., 160, 256
Drewett, R., 25, 93, 94, 100, 101, 104, 114, 142, 143, 162, 258
Dunning, J. H., 79, 200, 256

Echenique, M., 232, 234, 256
Economist Intelligence Unit, 200, 256
Eddison, A., 257
England, J. R., 106, 232, 252
Evans, A. W., 164, 198, 254, 257
Eversley, D. E. C., 22, 139, 205
Expenditure Committee, 257

Faludi, A., 257
Fitch, B., 46, 257
Florence, P. S., 19, 257
Flynn, P., 120, 255
Foot, D. H. S., 232, 255
Forrester, J. W., 210, 215, 257
Foster, C. D., 257
Friedman, M., 31, 257
Frost, M. E., 257

Galbraith, J. K., 45, 229, 257
Geddes, P., 149, 257
George, K. D., 19, 257
Gleave, D., 77, 100, 255
Gibson, N. E., 66, 257
Glyn, A., 2, 4, 7, 18, 257
Goddard, J., 94, 100, 256, 257
Godson, V., 114, 255
Goldberg, M. A., 200, 257
Goodman, R., 257
Gottmann, J., 257
Gower Economic Publications, 108, 258
Gracey, H., 25, 93, 94, 100, 101, 104, 114, 142, 143, 162, 258
Gray, H., 258
Green, R. J., 258
Griffith, J. A. G., 135, 258

Hall, A. D., 87, 258
Haggett, P., 109, 198, 254
Hall, P. G., 2, 25, 28, 93, 94, 100, 101, 104, 114, 142, 143, 234
Hamilton, F. E. I., 200, 258
Hammond, E., 258
Hansen, N., 258
Harris, N., 258
Harrod, R. F., 37, 258
Hart, P. E., 258
Hartwick, P. G., 258
Harvey, D., 47, 115, 173, 208, 209, 258
Haworth, C. T., 263
Hayward, J., 258
Heap, D., 258
Heclo, H., 258
Hepworth, N. P., 131, 139, 140, 258
Herbert, J., 258
Hicks, J. R., 258
Hill, M., 238, 258
Hird, C., 258
Hobsbawm, E. J., 15, 16, 258
Hoch, I., 258
Holland, S., 18, 25, 58, 72, 259
Holliday, J., 259
Hoover, E. M., 199, 259
Hoskins, W. G., 143, 259
Howard, E., 148, 259
Hoyt, H., 199, 201, 259
Hudson Institute Europe, 2, 4, 259
Hunt, E. K., 13, 259
Hyman, G., 198, 251, 259
Hymer, S., 20, 263

Ilersic, A. R., 259
Institute of Economic Affairs, 170, 259
Isard, W., 139, 192, 259

Jacobs, J., 259
Jaffe, M., 225, 259
Jarman, T. L., 256
Johnson, K. M., 200, 254
Johnson, M. B., 259
Jones, G. P., 259

Kain, J. F., 216, 261
Kaldor, N., 14, 259
Karlquist, A., 264
Keeble, D., 105, 260
Kennedy, M. C., 51, 54, 259
Kettle, P., 237, 260
Keynes, J. M., 3, 16, 24, 259
King, M. A., 7, 60, 259
Klir, J., 88, 259
Kohn, C. F., 200, 261
Kozlowski, J., 126, 259
Krueckenberg, D. A., 215, 229, 259

Labour Party, 72, 259
Labour Research Department, 23, 259
Lambert, C., 259
Lampard, E. E., 193, 259
Landes, D. S., 260
Lane, R., 216, 260
Langford, T. W., 259
Langley, P., 260
Layard, R., 235, 236, 260
Layfield Report, 132, 133, 135, 260
Lean, W., 164, 260
Le Corbusier, 260
Lecomber, J. R. C., 71, 252, 260
Lee, C., 231, 260
Lee, C. H., 260
Lee, D. B., 212, 228, 229, 233, 260
Leicester City Council, 154, 260
Leontief, W., 187, 260
Levinson, C., 20, 260
Lewis, E. W., 105, 253, 260
Lichfield, N., 237, 260
Liggins, D., 260
Lindsay, W., 256
Little, I. M. D., 35, 49, 260
Livingstone, J. M., 260
Local Government Financial Statistics, 131, 133, 260
Location of Offices Bureau, 108, 200, 260
Losch, A., 193, 198, 260
Lowe, A., 265
Lowry, I. S., 230, 260

McCormick, B. J., 145, 186, 260
McCrone, G., 77, 260
McDonald, C., 225, 253
McLoughlin, J. B. 164, 210, 222, 260
Mackinder, I. H., 220, 260
Maggs, K. R. A., 255, 260
Manners, G., 76, 254, 260
Marshall, A., 175, 189, 260
Masser, I., 212, 261
Massey, D. B., 225, 232, 252, 261
Mathias, P., 7, 261
Mattick, P., 18, 39, 261
Maud Committee, 135, 261
Mayer, H. M., 200, 261
Meade, J. E., 58, 261
Meadows, D. H., 261
Medio, A., 261
Merseyside Metropolitan County Council, 112, 225, 261
Meszaros, I., 261
Meyer, J. R., 216, 261
Miliband, R., 46, 261
Mills, E., 261
Mills, E. S., 109, 261
Minns, R., 139, 261
Mishan, J., 3, 261
Moore, B., 74, 75, 261
Morgan, R., 74, 251
Morishima, M., 16, 210, 261
Morrison, W. I., 106, 208, 210, 261, 264
Moser, C. A., 104, 261
Moses, L., 261
Mouzelis, N. P., 3, 242, 261
Mowat, C. L., 31, 261
Mumford, L., 2, 261
Murata, Y., 16, 261
Musgrave, R. A., 37, 261
Muth, R., 199, 261
Myrdal, G. M., 262

Nanton, V., 262
National Economic Development Office, 262
National Industrial Conference Board, 262
National Institute of Economic and Social Research, 262
Nelson, H. J., 105, 262
Nettleford, J. S., 262
Neuberger, H. N., 217, 262
Neumann, J. von, 185, 262
Neutze, M., 170, 194, 262
Nevitt, A. A., 133, 262
North, G. A., 22, 106, 262
Notts-Derby Sub Regional Planning Unit, 154, 262

O'Connor, J., 34, 37, 39, 44
Organization of Economic Co-operation and Development, 2, 5, 262
Orr, S. C., 262

Pahl, R. E., 23, 47, 208
Palmer, D., 106, 227
Palmer, J. A. D., 145, 262

Parsons, G., 105, 262
Parsons, T., 206, 210, 262
Peacock, A., 262
Peaker, A., 19, 262
Pen, J., 23, 262
Perloff, H. S., 192, 201, 262
Pigou, A. C., 32, 36, 262
Planning Research Applications Group, 232, 262
Political and Economic Planning, 262
Poulantzas, N., 46, 263
Prais, S. J., 263
Pratten, C. F., 185, 263
Pred, A., 112, 263
Prest, A. R., 21, 27, 58, 236, 263
Price, C., 252, 263
Prior, M., 265
Pryor, F., 45, 263

Quandt, R. E., 263
Quarmby, D. E., 263

Rasmussen, D. W., 263
Rees, J., 115, 263
Rees, M., 57, 263
Reilly, W. J., 197, 263
Rhodes, J., 74, 261
Richardson, H. W., 193, 202, 263
Roberts, M., 212, 263
Robinson, J., 263
Rodgers, A., 263
Rodgers, B., 260
Roll, E., 175, 177, 263
Rostow, W. W., 3, 214, 263
Round, J. I., 263
Rowthorn, R., 20, 263
Roy, D., 263

Samuelson, P. A., 80, 174, 176, 181, 182, 186, 255, 263
Schumacher, E. F., 3, 264
Schwartz, T. G., 13, 259
Scitovsky, T. A., 193, 237, 264
Self, P., 264
Senior, D., 114, 264
Shanks, M., 12, 264
Shapcott, M., 254
Sharpe, L. J., 262
Shepherd, W. G., 19, 264
Sherman, H. J., 184, 187, 264
Shonfield, A., 58, 69, 264
Silbertson, A., 190, 253, 264
Silvers, A. L., 215, 229, 259
Simmie, J. M., 23, 47, 204, 208, 210, 264
Smith, P., 187, 202, 261, 264
Smith, T., 67, 264
Snickers, F., 264
Solesbury, W., 26, 158
Solow, R. J., 255
South Hampshire County Council, 1, 2, 160
Spence, N., 94, 100, 256
Staffordshire County Structure Plan, 225, 264
Stalley, M., 264
Steadman, P., 254
Stewart, M., 264
Stone, P. A., 96, 264
Stone, R., 6, 264
Swanson, J. A., 193, 265
Sweezey, P., 12, 18, 252

Thomas, R., 25, 93, 94, 100, 101, 103, 114, 142, 143, 162, 258
Thompson, W. R., 193, 264
Thomson, R., 160, 256
Thornley, J., 139, 160, 261
Thünen, J. von, 178, 193
Tiebout, C. M., 202, 264
Titmus, R. M., 151, 265
Townroe, P. M., 265
Turner, G., 265
Turvey, R., 58, 198, 236, 263

Unwin, R., 265
Uthwatt Report, 140, 265
Utton, M. A., 265

Valach, M., 88, 259
Vernon, R., 199, 200, 259, 265

Wabe, J. S., 200, 265
Warneryd, D. O., 106, 265
Warren, B., 265
Warren, K., 260, 265
Watson, M., 258, 265
Webber, R., 120, 127, 255, 265
Weber, A., 193, 265
Weimar, A. M., 265
Weiss, S. J., 202, 265
Westaway, E. J., 105, 265
Whitbread, M., 237, 260, 265
Wildavsky, A., 2, 258
Williamson, H. F., 261
Williamson, J. G., 193, 265
Willis, J., 265
Wilson, A. G., 210, 229, 265
Wilson, J. Q., 265
Wingo, L., 192, 199, 201, 254, 262, 265
Wiseman, J., 262
Wohl, M., 216, 261
Worswick, G. D. N., 265

Yamey, B. S., 265
Young, S., 265

Subject index

Action area plans, 159
Activity analysis, 9, 26, 79ff., 114ff.
Advance factories, 76
Aggregate production function, 186, 203
Agriculture, 15, 93
Assignment of trips, 216

Barriers, to development, 13ff.; to public sector, 1, 13ff., 58; *see* Margin
Bid rent, 200
Boom, 49
Boundary (*see* margin); of urban area 94, 114
Bureaucracy, 31, chapter 7
Budget, 48; balanced, 38; history, 64, 66, 68

Capital intensive industry, 20, 21, 60, 78
Capitalist economy, 1, 45
Car ownership, 154, 216
Central area redevelopment, 27
Central government, control of national economy, 52, 59, 60, 62, 157; reorganisation, 68; road investment, 215
Central place theory, 104, 197ff.
Centralization, of economy, 1; of metropolis, 102; urban, 100ff.
Channelling the forces, 41, 148, 155, 165, 246, 247
Charity, 69
City, large, 93; region, 93
Collaboration, 218
Commodity, 9, 82, 114, 115; buildings, 26
Commuting boundary, 94, 104
Community development project, 106, 107
Community Land Act, 169
Comparative advantage, 191
Compensation principle, 32
Competition, avoidance of, 45; between local authorities, 140; for land, 25, 92, 120ff.; in market economy, 6ff., 18ff.; perfect, 33; small firms, 19; urban development, 103
Component analysis, 104
Comprehensive redevelopment areas, 155; central areas, 27
Compulsory purchase, 155
Concentration, firms, 12; national economy, 1, 72; population in urban centres, 25
Conservatives, 69, 71, 72
Cost benefit analysis, 190, 214, 235ff.
Costs of production, 9, 10, 29, 34, 200; competition, 12; decentralization, 111; households, 111; large firms, 20; national accounts, 80; restructuring, 17
Consultants, planning, 222ff.; transport studies, 219ff.
Consumption, controls, 48; demand, 35; detailed models, 230, 234, 238; 'end' of production, 113; of products, 9; households, 111; preferences, 35; propensity, 36, 37; roads, 213, 217; 'schizophrenia', 91, 113, 117; spending, 5, 80, 91
Contradictions, in planning, chapter 4, chapter 5, 212; in state, chapter 2, 240
Control, over urban areas, 107; system, 88
Conurbation, 95
Convergence, 45
Counties, 154; metropolitan, 130; non metropolitan, 130
County map, 156
Corporate planning, 135
Credit, 64
Crises, 3
Cybernetics, 149, 229; *see* systems analysis
Cyclic planning process, 225

Debt, local authority, 135; national, 42
Decentralization, 147, 125; nineteenth-century private enterprise, 146; *see also* urban decentralization
Decline, long term, 7; national economy, 7ff.; manufacturing, 21
Decreasing returns, 179
Demand curve, 179; cost benefit analysis, 236
Density, 97
Department of Economic Affairs, 68, 70
Department of Environment, 215; chief planner, 222; structure plans, 224; transport, 215
Development, definition of, 156
Development control, 157, 164
Development potential, 213, 227, 235
Development tax, 166, 167
Direct planning, 78
Disaggregation, consumer behaviour, 234
Discontinuance order, 157
Distribution of trips, 216
District plans, 159
Districts, 154; metropolitan, 130; non metropolitan, 130
Division of labour, 10, 25, 67, 93, 105, 108

Dynamic simulation, as social theory, 210

East Sussex structure plan, 223
Economic base, 103, 201
Employment, 36, 38; and urban development, 91ff.
Engineering approach, transport planning, 216
Equality, 206; of opportunity, 206
Equilibrium, 33
Exploitation, labour theory of value, 175; monopoly profits, 186; neoclassical theory, 35
Extensive change, 14
Externalities, and internal economies, 194; deviation from perfect market, 34; internalization of, 195; market, 194; non market, 194; road traffic, 217

Factorization, of a system, 87
Fashions in planning, 213; Royal Town Planning Institute, 221; techniques, 228
Feedback, 88
Feudalism, 15, 112
Final demand, 80
Finance, and industry, 18; capital, 2, 7, 11, 17, 125, 132; sector, 27, 66, 67; urban development, 92, 110
Firms, large, 94, 105, 106, 111; multinational, 16, 19; organization of, 105
Fluctuations, 52
Forecasts, 51
Fractions of the ruling class, 46
Functionalist theory, 206; and systems analysis, 210; of city, 103

Game theory, 185
Garden city, 148; public ownership of land, 149
General development order, 156, 157
General equilibrium, 181
Generalist education of planners, 222
Generation of trips, 216
Goals achievement matrix, 238
Gross National Product, 6, 80ff.
Growth, and redistribution, 206; comparisons, 5, 6, 23; investment, 7; population, 12; theory, 37; urban and regional, 97ff., 201ff.; transport planning, 214

Hexagonal market areas, 197
Hierarchy, central places, 197; urban areas, 95, 104, 105
Households, 80ff.; as production activity, 117ff.; home/work split, 91
Housing, acts, 147; spending, 59, 60; substandard, 26, 27

Immobility of factors of production, 196
Imperfect market; in space, 203; ultra radical theory, 208
Imports, 55
Income, distribution, 35; flow, 80; inequality, 23
Indicative planning, 68, 69
Indifference, curve, 183; of consumers to suppliers at different locations, 197; of landowners to users of land at zone boundaries, 200
Indivisibilities, 195
Industrial location, 196
Inequality, 23
Industrial Reorganisation Corporation, 70
Information system, 225
Instruments, 48ff.; regional, 74
Integration of urban areas into national economy, 93
Intensive, change, 13ff.; use of land, 25
Inflation, 52, 53, 65; and growth, 6; 'new', 66
Inner area, 101, 127
Imperial Chemical Industries, 106
Input output analysis, 188; linkages, 9
Institutional reform, 68
Integration of economy, 74, 111
Interdependence, 93
Interest rate, 50
Interpretation of models, 232
Investment, 9, 16ff., 63; and profit, 18; and saving, 36, 37; direct, 48; fluctuations, 54; grants, 74, 77; in urban development, 26, 27, 106; social, 29

Justification, let present trends continue, 233

Keynesian economics, 24, 29, 35ff., 174, 201

Labour, female, 113; market, 21; pool, 90, 94, 110ff.; wage and salary earners, 1
Labour party, 31; in government, 69, 72
Labour theory of value, 175; and prices, 177; activity of, 'man', 111, 177
Laissez-faire, 69, 211
Land, commission, 167; public ownership, 140; tenure reform, 167, 168; use, 25, 93, 96, 126
Leading sectors, cars, 214
Leicester and Leicestershire sub regional studies, 154
Limits to state intervention, 30

Liquidity trap, 37
Local authority expenditure, stabilizing effect of, 139
Local government, 130; agents of central government, 135; competition between authorities, 140; finance, 132; size, 57
Local plans, 159
Local state, 128; capital spending, 130; debt, 133; education, 131, 131; housing expenditure, 130, 139; trading, 130, 139
London, 146; airport study, 237; Greater London Development Plan, 215; transport study, 218; trips to central area, 217
Lowry model, 228, 230ff.

Macro economics, 190ff., 191; micro economics, 191, 203; *see* Keynesian economics
Mainstream economics, 174, 175
'Manual adjustments', 217, 228; 'massaging', 229, 234; third London airport, 237
Manufacturing, 21
Margin, 177, 200, 201, 203; inner area studies, 223; planning operates at the, 208; social areas, 107; the poor, 206
Market area, 197; Lowry model, 232, 233
Market economy, 1, 3, 6ff.; 'mixed' Economy, 8; View of state, 31
Marxist analysis, 41ff., 49ff., 207
Mature economy, 2, 3
Maximization, profit and utility, 33
Megalopolis, 25, 93, 95
Merseyside, 225
Micro economics, 174; breakdown of assumptions, 190; cost benefit analysis, 235; imperfect market, 189; land use models, 230; misapplication, 217; profit, 189; underlying pluralist theory, 205; welfare theory of state, 32
Migration, 77, 99ff.
Mixed economy, 8, 41
Mobility of resources, 33
Modal split, 216
Models, transport, 219
Money, 50, 53, 63
Monopoly, 184, 185; convergence theory, 46; deviation from perfect market, 34; power of the state, 208; sector of the economy, 19
Multinational firms, 20, 46, 79; *see also* firms
Multiplier, 37, 202; illusion, 40, 55; non local, 112

National accounts, 79ff.
National Economic Development Office, 68, 70
National Enterprise Board, 72, 73
National income, flow of, 9, 80ff.
National plan, 51, 71
Nationalization, 31; interference, 64; public corporation, 57
Natural increase, population, 99
Necessary, production, 12; social provision, 69
Negative response, 155
Non conforming use, 157
Notts-Derby sub regional study, 233

Offices, 107ff.
Oligopoly, 184ff.
Open space, 96
Opportunity costs, 122, 188, 195
Oppressive nature of the state, 30
Organization, 105
Overdeveloped economy, 2, 74
Over loaded road networks, 216ff.
Over production, 37
Overseas competition, 16

Planning Advisory Group Report, 158
Pareto criterion, 32, 235
Partial equilibrium, 187; demand curves, 237; versus general equilibrium, 182, 186
Paternalism, 31
PESC, 56
Physics, 34
Planning and profit, 11
Planning, experiment, 65, 67; fashions, 173
Planning authorities, 154
Planning considerations, 164
Planning permission, 31
Planning Research Applications Group (PRAG), 104, 121, 220, 234
Planning system, cost of, 223
Plant size, 22
Pluralism, 31, 112
Policy resolutions, of local authority, 158
Pluralism, 31, 112, 119
Population growth, 154
Poverty, social policy, 24, 69
Post industrial society, 105
Post war acts, 151
Prices, 177; labour theory of value, 177; marginal cost, 181; utility, 177
Private sector, driving force for the economy, 11, 29; goods, 33; regional policy, 75

Production, 9, 10, 184, 195, 198; social, 113; urban development, 93, 116ff.
Productivity, 13, 14
Profit, 85, decline, 7; difference between sectors, 20; motor generator, 40; payment for services, 183
Programme planning and budgeting, 135
Property companies, 26, 67
Public health legislation, 28, 31
Public sector, 24, chapter 2; borrowing, 41; danger point, 57; finance, 37; goods, 34; limited by private sector, 29; 'not-for-profit' sector, 42, 88; size, 45; spending, 25, 37; finance, 37

Quasi government agencies, 68

Rank size, 97
Rates, 132; destabilizing effect of, 139; rateable value, 133
Rationality, 213
Rationality of city structure, 101
Rate support grant, 132
Reforming legislation, 31, 207
Region, definition, 115
Regional disparity, 74
Regional employment premium, 74
Regional planning, 73ff.
Regional planning councils, 68
Regional policy, 30, 73ff.; cost, 75
Regions of the UK, 76
Rent, 92, 121ff.; differential, 190
Report of survey, 159
Residential location, concentric zones, 198; micro economic theory, 199; sector theory, 199
Restraint of traffic, 217
Restructuring, 13, 14, 16, 17, 21, 28
Royal Town Planning Institute, 213; control of expertise, 220ff.; examinations, 221

Saving, 36, 37
Say's Law, 36
Scale, economies, 10, 14, 22; increasing returns, 34; urban areas, 101
Sector theory of residential location, 199
Selectivity, 77
Separation of powers, 30, 218, 243
Services, 21
Shake out, 67
'Short-term' fetish, 67
Sieve maps, 227
Spatial monopoly, 197
Social consumption, 29, 40, 58, 59
Social democracy, 31
Social expenses, 40, 58, 59
Social investment, 39; roads, 59, 214
Social structure, 23; urban, 112
Sociology, 204, 205
South Hampshire structure plan, 160ff., 223, 234
Specialization, 93; cities and regions, 196
Spillover effect, 34; stabilization policies, 29, 48ff., 59ff.
Stable equilibrium, 33
Stages, in structure planning, 224; of production, 10
Standard metropolitan labour area, 94ff.
Standard of living, 4, 5, 47
State, committee of bourgeoisie, 31; contradiction in, 30; relative independence of, 207, 46; oppressive, 47; regulations of spillovers, 34; size, 57, 58; sounding box, 46; unlimited, 32, 47
State management, 48
Steering the economy, 48; *see* stabilization policies
Stocks, 55
Structure, economic, 68
Structure plans, 3, 4, 154, 159; content, 159; examination question, 221; methods for, 223ff.; shortcomings, 160; teething troubles, 159
Subject plans, 159
Substitution, 183, 201
Supply curve, 179, 180; ultra radical theory, 208
Surplus value, 176
Survey-analysis-plan, 149
Systems analysis, 86ff.; functionalist sociology, 210; progress of man, 229; training of planners, 220

Tautology, in planning models; 229 Lowry models, 233
Tax, 37, 50, 59, 60, 63, 74
Technical progress, 13
Techniques for planning, books on, 212; development potential, 227; elementary, 226; forecasting, 227; neutral, 210; sieve maps, 227
Teesside, 22, 78, 106
Tertiary sector, 21
Theory, 173; counter revolutionary, 174; explanation of urban development, 173; jusfication of planning, 173; revolutionary, 174; *status quo*, 174
Three estates, 29, 79, 120
Threshold analysis, 126
Time, economy of, 114
Town map, 156
Town planning act, 15; 1947, 152ff.; 1968, 155

Town planning measures, 1900-1935, 147
Trade, 191, 65
Trading, local authorities, 130, 139ff.
Transfer payment, 30, 44, 47, 57, 59, 60, 122, 183, 201; rent, 200
Transformation problem, 177
Transport costs, rent, 123
Transport planning, 214ff; spending, 59; studies, 213ff.
Transport plans and programmes, 215
Treasury, 30, 48ff., 68, 73

Ultra radicals, 31, 78, 112
Ultra vires, 139
Under consumption, 44ff.
Unemployment, 34, 52, 61, 77
Urban, centralization, 100ff.; decentralization, 96, 100ff.; definition, 114; growth, 96; preliminary definition, 14; system, 90ff.; urbanization, 47
Urban planning, 158; books, 27ff.; containment, 162; decentralization, 91; effects, 162ff.; growth, 91; land utilization, 163; land values, 164ff.; market economy, 158; U.K. system, 2
Use classes order, 156ff.
Uthwatt committee, 166
Utility, 33, 90, 176; diminishing, 178; inter-personal comparison, 35; labour theory of value, 189; marginal, 178ff.

Value, added, 20; of assets, 26ff.
Victorian, housing legacy, 27; social provision, 119

Wartime reports, 151ff.
Wages, 85; payment for services, 183; standard of living, 4ff.; unions, 16
Wealth, 23
Welfare, economics, 32, ff.; new and old, 35
Work, 113

Zoning, concentric, 198; inner area problem, 102; land market, 92; Lowry model, 233; product of market, 127; rent, 122ff.

For Product Safety Concerns and Information please contact our EU
representative GPSR@taylorandfrancis.com Taylor & Francis Verlag GmbH,
Kaufingerstraße 24, 80331 München, Germany

Printed and bound by CPI Group (UK) Ltd, Croydon, CR0 4YY
11/04/2025
01843977-0006